Primary and Secondary Metabolism of Plant Cell Cultures I
K. H. Neumann, W. Barz, and E. Reinhard (Eds.) 1985

W. G. W. Kurz (Ed.)

Primary and Secondary Metabolism of Plant Cell Cultures II

With 117 Figures

Springer-Verlag Berlin Heidelberg New York
London Paris Tokyo Hong Kong

Dr. Wolfgang G. W. Kurz
Plant Biotechnology Institute
National Research Council of Canada
Saskatoon
Saskatchewan S7N OW9
Canada

ISBN-13: 978-3-642-74553-9 **e-ISBN-13: 978-3-642-74551-5**
DOI: 10.1007/978-3-642-74551-5

Library of Congress Cataloging-in-Publication Data. Primary and secondary metabolism of plant cell cultures II / W. G. W. Kurz, ed. p. cm. The second of a planned series of meetings on Primary and secondary metabolism of plant cell cultures, held in 1988 at the Plant Biotechnology Institute of the National Research Council of Canada in Saskatoon, Saskatchewan. Includes index. ISBN 0-387-50861-9 (alk. paper) 1. Plant cell culture–Congresses. 2. Plants–Metabolism–Congresses. 3. Metabolism, Secondary–Regulation–Congresses. I. Kurz, W. G. W. (Wolfgang G. W.), 1933- . QK725.P773 1989 582.087–dc 20

Softcover reprint of the hardcover 1st edition 1989

Typesetting: International Typesetters Inc., Makati, Philippines
Printing: Druckhaus Beltz, 6944 Hemsbach/Bergstraße
Binding: J. Schäffer, Grünstadt
2131/3145-543210 – Printed on acid-free paper

Preface

At the end of the initial meeting on Primary and Secondary Metabolism of Plant Cell Cultures at Schloss Rauischholzhausen, it was decided to convene similar events on a regular basis midway between the International Congress for Plant Tissue and Cell Culture. We felt it was necessary to bring representatives of the research teams working in this field together to assess progress as well as to discuss future directions.

The Plant Biotechnology Institute of the National Research Council of Canada in Saskatoon, Saskatchewan was chosen to be the site for the second symposium held in 1988. The theme of this meeting was "The Regulation of Primary and Secondary Metabolism in Plant Cells" and judging by the contributions published in this volume it was a timely and actual one indeed.

I would like to thank my organizing committee and my research group for their efforts, the National Research Council of Canada for the financial assistance and the participants for their contributions which all contributed to make this symposium a success. A special thank you to Mrs. L. Blashill and R. Gallays for their assistance in hosting this event. Last but not least I would like to thank our publisher, Springer-Verlag, Heidelberg for their decision to publish the proceedings of this and future symposia.

Saskatoon, Spring 1989 W. G. W. Kurz

Contents

Part IV Gene Expression in Transformed and Stressed Plant Cells

Part V Round Table Discussion

Part I
Interaction Between Primary and Secondary Metabolism

Initiation and Characterization of Photoautotrophic Suspension Cultures

J. M. Widholm[1]

1 Introduction

A review of the literature shows that photoautotrophic cultures of the following 16 different species have been established: *Nicotiana tabacum* (Bergmann 1967, 1968), *Ruta graveolens* (Corduan 1970), *Chenopodium rubrum* (Husemann and Barz 1977), *Cytisus scoparius* (Yamada and Sato 1978), *Peganum harmala* (Barz et al. 1980), *Spinacia oleracea* (Dalton 1980), *Datura stramonium* (Yasuda et al. 1980), *Hyoscyamus niger* (Yasuda et al. 1980), *Arachis hypogea* (Bender et al. 1981), *Asparagus officinalis* (Peel 1982), *Marchantia polymorpha* (Katoh 1983), *Glycine max* (Horn et al. 1983b), *Solanum tuberosum* (LaRosa et al. 1984), *Morinda lucida* (Igbavboa et al. 1985), *Catharanthus roseus* (Tyler et al. 1986), and *Euphorbia characias* (Hardy et al. 1987). While this number (16) is not small it should be pointed out that the establishment of photoautotrophic cultures which are convincingly photoautotrophic, i.e., cultures which can be grown for many subcultures in medium without sugar or other energy sources, is usually a time-consuming and often difficult task. Several of the cultures listed above are grown on agar-containing medium which can complicate the definition since McHale (1985) has shown that agar or agarose-containing medium without sucrose can support the growth of green *Nicotiana tabacum* cells incubated in light in CO_2-free air.

More in-depth descriptions of photoautotrophic plant cultures and their characteristics are found in the reviews of Horn and Widholm (1984), Husemann (1985), and Neumann and Bender (1987).

In the following discussion I will describe the eight different photoautotrophic suspension cultures (six different species) which we have initiated and are presently maintaining and will summarize how this information helps us understand basic questions about photoautotrophy.

2 *Glycine max* (Corsoy and PI 437833) Photoautotrophic Cultures

Our initial work on photoautotrophic cultures developed the *G. max* (soybean) suspension culture of the cv. Corsoy using rubber stoppered Erlenmeyer flasks with syringe needles inserted to allow 5% CO_2 in air to be blown into the flask (Horn et al. 1983b). This culture which was denoted SB-P originally had a doubling time of

[1]University of Illinois, Department of Agronomy, 1102 South Goodwin, Urbana, IL 61801, USA

Primary and Secondary Metabolism of Plant Cell Cultures, ed. by W.G.W. Kurz

about 4 days and chlorophyll (Chl) levels near 400 $\mu g\ g^{-1}$ FW. The growth required light and elevated CO_2 (1% or above) and was inhibited by DCMU [3-(3,4-dichlorophenyl)-1,1-dimethylurea] concentrations which did not inhibit heterotrophic growth. The peak $^{14}CO_2$ fixation rate was near 80 $\mu mol\ mg^{-1}$ Chl h^{-1} late in the culture period when dark respiration was lowest.

The SB-P cells were characterized again after about an additional 4 years in culture (Rogers et al. 1987). The medium was simplified by removal of the complex vitamin mixture leaving only 0.1 mg l^{-1} thiamine HC1. The growth rate, small cell clump size, and elevated CO_2 requirement had not changed but the Chl content had increased to between 1500 and 2000 $\mu g\ g^{-1}$ FW. The peak $^{14}CO_2$ fixation rate in the light was 66 $\mu mol\ mg^{-1}$ Chl h^{-1} while the dark fixation rate was never more than 10% of the rate in the light. The cells evolved up to 80 $\mu mol\ O_2\ mg^{-1}$ Chl h^{-1} in light and the dark respiration rate was usually near 10 $\mu mol\ CO_2\ mg^{-1}$ Chl h^{-1}. While cell growth was linear for the 21-day batch culture period the Chl level declined somewhat by day 4 at a time when dark respiration increased and net CO_2 fixation declined as measured by gas exchange by infrared gas analysis (IRGA). This time also corresponds with a peaking of the soluble protein content indicating a rapid synthesis period which would increase respiration and in turn decrease net photosynthesis.

The SB-P cells always showed low positive net CO_2 fixation by IRGA except on day 4 when there was net CO_2 evolution. These cells also had a measured CO_2 compensation concentration of about 230 $\mu l\ l^{-1}$ which is below the ambient CO_2 level of about 350 $\mu l\ l^{-1}$. This raises the question as to why the cells will not survive under ambient CO_2 levels. Part of the reason could be that the already low total ribulose-1,5-bisphosphate carboxylase (RuBPcase) activity of the SB-P cells (60 to 90 $\mu mol\ CO_2\ mg^{-1}$ Chl h^{-1}) which is only about 50% active when measured as initial activity subsequently becomes only about 14% active after 24-h incubation under ambient CO_2 conditions. The RuBPcase activation levels are determined by measuring activity in rapidly prepared cell extracts (initial activity) and then activating the RuBPcase with saturating Mg^{2+} and HCO_3^- levels before measuring the activity again (total activity). Leaves of most plants contain RuBPcase which is about 90% activated under high light.

The phosphoenolpyruvate carboxylase (PEPcase) activity peaks near day 3 in the culture cycle near 85 $\mu mol\ CO_2\ mg^{-1}$ Chl h^{-1} and then declines to about 25 by day 21. These changes in enzyme activities, protein levels, and CO_2 exchange would indicate that early in the culture cycle, especially, the cultured cells are like a developing leaf which has relatively low RuBPcase, high PEPcase, and high dark respiration (Kisaki et al. 1973, Aoyagi and Bassham 1986).

A second photoautotrophic soybean suspension culture was initiated from a different genotype (PI 437833) to determine whether the characteristics of two different cultures would be similar (Rogers and Widholm 1988). Friable green callus was initiated from stem tissue and subsequently a suspension was obtained in 3% sucrose-containing medium. When these cells were cultured in medium with 1% sucrose, many cells became bleached and died but after 42 days a culture composed predominantly of green, viable cell aggregates was obtained. This culture was maintained on 1% sucrose for a total of 15 transfers over 33 weeks after which the cells were transferred to medium devoid of sucrose with a 5% CO_2

atmosphere. Here again a selection process occurred, but a viable photoautotrophic culture was obtained from the surviving cells.

The new line, which was named SB1-P, acquired almost all of the same characteristics of the SB-P cells within the 14-month initiation period. These characteristics included growth rate, cell (30 μm diameter) and clump size (150 to 190 μm diameter), Chl content, RuBPcase and PEPcase levels, $^{14}CO_2$ fixation rates, O_2 evolution rates, dark respiration, and oxygen inhibition of photosynthesis.

From this study we can conclude that photoautotrophic cultures of the same species will generally have similar characteristics which might be expected since photoautotrophic cultures of different species also have similar characteristics.

3 *Datura innoxia* and *Nicotiana tabacum* + *Nicotiana glutinosa* Fusion Hybrid Photoautotrophic Cultures

We have also developed photoautotrophic suspension cultures of *Datura innoxia* and a *Nicotiana tabacum* + *N. glutinosa* fusion hybrid T3g1C (Horn et al. 1983a) by gradual adaptation to medium without a carbon source (Xu et al. 1988). The original green suspensions were grown on 3% soluble potato starch and then 1% starch as intermediate steps before placing in medium without any carbohydrate with a 5% CO_2 atmosphere. The present doubling times of these cultures, denoted DAT-P for *D. innoxia* and NTG-P for the fusion hybrid, is about 14 days and the Chl contents up to about 200 μg g^{-1} FW. The cultures required elevated CO_2 (5%) and light for growth. They show high dark respiration (about 50 μmol CO_2 mg^{-1} Chl h^{-1}) and have near zero or negative CO_2 uptake in the light in 21% O_2 and air CO_2 levels by IRGA. The light-induced net CO_2 uptake is enhanced by lowering the O_2 level to 2%. The $^{14}CO_2$ fixation in the light is near 30 to 40 μmol mg^{-1} Chl h^{-1} or more for both lines. The PEPcase to total RuBPcase activity ratios were variable but were usually near 1:1 with activity levels of from about 30 to 100 μmol CO_2 mg^{-1} Chl h^{-1}. The RuBPcase activation level was only about 30% in freshly prepared extracts.

4 *Amaranthus cruentus* and *Amaranthus powellii* Photoautotrophic Cultures

Until now there have been no reports of photoautotrophy in cell cultures initiated from any C_4 plant species. We (Xu et al. 1988) have initiated photoautotrophic cultures from *A. cruentus* and *A. powellii*. These cultures were initiated by the 3% sucrose, 3% soluble starch, 1% starch to no carbohydrate sequence used for the DAT-P and NTG-P cultures above. The respective cultures have been denoted ACR-P and APO-P. The strains contain about 200 μg Chl g^{-1} FW, have doubling times near 14 days and require elevated CO_2 and light for growth. The cultured cells did not have elevated PEPcase levels as one might have expected with cultures of C_4 species. The PEPcase levels were in the range of 12 to 22 μmol CO_2 mg^{-1} Chl h^{-1}

while the RuBPcase activities usually ranged from 40 to 100 μmol CO_2 mg^{-1} Chl h^{-1}. The RuBPcase activation levels were near 50% when grown in 5% CO_2.

The $^{14}CO_2$ fixation levels in the light of these two cultures ranged from 9 to 38 μmol CO_2 mg^{-1} Chl h^{-1} while the dark rates were about 10% of this. Dark respiration rates varied from 9 to 40 μmol CO_2 mg^{-1} Chl h^{-1} while CO_2 exchange under air conditions of CO_2 and O_2 was usually negative.

These results show that the cultures from two C_4 *Amaranthus* species have characteristics similar to those of cultures initiated from the other C_3 species described here and by others, and did not have elevated PEPcase levels.

5 *Gossypium hirsutum* (Cotton) Photoautotrophic Cultures at 5% and Ambient CO_2 Levels

A callus and subsequently a suspension culture were initiated from hypocotyl and cotyledon tissue from a cotton seedling of cv. Stoneville 825 (Blair et al. 1988). Heterotrophic cells growing with 3% sucrose were grown with 1% sucrose and continuous light to produce a mixotrophic suspension. These cells were then put under photoautotrophic conditions (no sucrose and 5% CO_2 atmosphere) where intensive selection occurred until a stably growing culture was obtained (denoted COT-P). After about 2 years of continuous photoautotrophic culture the COT-P cells can have a minimal doubling time of about 4 days and contain about 500 to 600 μg Chl g^{-1} FW. The cells required high CO_2 and light for growth and the growth was inhibited by the photosynthetic electron transport inhibitor DCMU.

Characterization of the CO_2 exchange by IRGA showed the dark respiration to be about 24 to 44 μmol CO_2 mg^{-1} Chl h^{-1} and the exchange in the light at 0.035% CO_2 and 2% O_2 to be near zero. Increasing the O_2 level to 21 or 50% increased the net CO_2 loss in the light. The total activated RuBPcase activity was about 63 μmol CO_2 mg^{-1} Chl h^{-1} and was about 70% activated when initially prepared. The PEPcase activity was sometimes near that of the activated RuBPcase (19 to 76 μmol CO_2 mg^{-1} Chl h^{-1}).

We (Xu et al. 1988) have been able to adapt the *G. hirsutum* suspension culture (COT-P) originally growing photoautotrophically in 5% CO_2 to grow under ambient CO_2 levels of the culture room which are about 0.06 to 0.07%. This culture, denoted COT-PA, has been growing continuously for about 2 years now. The initial transition was made by incubating the COT-P cells under ambient CO_2 conditions until the cells were visually mostly dead and then the 5% CO_2 atmosphere was reinstated to rescue the live cells. After several cycles, cells were selected which would survive the ambient CO_2 conditions.

This would appear to be the first line other than the *Arachis hypogea* cells of Bender et al. (1981) which can carry out sustained growth under ambient CO_2 conditions. The *A. hypogea* strain is maintained on agar-solidified medium, however, which would indicate that some of their carbon is derived from compounds present in the medium (McHale 1985). This line can, however, apparently grow under ambient CO_2 levels in liquid medium (K. H. Neumann, personal communication).

Recent characterization studies with the COT-PA cells (C. Roeske, W. L. Ogren, J. M. Widholm, unpublished) show that the doubling time on ambient CO_2 levels is near 10 days and the Chl content is about 600 to 800 $\mu g\ g^{-1}$ FW. The dark respiration rates vary from 8 to 36 $\mu mol\ CO_2\ mg^{-1}\ Chl\ h^{-1}$. When the CO_2 fixation enzyme activities were measured, the PEPcase level was 20 $\mu mol\ CO_2\ mg^{-1}\ Chl\ h^{-1}$ and the RuBPcase about 54 when measured in extracts using conventional methods (Rogers and Widholm 1987, Blair et al. 1988, Xu et al. 1988). The conventional method involved taking cells directly from flasks under the incident low light of the shaker (about 300 $\mu E\ m^{-2}s^{-1}$), grinding in liquid N_2 or in a homogenizer with a buffer without a protease inhibitor. Much higher RuBPcase activities were obtained from COT-PA and three other lines, as discussed in the next section, if the cells were initially incubated under higher light conditions before extracting the enzyme and if the protease inhibitors, phenylmethylsulfonyl fluoride (PMSF), antipain, and leupeptin, were included in the grinding buffer. In the case of the COT-PA cells initial activities of 151 $\mu mol\ CO_2\ mg^{-1}\ Chl\ h^{-1}$ were obtained with a Mg^{2+} and HCO_3^- total activated level of 162. These levels are three times those obtained previously with these cells and also shows a 93% in vivo activation level versus about 60% found before (Table 1).

When the $^{14}CO_2$ fixation products were analyzed after a 2-min pulse with the COT-PA cells, the label was distributed in sugar monophosphates (39%), sugar bisphosphates (8%), malate (9%), organic acids (trace), sugars (10%), acid phosphates (e.g., phosphoglycerate, 7%), amino acids (largely glycine, aspartate, and alanine, 23%), and starch (5%). Following a chase of up to 30 min, the label in the sugar phosphates decreases and that in starch increases. These labeling patterns are similar to those of normal C_3 photosynthesis and fit with the RuBPcase and PEPcase activities found in these cells.

6 Additional Characteristics of the Photoautotrophic Lines Described Here

All of our photoautotrophic cultures were originally grown on more complex media with the carbon-containing compounds, several vitamins, FeEDTA, and usually 5 mM Hepes buffer. However, it was initially shown that the soybean lines (SB-P, SB1-P) could grow just as rapidly without the Hepes and with only 0.1 mg l^{-1} thiamine HCl, FeEDTA chelate, and the growth regulators as carbon-containing compounds in the medium (Rogers et al. 1987; Rogers and Widholm 1988). All of the other cultures are likewise now being grown in simplified media with only the growth regulators, FeEDTA and thiamine (C. Goldstein and J. M. Widholm, unpublished).

Since the Chl levels continue to change with time (usually increase) and the Chl values given in Table 1 were obtained about 1 year ago, except for COT-PA, new measurements were made recently (C. Goldstein and J. M. Widholm, unpublished). The largest increases were in APO-P and ACR-P where the levels are now 1700 and 875 μg Chl g^{-1} FW, respectively. Little change was seen with the SB-P, SB1-P, DAT-P, NTG-P, and COT-P lines, however.

Table 1. Generalized characteristics of the photoautotrophic cultures initiated in this laboratory. All cultures are grown in 5% CO_2 except for COT-PA which is grown in ambient CO_2

Cell line	Doubling time (days)	Chl content ($\mu g\ g^{-1}$ FW)	Dark respiration	$^{14}CO_2$ Fixation Dark	$^{14}CO_2$ Fixation Light	PEPcase activity	RuBPcase activity	RuBPcase activation in vivo (%)
			($\mu mol\ CO_2\ mg^{-1}$ Chl h^{-1})					
SB-P (Soybean)	4	1500–2000	10	6	66	25– 85	60– 90	50
SB1-P (Soybean)	4.5–5.0	1500–1700	10	3.6	50–60	40–110	50– 80	40
DAT-P (*D. innoxia*)	14	200	50	5	30– 40	30–40	30–100	29
NTG-P (*N. tabacum* + *N. glutinosa* fusion hybrid)	14	200	50	10	30–70	30–90	30–100	35
ACR-P (*A. cruentus*)	14	200	17–34	1.3–3.9	13–38	21	40–100	46
APO-P (*A. powellii*)	14	200	9–40	1.8–6.1	9–28	13	20– 80	42
COT-P (Cotton)	4	500– 600	24–44	6	24	19–76	63	70
COT-PA (Cotton)	10	600– 800	8–36	4	64	20	54	60

The high CO_2 requirement for growth by most photoautotrophic cultures might be explained by the absence of carbonic anhydrase which could reduce the availability of CO_2 as a substrate for fixation. We (C. Roeske, W. L. Ogren, J. M. Widholm, unpublished) found that the COT-PA line which does grow under ambient CO_2 conditions has no carbonic anhydrase detectable by the veronal buffer assay method or by Western blotting using antibodies to carbonic anhydrase from spinach. In contrast, the soybean lines growing on 5% CO_2, SB-P, and SB1-P do contain from 20 to 77% of the spinach leaf levels of carbonic anhydrase as measured by enzyme assay and immunological methods. No enzyme was detected in COT-P, ACR-P, or APO-P. The spinach antibody does cross-react with a protein from leaves of cotton and soybean. Thus, the presence of carbonic anhydrase does not determine the CO_2 requirements of these photoautotrophic cultures.

As described earlier in the COT-PA cell description, higher RuBPcase total activity and in vivo activation levels could be obtained if the cells were collected and illuminated with bright light before enzyme extraction and if the extraction buffer contained protease inhibitors. This was also found to be true with SB-P, SB1-P, and COT-P cells where the respective total RuBPcase activities were increased to 156, 262, and 117 μmol CO_2 mg^{-1} Chl h^{-1} (C. Roeske, W. L. Ogren and J. M. Widholm, unpublished). These levels are two or more times those measured previously (Table 1) so we can assume that the previous assays may show the actual low activation state of the enzyme in vivo, but the total enzyme level was underestimated due to proteolytic degradation. Thus, all of the cell lines described here or elsewhere may have higher RuBPcase levels than actually measured unless protease inhibitors are used in the extraction medium. The higher measured levels are still not as high as leaves, like growth chamber-grown spinach, which had RuBPcase activity of 500 μmol CO_2 mg^{-1} Chl h^{-1} (Blair et al. 1988) or field-grown soybeans with over 600 μmol CO_2 mg^{-1} Chl h^{-1} (Vu et al. 1986).

The fluorescence emission spectra of the photoautotrophic cultures (SB-P, DAT-P, COT-P, COT-PA) were determined using light of about 400 to 600 nm for excitation (Xu et al. 1988). The fluorescence emission spectra obtained at room temperature showed a broad peak at 685 nm for all four strains which should originate from the antenna Chl *a* molecules of photosystem II (Murata and Satoh 1986). The fluorescence emission spectra obtained at 77 K consisted of three peaks at 685, 695, and 740 nm. Both the 685 and 695 nm bands should originate in the antenna Chl *a* molecules of photosystem II. The peak at 740 nm should originate in the antenna Chl *a* of photosystem I. These spectra are similar to those of green algae and isolated chloroplasts (Briantais et al. 1986) indicating that these photoautotrophic cultures have antenna systems similar to those of other green plants.

7 Discussion

In general, it is felt that photoautotrophic cultures have high rates of dark respiration which may be a consequence of their need for growth and cell division in contrast to mature leaf cells which are no longer growing or dividing. The dark respiration rates of the cultures described here range from a low of 10 μmol CO_2

mg^{-1} Chl h^{-1} for the SB-P and SBl-P strains to about 50 for DAT-P and NTG-P. If calculated on a FW basis, however, the values range from about 5 μmol CO_2 g^{-1} FW h^{-1} for the ACR-P and APO-P cells to about 15 to 19 for the COT-P, COT-PA, SB-P, and SB1-P cells. The values for DAT-P and NTG-P are 10 μmol CO_2 g^{-1} FW h^{-1}.

Kisaki et al. (1973) found that the dark respiration rates of mature leaves of greenhouse-grown tobacco (*N. tabacum*) were about 5 to 8 μmol CO_2 mg^{-1} Chl h^{-1} and about 10 μmol g^{-1} FW. In studies with mature leaves from growth chamber-grown soybean plants Forrester et al. (1966) found the dark respiration to be no more than 14 μmol CO_2 g^{-1} FW with the assumption that the Chl level is about 3 mg g^{-1} FW. The dark respiration rates of both growth chamber-grown *Helianthus annuus* (sunflower, Sharp et al. 1984) and *Triticum aestivum* (wheat, Azcon-Bieto and Osmond 1983) are near 1 μmol CO_2 m^{-2} s^{-1} which can be converted to a Chl basis of about 12 μmol CO_2 mg^{-1} Chl h^{-1} assuming a Chl content of about 300 mg m^{-2}. The value for these two species would be about 36 μmol CO_2 g^{-1} FW assuming a value of 100 g FW m^{-2}. Thus, the dark respiration in the mature leaves of these four species ranges from about 5 to 12 μmol CO_2 mg^{-1} Chl h^{-1} or 10 to 36 μmol CO_2 g^{-1} FW.

Thus, if we compare the photoautotrophic culture dark respiration with that of mature leaves on a Chl basis, the cultures are higher, i.e., 10 to 50 μmol CO_2 mg^{-1} Chl h^{-1} versus 5 to 12. The SB-P and SB1-P cells which have Chl levels near that of leaves do have dark respiration rates (about 10) like those of mature leaves (5 to 12). Likewise when the tissue culture dark respiration rates are put on a FW basis, the values (5 to 19) are very similar to those of leaves (10 to 36). Thus, we can conclude that with the cultures described here the dark respiration rates are similar to those of mature leaves if calculated on a FW basis. The dark respiration values are higher if calculated on a Chl basis especially if the Chl levels are much lower than leaves, which is usually the case, except for the soybean lines (SB-P, SB1-P).

Studies of the effect of 21 and 2% O_2 levels on net CO_2 exchange measured by IRGA with CO_2 concentrations near 350 μl l^{-1} has shown that increased O_2 concentrations decrease the net CO_2 fixation greatly in all eight photoautotrophic cultures (Rogers et al. 1987, Rogers and Widholm 1988, Xu et al. 1988). This indicates that ribulose-1,5-bisphosphate carboxylase/oxygenase is fixing a large proportion of the CO_2 in these cells and that O_2 is able to affect this net fixation either directly or indirectly, as expected (Ogren 1984).

While the COT-PA cells grow under ambient CO_2 conditions, there are no clear differences between the characteristics of this line and those of the other cultures to explain this ability. Likewise the SB-P and SB1-P cells have a CO_2 compensation concentration of about 0.023% which is below the normal ambient CO_2 level of about 0.035% and the 0.06 and 0.07% CO_2 levels found in the culture room. These cells will not, however, survive in the culture room without elevated CO_2 levels (1% or higher).

One reason for lower photosynthesis in most of the cultures is that the low light levels which enter the flask to illuminate the cells may not be high enough to activate RuBPcase completely. Preliminary studies show that about 50% activation occurs at 50 μE $m^{-2}s^{-1}$ (C. Roeske, W. L. Ogren and J. M. Widholm, unpublished). The flasks are illuminated with about 250 μE $m^{-2}s^{-1}$ but the amount of light at the cell surface would be much less due to shading by the flask and other cells which could help explain the low activation levels found in vivo (29 to 70%, Table 1). Preliminary attempts to increase the light intensity have not proven satisfactory.

8 Summary

These studies show that cells of a number of plant species can be adapted to grow photoautotrophically and one cotton line (COT-PA) can even grow under an ambient CO_2 atmosphere. The characterization studies have not yet, however, explained why this cotton line and not the others can survive the low CO_2 conditions. Carbonic anhydrase was eliminated as the cause since activity is found only in the soybean lines (SB-P and SB1-P) which will not grow under ambient CO_2 conditions. These two soybean lines, initiated from different genotypes at different times, have very similar characteristics. The use of protease inhibitors has increased the measured RuBPcase activity allowing these cultures to be more leaflike in appearance, as far as CO_2 fixation enzyme levels are concerned.

The cultures of the two C_4 species (*Amaranthus*), ACR-P and APO-P, do not have elevated PEPcase activities relative to those of the cultures from C_3 species. While dark respiration is higher than mature leaves in cultures with low Chl levels when calculated on a Chl basis, the dark respiration is not higher when calculated on a fresh weight basis.

Studies of the fluorescence emission spectra show that the Chl antenna systems are similar to those of green algae and isolated chloroplasts and in contrast to studies with leaves, little distortion of the spectra was observed.

These cultures now provide ample material for numerous physiological, biochemical, molecular, and genetic studies.

Acknowledgments. The unpublished work was supported by funds from the Illinois Agricultural Experiment Station, the SOHIO-UI Center of Excellence in Crop Molecular Biology and Genetic Engineering, and the McKnight Interdisciplinary Photosynthesis Research Program.

References

Aoyagi K, Bassham JA (1986) Appearance and accumulation of C_4 carbon pathway enzymes in developing wheat leaves. Plant Physiol 80:334–340

Azcon-Bieto J, Osmond CB (1983) Relationship between photosynthesis and respiration: the effect of carbohydrate status on the rate of CO_2 production by respiration in darkened and illuminated wheat leaves. Plant Physiol 71:574–581

Barz W, Herzbeck H, Hüsemann W, Schneiders G, Mangold HK (1980) Alkaloids and lipids of heterotrophic, photomixotrophic and photoautotrophic cell suspension cultures of *Peganum harmala*. Planta Medica 40:137–148

Bender L, Kumar A, Neumann KH (1981) Photoautotrophic pflanzliche Gewebekulturen in Laborfermentern. In: Lafferty RM (ed) Fermentation. Springer Berlin Heidelberg, New York, pp 193–203

Bergmann L (1967) Wachstum grüner Suspensionskulturen von *Nicotiana tabacum* var. "Samsun" mit CO_2 als Kohlenstoffquelle. Planta 74:243–249

Bergmann L (1968) Photosynthesis and growth of suspension cultures of *Nicotiana tabacum* with CO_2 as carbon source. In: Gautheret MRJ, Hirth ML (eds) Les Cultures de Tissus de Plantes. Centre Nat Rech Scien, Paris, pp 213–221

Blair LC, Chastain CJ, Widholm JM (1988) Initiation and characterization of a cotton (*Gossypium hirsutum* L.) photoautotrophic cell suspension culture. Plant Cell Reports 7:266–269

Briantais JM, Vernotte C, Krause GH, Weis E (1986) Chlorophyll a fluorescence of higher plants: chloroplasts and leaves. In: Govindjee, Amesz J, Fork DC (eds) Light emission by plants and bacteria. Academic Press, Orlando, pp 539–586

Corduan G (1970) Autotrophe Gewebekulturen von *Ruta graveolens* und deren $^{14}CO_2$-Markierungsprodukte. Planta 91:291–301

Dalton CC (1980) Photoautotrophy of spinach cells in continuous culture: photosynthetic development and sustained photoautotrophic growth. J Exp Bot 31:791–804

Forrester ML, Krotkov G, Nelson CD (1966) Effect of oxygen on photosynthesis, photorespiration and respiration in detached leaves. I. Soybean. Plant Physiol 41:422–427

Hardy T, Chaumont D, Brunel L, Gudin C (1987) Photoautotrophic suspension cultures I – obtention of photoautotrophic cultures from *Euphorbia characias* L. J Plant Physiol 128:11–19

Horn ME, Kameya T, Brotherton JE, Widholm JM (1983a) The use of amino acid analog resistance and plant regeneration ability to select somatic hybrids between *Nicotiana tabacum* and *N. glutinosa*. Mol Gen Genet 192:235–240

Horn ME, Sherrard JH, Widholm JM (1983b) Photoautotrophic growth of soybean cells in suspension culture. I. Establishment of photoautotrophic cultures. Plant Physiol 72:426–429

Horn ME, Widholm JM (1984) Aspects of photosynthetic plant tissue cultures. In: Collins GB, Petolino JG (eds) Applications of genetic engineering to crop improvement. Martinus Nijhoff/Dr. W. Junk, Boston, pp 113–161

Hüsemann W (1985) Photoautotrophic growth of cells in culture. In: Vasil IK (ed) Cell culture and somatic cell genetics of plants, Vol II. Academic Press, Orlando, pp 213–252

Hüsemann W, Barz W (1977) Photoautotrophic growth and photosynthesis in cell suspension cultures of *Chenopodium rubrum*. Physiol Plant 40:77–81

Igbavboa U, Sieweke HJ, Leistner E, Roewer I, Hüsemann W, Barz W (1985) Alternative formation of anthroquinones and lipoquinones in heterotrophic and photoautotrophic cell suspension cultures of *Morinda lucida* Benth. Planta 166:537–544

Katoh K (1983) Kinetics of photoautotrophic growth of *Marchantia polymorpha* cells in suspension culture. Physiol Plant 59:242–248

Kisaki T, Hirabayashi S, Yano N (1973) Effect of the age of tobacco leaves on photosynthesis and photorespiration. Plant Cell Physiol 14:505–514

LaRosa PC, Hasegawa PM, Bressan RA (1984) Photoautotrophic potato cells: transition from heterotrophic to autotrophic growth. Physiol Plant 61:279–286

McHale NA (1985) Conditions for strict autotrophic culture of tobacco callus. Plant Physiol 77:240–242

Murata N, Satoh K (1986) Absorption and fluorescence emission by intact cells, chloroplasts, and chlorophyll-protein complexes. In: Govindjee, Amesz J, Fork DC (eds) Light emission by plants and bacteria. Academic Press, Orlando, pp 137–160

Neumann KH, Bender L (1987) Photosynthesis in cell and tissue culture systems. In: Green CE, Somers DA, Hackett WP, Biesboer DD Plant tissue and cell culture. Alan R. Liss, Inc., New York, pp 151–165

Ogren WL (1984) Photorespiration: pathways, regulation and modification. Ann Rev Plant Physiol 35:415–442

Peel E (1982) Photoautotrophic growth of suspension cultures of *Asparagus officinalis* L. cells in turbidostats. Plant Sci Letters 24:147–155

Rogers SMD, Ogren WL, Widholm JM (1987) Photosynthetic characteristics of a photoautotrophic cell suspension culture of soybean. Plant Physiol 84:1451–1456

Rogers SMD, Widholm JM (1988) Comparison of photosynthetic characteristics of two photoautotrophic cell suspension cultures of soybean. Plant Sci 56:69–74

Sharp RE, Matthews MA, Boyer JS (1984) Kok effect and the quantum yield of photosynthesis, light partially inhibits dark respiration. Plant Physiol 75:95–101

Tyler RT, Kurz WGW, Panchuk BD (1986) Photoautotrophic cell suspension cultures of periwinkle (*Catharanthus roseus* (L.) G. Don): transition from heterotrophic to photoautotrophic growth. Plant Cell Reports 3:195–198

Vu CJV, Bowes G, Allen LH (1986) Effects of light and elevated atmospheric CO_2 on the ribulose bisphosphate carboxylose activity and rubulose bisphosphate level in soybean leaves. Plant Sci 44:119–123

Xu C, Blair LC, Rogers SMD, Govindjee, Widholm JM (1988) Characteristics of five new photoautotrophic suspension cultures including two *Amaranthus* species and a cotton strain growing on ambient CO_2 levels. Plant Physiol 88:1297–1302

Yamada Y, Sato F (1978) The photoautotrophic culture of chlorophyllous cells. Plant Cell Physiol 19:691–699

Yasuda T, Hashimoto T, Sato F, Yamada Y (1980) An efficient method of selecting photoautotrophic cells from cultured heterogeneous cells. Plant Cell Physiol 21:929–932

Inputs and Outputs: Primary Substrates and Secondary Metabolism

R.C. Cresswell, M.W. Fowler, A. Stafford, and G. Stepan-Sarkissian[1]

1 Introduction

Our objectives in the following discussion are to address the question of where the carbon and nitrogen required to synthesize secondary metabolites arise in cultured cells, and how competing biosynthetic pathways may affect the flux of these primary substrates. This chapter is not intended as a review, rather as a means of focussing attention on the poorly understood interrelationships between primary and secondary metabolism.

2 Interaction of the Cultured Cell with its Environment

2.1 The Carbon Source

In the intact plant, CO_2 is assimilated to sucrose as the main translocatable carbon source. In this respect, a sucrose-fed cell culture is *probably* similar to the whole plant in terms of its primary metabolic pathways, although the feeding should be at a concentration similar to that of the whole plant. Sucrose is present in the phloem and cell sap at levels of around 10–25%, considerably higher than conventional concentrations in tissue culture media.

It has been demonstrated that sucrose administered externally to plant cell suspensions is usually rapidly hydrolyzed to glucose and fructose which are then taken up by a passive or active process, the mechanism probably depending on the species. The invertases responsible for sucrose hydrolysis may reside in the external medium, the cell wall or the plasmalemma. Glucose is always more rapidly removed from the medium than fructose, but uptake profiles of the two monosaccharides are correlated (Stepan-Sarkissian and Fowler 1986).

Sucrose can be envisaged as merely an alternative means of supplying glucose. However, reported differences in the effects of sucrose and glucose upon culture growth and secondary metabolism (Fowler and Stepan-Sarkissian 1985) indicate that their mode of uptake and utilization is regulated in a rather different fashion. Reports of the utilization of other glucose-containing dissaccharides should also be considered in this context; both maltose and lactose can be used as an energy source

[1]Wolfson Institute of Biotechnology, The University of Sheffield, Western Bank, Sheffield, S10 2TN, UK

Primary and Secondary Metabolism of Plant Cell Cultures, ed. by W.G.W. Kurz

by certain cell lines such as *Daucus carota* (Verma and Dougall 1977) and *Medicago sativa* (Chaubet and Pareilleux 1982) respectively, though generally less efficiently than sucrose, and of course the primary requisite for their use is the availability of the appropriate hydrolytic enzyme to the external sugar, i.e. maltase or β-galactosidase respectively. Reports of the effects of these alternative carbon sources on secondary metabolism are scarce, but generally they appear to have a detrimental effect. Once glucose has passed through the plasmalemma into the cytoplasm, it can enter primary metabolism via hexokinase as glucose-6-phosphate and from there into glycolysis or the pentose phosphate pathway (Fig. 1). While the fate of glucose is known, the fate of fructose after uptake from the medium can only be speculated. The fact that some cell lines have been shown to grow on fructose for one growth cycle only could be interpreted as evidence for a glucose-mediated fructose-uptake mechanism. Some cell lines can survive indefinitely on a fructose-only medium (Stepan-Sarkissian and Grey, unpublished) however, and are therefore capable of using this sugar as their sole energy source.

The mode of entry of fructose into primary metabolism is therefore unclear for plant cells; it would appear that under most conditions the enzymic capacity for handling fructose is low, though at what point the limiting step takes effect is arguable. It may be, for example, at the point of entry of fructose into the cell, or at the level of fructokinase activity/expression.

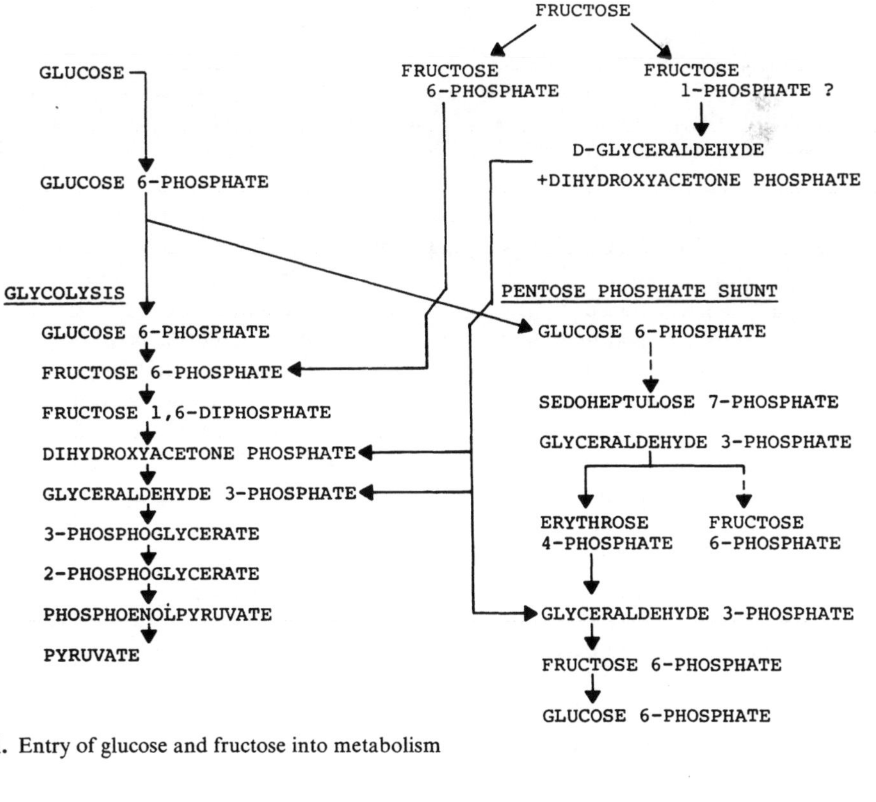

Fig. 1. Entry of glucose and fructose into metabolism

With respect to the fructokinase, two intermediates are possible. Hexokinases, in addition to phosphorylating glucose, will catalyze the formation of fructose-6-phosphate. An alternative enzyme, fructose-1-kinase may also be postulated. This fructokinase operates in liver and catalyzes the phosphorylation of fructose at carbon atom 1, yielding fructose-1-phosphate which can then be cleaved by an aldolase to yield glyceraldehyde and dihydroxyacetone phosphate. The phosphorylated glyceraldehyde and DHAP can then enter into either glycolysis or the pentose phosphate pathway. It is at this point relevant to consider what effect, if any, the production of fructose-1-phosphate as opposed to fructose-6-phosphate as an intermediate may have upon secondary metabolism.

The initial stages of most secondary biosynthetic pathways rely heavily upon the provision of reducing equivalents, which may be furnished via the initial oxidative steps of the pentose phosphate pathway, producing NADPH, or via glycolysis, producing NADH. The point of entry of fructose-1-phosphate into primary metabolism would be either glycolysis or the non-oxidative stage of the pentose phosphate pathway. Fructose-6-phosphate would have the alternatives of entry directly into glycolysis, and into the pentose phosphate pathway or starch synthesis via glucose-6-phosphate. Perhaps in this respect then, fructose-6-phosphate would be the preferred intermediate for the generation of reducing power, in the form of NADPH. It would be of interest to examine the role, if any, of a fructose-1-phosphate intermediate in plant cell metabolism.

We have found that certain cell lines are capable of indefinite growth on a fructose-only medium. In an experiment directed at enhancing glucosidase levels, potato (*Solanum tuberosum*) and Evening Primrose (*Oenothera biennis*) cell suspension cultures were grown on 2% fructose for 11 and 20 subcultures respectively. A comparison with 2% sucrose-grown cultures of the same cell lines showed that significant alterations in α- and β-glucosidase activities occurred even though growth was relatively unaffected. With β-glucosidase in potato, a significant increase in activity in the presence of 2% fructose compared with 2% sucrose was observed, indicating that the specific type of monosaccharide was the primary effector (Fig. 2). With α-glucosidase in Evening Primrose, only a concentration-linked endurance of activity was noted (Fig. 3). These results concur with reports that glucose is a repressor of glucosidase activity (Hughes and Ayre 1976). Though these assays were of non-specific-α- or β-glucosidase activities using the synthetic substrate *p*-nitrophenylglucose, the data suggest that more specific effects upon the activities of β-glucosidases involved in secondary biosynthetic pathways may be expected. Though the enzymic activity of most glycosidases is reversible, hydrolysis is generally the favoured reaction, transferase enzymes probably being responsible for the biosynthesis of glycosides (Hosel 1981). Most naturally occurring plant heterosides are β-linked. The specificities of the β-glycosides are variable but most appear to be fairly specific and have roles, for example, in the deactivation and activation of plant hormones (such as ABA, 2,4-D and gibberellins), and the hydrolysis of cyanogenic glycosides, flavonoid, steroid or monoterpenoid glycosides, and secondary biosynthetic intermediates such as s rictosidine.

The significance of the observed relative increase in glucosidase activity in relation to secondary metabolism can also only be speculated. Glycosylated secondary products differ from aglycones in two main properties, that is they

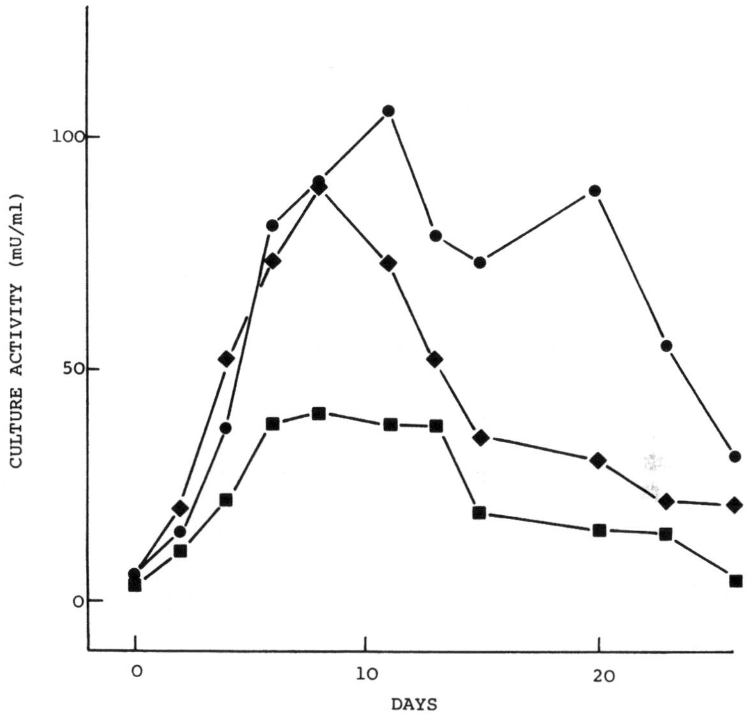

Fig. 2. β-Glucosidase activity in *Solanum tuberosum* cell suspensions. Cultures were grown on M and S medium (Murashige and Skoog 1962) containing 2% sucrose (■), 2% fructose (◆) and 4% fructose (●). Prior to the experiment fructose-adapted cells were cultured for 11 growth cycles on 2 % and 4% fructose media. Cultures were dark-grown in 100 ml volumes at 25°C on a gyrorotatory shaker at 150 rpm. β-Glucosidase assays were performed as described by Hughes and Ayre (1976), and the data shown are combined cell and medium activities

exhibit enhanced water solubility and have decreased chemical reactivity. They are therefore more easily accumulated in the vacuole and less cytotoxic (Matile 1987). The simplest conclusion that could be drawn from the observed enhancement of glucosidase activity by fructose feeding is that any storage of secondary products as glycosides would be reduced, the kinetics favouring hydrolysis, especially when intracellular glucose concentrations are low as would be expected in the later stages of culture growth. A reduction in the potential of cells to store secondary products as glycosides is likely to result in a parallel reduction in the potential for secondary product accumulation if the products are cytotoxic in their aglycone form (Fig. 4).

The importance of secondary product glucosides and glucosidases is, however, further complicated by the numerous demonstrations of spatial separation of glycosides from the enzyme that produces the active component, as with e.g. *Sorghum* leaves, where dhurrin is located in the sap of epidermal cells, whereas the two enzymes responsible for the liberation of cyanide, β-glucosidase and hydroxynitrile lyase, are located in the chloroplasts and cytosol respectively, of mesophyll cells (Thayer and Conn 1981).

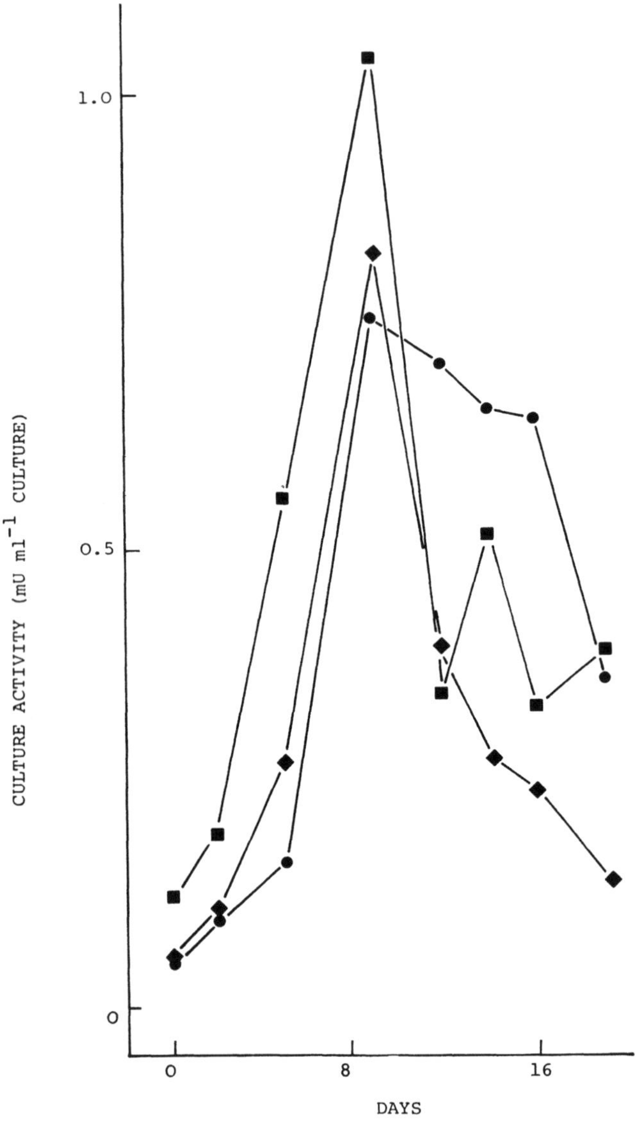

Fig. 3. α-Glucosidase activity in *Oenothera biennis*. Cultures were grown on M and S medium (Murashige and Skoog 1962) containing 2% sucrose (■), 2% fructose (◆) and 4% fructose (●). Prior to experiments fructose-adapted cells were cultured for 20 growth cycles on 2 and 4% fructose media. Cultures were grown in continuous light (200 lux) in 100 ml volumes at 25°C on a gyrorotatory shaker at 150 rpm. α-Glucosidase assays were performed as described by Khan and Eaton (1967), and the data shown are combined cell and medium activities

The effects of growing secondary product accumulating plant cells upon alternative carbon sources has been quite widely explored. As a general rule, sucrose tends to be the most efficacious for secondary metabolic production (Mizukami et al. 1977; Fowler and Stepan-Sarkissian 1985), but precisely how the effect is mediated is unknown. A further point which arises from these observations concerns the situation in sucrose media, where fructose might be expected to be relatively more available after glucose utilization as carbon substrate, later in the linear growth phase. During this time, glycosylation of secondary products may, therefore, be repressed, though at this point they are normally seen to accumulate most rapidly.

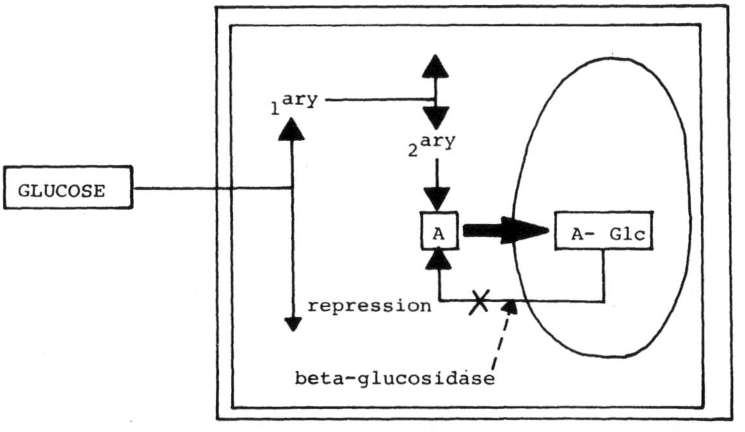

NET RESULT: PRODUCT ACCUMULATION

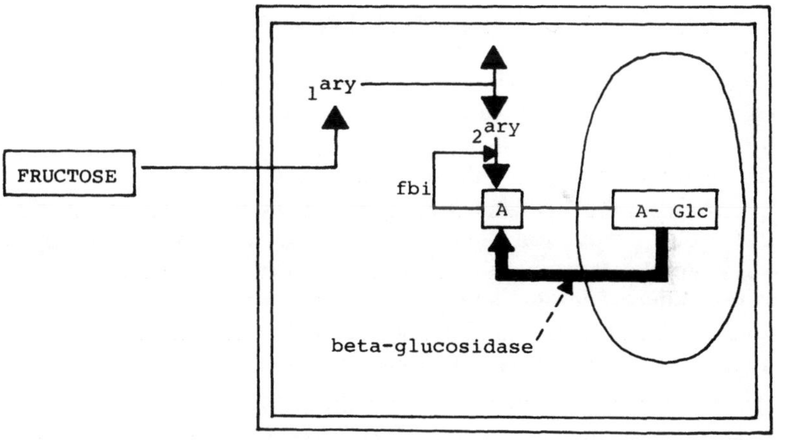

NET RESULT: LOWER LEVELS OF SECONDARY PRODUCT ACCUMULATION OR CELL DEATH

Fig. 4. Hypothesized effects of glucose and fructose on glucosidase activity and secondary product accumulation. *A* represents a secondary product aglycone which is stored normally in the vacuole as a glucoside (*A-Glc*); *Fbi* feedback inhibition

2.2 The Nitrogen Source

The supply of nitrogen to plant cells as nitrate imposes a high energy demand on the assimilation of nitrogen. The supply of nitrogen as ammonia is energetically more efficient than supplying the more oxidized form; the removal of NH_3 from culture medium does, however, lead to a rapid pH drop with the weak buffering usually present in tissue culture media. When NH_3 is present in tissue culture medium, together with other more oxidized forms such as NO_3^-, the NH_3 generally appears to be removed and utilized preferentially.

Other N-sources used alone in tissue culture media include urea, glutamine, glutamate, alanine or casein (Dougall 1980). It seems logical that organic nitrogen sources would satisfy not only nitrogen demand but also the carbon requirement. Amino acids have also been used to supplement other N-sources, and with respect to nitrate reductase activity, culture growth, and especially secondary product synthesis, their recorded effects are variable. It is likely that these organic nitrogen sources have complex and numerous effects, with possible roles as carbon sources, secondary product precursors and regulatory molecules. To extrapolate from *Claviceps* and ergot alkaloid production (Krupuiski et al. 1976), the addition of tryptophan to the fermentation produces a response resembling the phenomenon of enzyme induction. It has been noted for instance that tryptophan analogues, not incorporated into alkaloids, would also stimulate alkaloid production. In addition, tryptophan was required to be added to the culture early during the growth phase in order to stimulate alkaloid production, and has been found to induce secondary biosynthetic enzymes, such as dimethyallyl-tryptophan synthetase, on the ergot alkaloid pathway.

The effects of secondary product precursors in plant cell culture systems are generally diverse and often contradictory. There are two distinct methods of increasing the precursor supply within the cell; firstly, by addition to the medium, in which case the uptake mechanism may/may not be limiting (Table 1), and secondly by selecting for resistance to precursor analogues (or by treating with antimetabolites), in which case the intracellular level may be modified (Table 2).

Both of these approaches have yielded some examples where secondary metabolite production is increased. Frequently, however, no increases in secondary product accumulation are observed and there are several possible explanations for these findings. With precursor feeding experiments (Table 1), it is possible that the precursor is degraded and recycled or is not taken up. Alternatively, other regulatory mechanisms, the non-expression of later parts of the pathway, or

Table 1. Effects of precursor feeding on secondary metabolite accumulation in plant cell cultures

Precursor supplied	Secondary metabolite	Species	Effects	Reference
Tryptamine	Serotonin	*Peganum harmala*	Increased production in low-yielding strain	Sasse et al. (1982)
L-Tryptophan	Alkaloids	*Catharanthus roseus*	Mixed: a) Growth and alkaloid production inhibited b) Increased alkaloid yields	Doller (1978)
L-Tryptophan	Carboline alkaloids	*Cinchona ledgeriana*	Production of carbolines Artefacts of a non-enzymic reaction	Harkes et al. (1986)

Table 2. Effects of precursor overaccumulation in variant cell lines on secondary metabolite accumulation

Precursor overaccumulated	Secondary metabolite	Species and variance[a]	Effects	Reference
Tryptamine	Indole alkaloids	*C. roseus* 4MT-resistant	Accumulation of indole alkaloids in some lines but not in wild-type or some other 4MT-resistant lines	Sasse et al. (1983)
Tryptophan	Indole alkaloids	*C. roseus* 5MT-resistant	Frequently no overaccumulation of alkaloids	Stafford and Smith (1986) Cresswell (1986)
Phenylalanine	Phenolics	*N. tabacum* PFP-resistant	Higher accumulation of phenolics	Berlin (1979)

[a] MT = Methyltryptophan; PFP = parafluorophenylalanine.

absence of other key substrates may prevent conversion of the precursor into secondary metabolites.

Variant cell lines have been selected for tolerance of antimetabolites (especially amino acid analogues) in attempts to yield cultures which overaccumulate precursors such as L-tryptophan and may also, therefore, overaccumulate secondary metabolites. There are some examples where this approach has been successful (Table 2) but frequently no overaccumulation of secondary metabolites is observed in such lines. A primary explanation for this is that the selected cell line may be tolerant by virtue of mechanisms which do not result in overaccumulation of the precursor amino acids. Alternatively, other substrates may be absent, or other regulatory steps further down the biosynthetic pathway may prevent conversion of the overaccumulated precursor into secondary metabolites.

3 Culture Growth and Secondary Metabolism

With respect to energy demand, the culture cycle falls into three phases with arbitrary limits (Fig. 5). In the first phase (A), loading of carbon and nitrogen pools occurs with rapid uptake of sugars and nitrogen from the medium. Oxidation is at a minimum, and monosaccharides are stored as starch (in most species). In the second phase (B), cell division occurs, with oxidation of incoming carbon being the major direction of flux via glucose-6-phosphate. At this point, cell respiration is high. During this phase, in carbon-limited media, some mobilization of carbohydrate reserves may also occur, but the major phase for this process is likely to be phase (C), when these reserves are remobilized and generated as carbon skeletons for both oxidation and the synthesis of structural components and secondary metabolites.

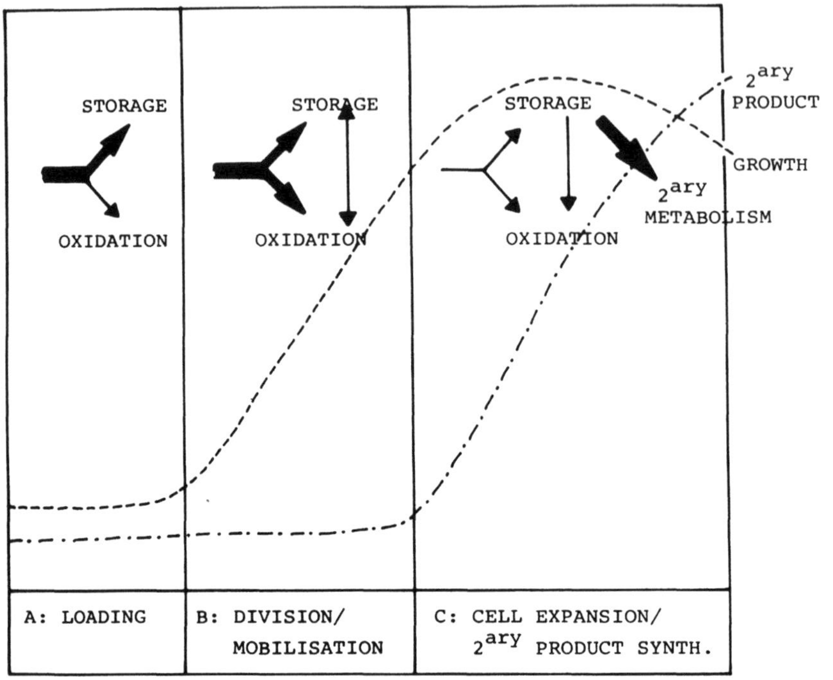

Fig. 5. Carbon flow (as illustrated by arrows) in relation to oxidation, storage and secondary product synthesis during three phases of a batch cell culture cycle

It is often suggested that there is an inverse relationship between growth rate and the rate of secondary product accumulation (Lindsey and Yeoman 1983). Certainly in some culture systems cell growth in terms of biomass increase and secondary product level appear to be more or less distinct phenomena. On one frequently quoted medium, known as M3 (Morris 1986), *C. roseus* cell suspensions accumulate monomeric indole alkaloids predominantly after cell division has ceased, suggesting a possible link between the non-dividing state and the channelling of substrate into secondary biosynthesis. Further support for this possibility is gained from the use of cell cycle inhibitors, such as colchicine, hydroxyurea and trigonelline which we have found to produce transient increases in the level of secondary products, correlated with a suppression of cell division (A. Stafford, unpublished).

Also with *C. roseus* cell cultures, we have observed a linear and direct relationship between the dry mass/cell and serpentine/cell when cultures growing on a hormone-free, 5% sucrose medium for 6 days were supplemented with a range of hormone treatments (Fig. 6). Considering the wide range of hormones used in the experiment, the data are indicative of a relatively tight control on the level of alkaloid that can be accumulated in a cell of a given mass (or vice versa). Suspensions of all treatments contained cells with large plastids, presumably starch grains, on this high carbohydrate medium. Here, there is evidence that in this case at least, the flux of primary precursors into storage and structural biosynthesis

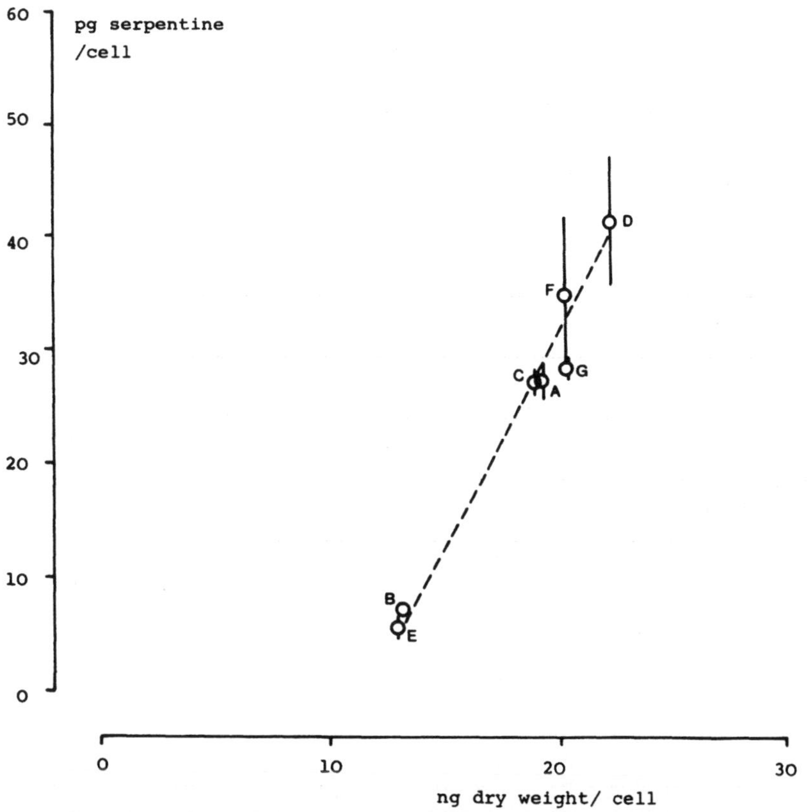

Fig. 6. Relationship between serpentine/cell and dry weight/cell in cell suspensions of *Catharanthus roseus* treated with a range of phytohormone supplements. Cells were grown for 6 days on hormone-free B5 medium (Gamborg et al. 1968) prior to the following additions being made: *A* No addition; *B* zeatin 0.1 mg l^{-1}, NAA 1 mg l^{-1}; *C* zeatin 0.1 mg l^{-1}, IAA 0.175 mg l^{-1}; *D* 2,4-D 1 mg l^{-1}, NAA 1 mg l^{-1}; *E* 2,4-D 1 mg l^{-1}, IAA 0.175 mg l^{-1}; *F* 6-BA 1.125 mg l^{-1}, NAA 1 mg l^{-1}; *G* 6-BA 1.125 mg l^{-1}, IAA 0.175 mg l^{-1}. The graph comprises data from cell numbers, dry weights and cellular serpentine contents sampled 96 h after phytohormone addition

versus secondary product biosynthesis may be stringently controlled perhaps via an early branch-point enzyme, as at this stage of the culture cycle, i.e. 6 days, it is unlikely that the maximum capacity for alkaloid storage/cell has been attained.

4 How Will the Presence/Absence of Differentiated Compartments in Plant Cell or Tissue Cultures Affect Substrate Utilization?

At the subcellular level, undifferentiated cultures may lack refinements in membrane or organelle organization required for the compartmentalization of metabolites and/or enzymes. Furthermore, the frequent inability of such cultures to differentiate, for example, glandular structures, for the accumulation of toxic secondary products, or laticifers for the expression of partial biosynthetic pathways is likely to cause an overall energy flow in the direction of primary metabolism or structural biosynthesis. It follows from this, that a lack of accumulating compartments does not necessarily infer a total absence of secondary biosynthetic potential; the enzymic machinery may indeed be present, only requiring product removal to become totally functional.

In cases where cellular differentiation is also a prerequisite for enzyme activity, as appears to be the case in, for example, morphinan alkaloid synthesis in *Papaver somniferum* cell cultures (Yoshikawa and Furuya 1985), no alteration in the levels or types of primary metabolites are likely to alter the flux through the morphinan alkaloid pathway.

It has been shown, however, that the addition of an accumulating compartment such as a lipophilic phase, e.g. Miglyol (a liquid triglyceride), or ion exchange resins, can alter not only the proportion of secondary product accumulating in the medium but also the overall level accumulated by the total culture (Mulder-Krieger et al. 1988). Examples are the recent use of polymeric adsorbents to stimulate anthraquinone and nicotine production by cell cultures of *Cinchona ledgeriana* (Robins and Rhodes 1986) and root cultures of *Nicotiana rustica* (Rhodes et al. 1986) respectively. The reasons for the enhanced levels may be various, and include:

1. The "protection" of volatile or easily degraded secondary metabolites from the aqueous, oxidative culture environment.
2. Reduction in feedback inhibition at the level of primary or secondary precursor formation: any evidence?
3. Derepression at the level of the genome.

The final part of this chapter concerns related experiments performed with hop cell suspensions (A. Stafford, unpublished). After an exhaustive investigation of medium designs and culture conditions for α-acid accumulation, two cell lines were subjected to a regime in which they were subcultured repeatedly into growth media containing Miglyol. This is a mixed triglyceride of C6, C8, C10 and C12 fatty acids. The reason for choosing this particular "compartment" was to protect any secreted product from extracellular and possibly enzymic oxidation and in addition to provide the cells with an intracellular site for α-acid accumulation if indeed they could take up and tolerate the undegraded triglyceride.

One cell line derived from "Northern Brewer" cultivar exhibited similar growth profiles in both the presence and absence of Miglyol. However, the second cell line derived from cultivar "Target" consistently displayed (over five subcultures) a greater dry weight yield on Miglyol than in its absence (Fig. 7). Using iodine to stain for starch, microscope examination revealed that these cells were packed full of starch grains. The Miglyol added to the medium of this cell line became "waxy" in appearance and at the point of each subculture consisted partially of solid lumps rather than the usual liquid spherical globules.

The observations could be explained as either a stress response of the cells to the lipophilic phase, or a utilization of a portion of the Miglyol via, for example, lipase degradation of the triglycerides liberating glycerol and free fatty acids. The

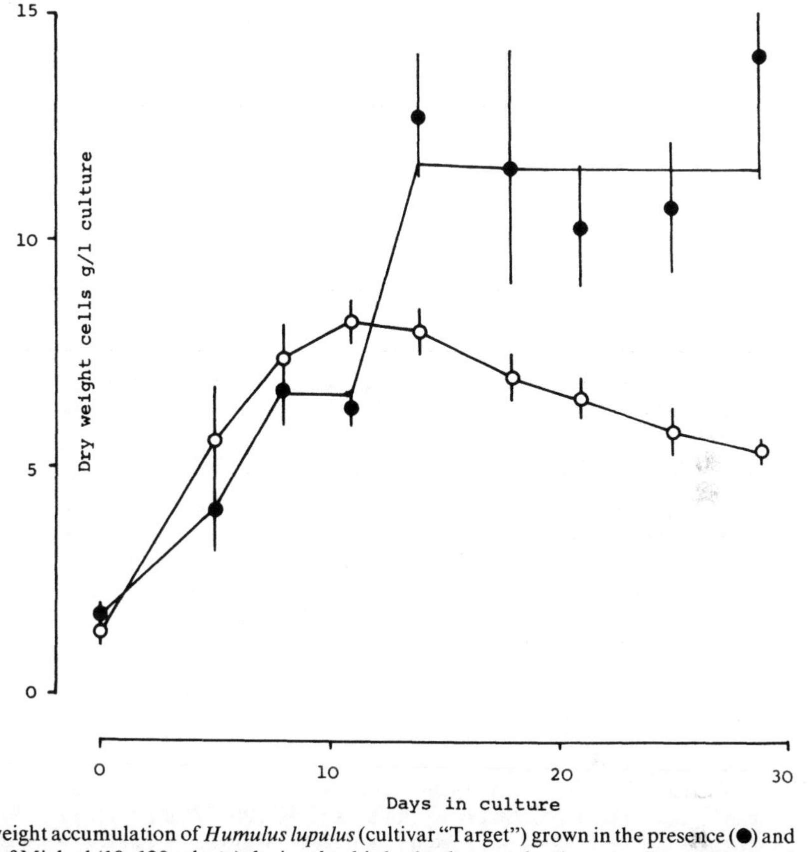

Fig. 7. Dry weight accumulation of *Humulus lupulus* (cultivar "Target") grown in the presence (●) and absence (○) of Miglyol (10: 120 ml v/v) during the third subculture under these conditions. The basal medium was B5 (Gamborg et al. 1968) containing 6-BA 1 mg l^{-1} and 2,4-D 1 mg l^{-1}

use of glycerol as a carbon source by plant cell cultures has been reported (Jones and Velicky 1980), and we are currently investigating the utilization of both glycerol and fatty acids by these cell lines.

Acknowledgements. Financial support for some of the data mentioned was provided by the Ministry of Agriculture, Fisheries and Food, Allied Breweries Ltd., Labatt Brewing Co. Ltd., Lupofresh Ltd. and Harwell A.E.R.E. Thanks go to Dr. D. Grey for providing previously unpublished data.

References

Berlin J (1979) Para-fluorophenylalanine resistant cell lines of tobacco. Z Pflanzenphysiol 97:309

Chaubet N, Pareilleux A (1982) Characterization of β-galactosidases of *Medicago sativa* suspension-cultured cells growing on lactose. Effect of the growth substrates on the activities. Z Pflanzenphysiol 106:401

Cresswell R (1986) Selection studies on *Catharanthus roseus*. In: Morris P, Scragg AH, Stafford A, Fowler MW (eds) Secondary metabolism in plant cell cultures. Cambridge Univ Press, pp 230–236

Doller G (1978) Influence of the medium on the production of serpentine by suspension cultures of *Catharanthus roseus* (L) G Don In: Alfermann W, Reinhard E (eds) Gesellschaft für Strahlen- und Umweltforschung, München, pp 109–117
Dougall DK (1980) Nutrition and metabolism. In: Staba EJ (ed) Plant tissue culture as a source of biochemicals. CRC, Boca Raton, Fla, pp 21–58
Fowler MW, Stepan-Sarkissian G (1985) Carbohydrate source, biomass productivity and natural product yield in cell suspension cultures. In: Neumann KH, Barz W, Reinhard E (eds) Primary and secondary metabolism of plant cell cultures. Springer, Berlin Heidelberg New York, pp 66–73
Gamborg OL, Miller RA, Ojima K (1968) Nutrient requirements of suspension cultures of soybean root cells. Exp Cell Res 50:151
Harkes PAA, De Jong PJ, Wijnsma R, Verpoorte R, Van Der Leer T (1986) Influence of production media on *Cinchona* cell cultures; spontaneous formation of B-carbolines from L-tryptophan. Plant Sci 47:71
Hosel W (1981) Glycosylation and glycosidases. In: Stumpf PK, Conn EE (eds) The biochemistry of plants, vol 7. Secondary plant products. Academic Press, New York London, pp 725–753
Hughes MA, Ayre L (1976) The production of β-glucosidase in cultured cells of *Trifolium repens* L. Plant Sci Lett 7:271
Jones A, Velicky IA (1980) Growth of plant cell suspension cultures on glycerol as a sole source of carbon and energy. Can J Bot 58:648
Khan NA, Eaton NR (1967) Purification and characterization of maltase and α-methyl glycosidase from yeast. Biochim Biophys Acta 146:173
Krupuiski VM, Robbers JE, Floss HG (1976) Physiological study of ergot: induction of alkaloid synthesis by tryptophan at the enzymatic level. J Bacteriol 125:158
Lindsey K, Yeoman MM (1983) The relationship between growth rate, differentiation and alkaloid accumulation in cell cultures. J Exp Bot 34:1055
Matile P (1987) The sap of plant cells. New Phytol 105:1
Mizukami H, Konoshima M, Tabata M (1977) Effect of nutritional factors on shikonin derivative formation in *Lithospermum* callus cultures. Phytochemistry 16:1183
Morris P (1986) Regulation of product synthesis in cell cultures of *Catharanthus roseus*. II. Comparison of production media. Planta Med 1986:77
Mulder-Krieger T, Verpoorte R, Baerheim Svendsen A, Scheffer JJC (1988) Production of essential oils and flavours in plant cell and tissue cultures. A review. Plant Cell Tissue Org Cult 13:85
Murashige T, Skoog F (1962) A revised medium for rapid growth and bioassays with tobacco tissue cultures. Physiol Plant 15:473
Rhodes MJC, Hilton M, Parr AJ, Hamill JD, Robins RJ (1986) Nicotine production by "hairy root" cultures of *Nicotiana rustica*: fermentation and product recovery. Biotech Lett 8:415
Robins RJ, Rhodes MJC (1986) The stimulation of anthroquinone production by *Cinchona ledgeriana* cultures with polymeric adsorbents. Appl Microbiol Biotechnol 24:35
Sasse F, Heckenberg U, Berlin J (1982) Accumulation of B-carboline alkaloids and serotonin by cell cultures of *Peganum harmala*. Z Pflanzenphysiol 105:315
Sasse F, Buchholz M, Berlin J (1983) Selection of cell lines with increased tryptophan decarboxylase activity. Z Naturforsch 38C: 916
Stafford A, Smith L (1986) Effect of modification of the primary precursor level by selection and feeding on indole alkaloid accumulation in *Catharanthus roseus*. In: Morris P, Scragg AH, Stafford A, Fowler MW (eds) Secondary metabolism in plant cell cultures. Cambridge Univ Press, pp 250–256
Stepan-Sarkissian G, Fowler MW (1986) The metabolism and utilisation of carbohydrates by suspension cultures of plant cells. In: Morgan MJ (ed) Carbohydrate metabolism in cultured cells. Plenum, New York London, pp 151–181
Thayer SS, Conn EE (1981) Subcellular localisation of dhurrin β-glycosidase and hydroxynitrile lyase in the mesophyll cells of *Sorghum* leaf blades. Plant Physiol 67:617
Verma DC, Dougall DK (1977) Influence of carbohydrates on quantitative aspects of growth and embryo formation in wild carrot suspension cultures. Plant Physiol 59:81
Yoshikawa T, Furuya T (1985) Morphinan alkaloid production by tissues differentiated from cultured cells of *Papaver somniferum* (1). Planta Med 1985:110

Photoautotrophic Culture: A Model System to Investigate the Regulation of Gene Expression in Cultured Plant Cells

F. Sato, S. Takeda, Y. Shigematsu, N. Koizumi, and Y. Yamada[1]

1 Introduction

Photoautotrophism is one of the outstanding characteristics carried out by chloroplasts in plant cells. This unique property is usually not expressed in cultured plant cells, because they are grown heterotrophically in the dark together with an organic carbon source. Thus, we need the selection of the cells which show high greening to establish the photoautotrophic (PA) cells, which have developed chloroplasts and can grow in the light without any addition of carbon sources [6,8,10,21]. But the development of chloroplasts in PA cells is still limited, as shown when we compared the characteristics of cultured cells with mesophyll cells of green leaves [7,9,22]. Little is known about the mechanisms controlling the chloroplast development especially among mesophyll cells, PA cells and other heterotrophic cells which cannot form fully differentiated chloroplasts. It is not clear whether these deficiencies are based on genetic mutations, modification of genes, or other regulatory mechanisms at the transcriptional or translational level.

Here, we report the characteristics of chloroplast development and of the polypeptides specifically accumulating in cultured green cells and in green leaves. Some analytical results on the chloroplast DNA and nuclear genome were obtained by restriction endonuclease treatment. Southern hybridization is also described and discussed in connection with photoautotrophism of plant cells.

2 Plant and Cell Materials

Photomixotrophic (PM) cultures of tobacco (*Nicotiana tabacum* cv. Samsun NN) were maintained in modified Linsmaier and Skoog liquid media containing 3% sucrose [21]. PA cells were maintained in the same media as those of PM cells without sucrose but with CO_2 enriched (1–2%) air using two-tier flasks as described previously [15]. Intact plants of tobacco (*Nicotiana tabacum* cv. Samsun NN) were grown in soil in a greenhouse for 2–3 months under natural illumination.

[1]Research Center for Cell and Tissue Culture, Faculty of Agriculture, Kyoto University, Kyoto 606, Japan

Primary and Secondary Metabolism of Plant Cell Cultures, ed. by W.G.W. Kurz

3 Changes in Chl Contents and Chloroplast Number During the Growth of Cultured Cells and During the Development of Leaves

Analyses revealed that the higher Chl contents of green leaves compared with cultured cells on a fresh weight basis were mainly due to the smaller cell volume of mesophyll cells than cultured cells (Fig. 1A,B). The Chl content per chloroplast was almost the same, while the chloroplast number per cell differed and accounted for the small differences between the two cell types tested. Our results indicate that the changes in Chl content of cells during growth were mainly due to changes in chloroplast number per cell in all cell types.

4 Changes in Activities of Photochemical Reactions During Growth

Our measurements of photosynthetic activity showed differences in the development of chloroplasts between leaves and cultured cells (Fig. 1C). The maximum activities of PSI and PSII in cultured cells were about half of those of green leaves on a Chl basis. Moreover, PSI and PSII activities of leaves changed simultaneously with the change of the Chl content on a fresh weight basis, while those of cultured cells did not closely follow the changes in Chl content. PA cells showed relatively constant activities of PSI and PSII. PM cells showed the maximum activities in the early exponential phase, and then the activities decreased.

5 Analysis of Thylakoid Membrane Polypeptides

LDS (Lithium dodesyl sulphate)-PAGE showed that the compositions of thylakoid polypeptides of cultured cells were almost similar to those of leaves at any growth phase, while the decreases in several polypeptides, such as α- and β-subunit of CF1 (coupling factor 1), LHC(I), LHC(II) (chlorophyll-binding polypeptides of the light-harvesting complex of PSI and PSII respectively), and polypeptides of the reaction center of PSI and PSII and so on, were obvious in cultured green cells, especially in PM cells. We concluded that the lower photosynthetic activity of cultured cells was mainly due to less expression of the thylakoid polypeptides. However, the changes in photosynthetic activities during growth were not explained by the changes in the amount of thylakoid polypeptides.

6 Analysis of Total Cellular Proteins by 2D-PAGE and NH_2-Terminal Amino Acid Sequencing

When we analyzed the total cellular proteins by SDS-PAGE, those proteins in cultured green cells also showed some similarity to those in mesophyll cells, but also some differences. Further separation of total cellular polypeptides was done by

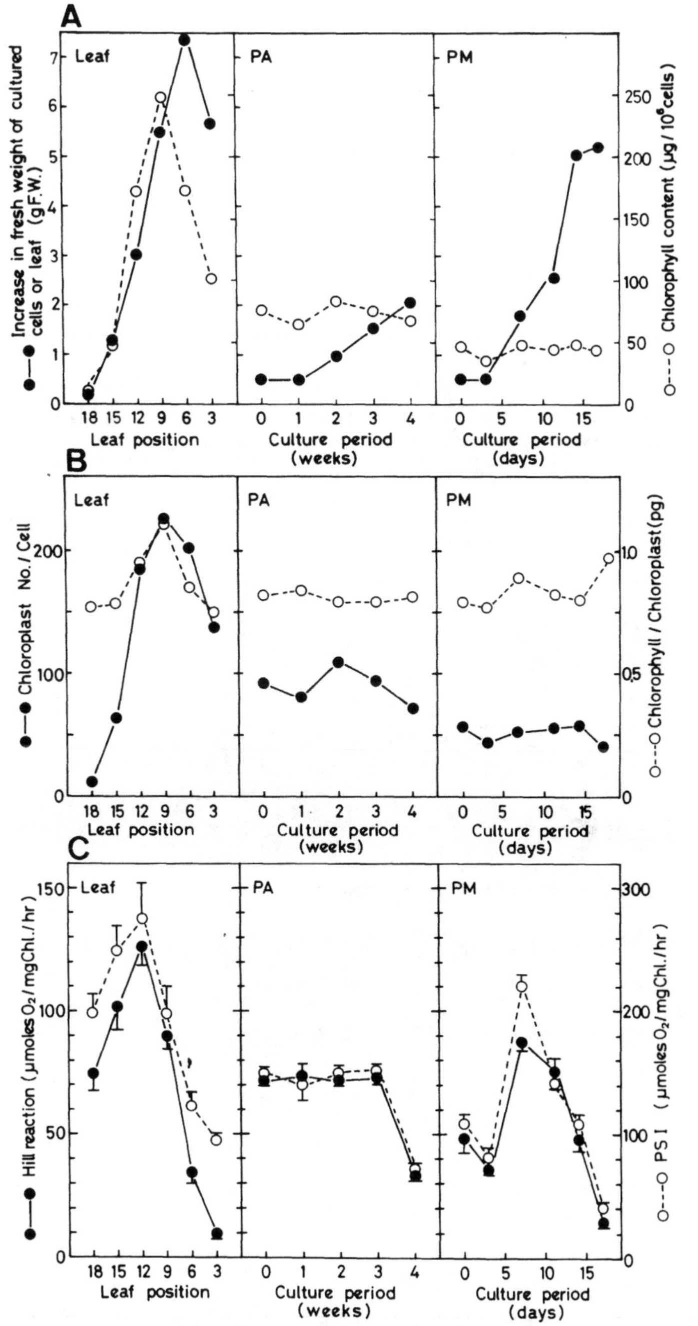

Fig. 1A-C. Characteristics of chloroplast development of cultured green cells and leaves of tobacco. **A** Increases in fresh weights and changes in Ch l contents per cell of leaves and suspension cultured cells of tobacco. Leaf position denotes the number of leaves counted from the primary leaf. **B** Changes in chloroplast number per cell and Ch l contents per chloroplast during the growth of leaves and cultured cells of tobacco. **C** Changes in photosynthetic activities during the growth of leaves and cultured cells of tobacco. *Bars* indicate the standard deviation of the activities measured

2D-PAGE..The separated proteins were electroblotted on polyvinyliden difluoride (PVDF) membranes [1,11], then visualized with Coomasie Brilliant Blue.

Visual comparison of the relative intensities of the protein spots allow us to identify the major proteins of which the amounts changed among cell types (Fig. 2). Protein L1 and L2 were present in a relatively larger amount in mesophyll cells than in cultured cells, while cultured cells accumulated protein P1-4 in relatively higher amounts than mesophyll cells. Protein L1 and L2 were identified with the large and small subunits of ribulose 1,5-bisphosphate carboxylase (RuBisCO) by the immunoblotting method using tobacco RuBisCO specific antiserum, respectively.

We sequenced some of the major proteins to identify the kind of protein predominantly accumulating in mesophyll and cultured cells. PVDF membrane-binding proteins were cut into small pieces and subjected directly to Edmann degradation to sequence NH_2 terminal amino acid with the Applied Biosystem, Model 477A sequencer, equipped with an on-line PTH analyzer using the regular sequence program.

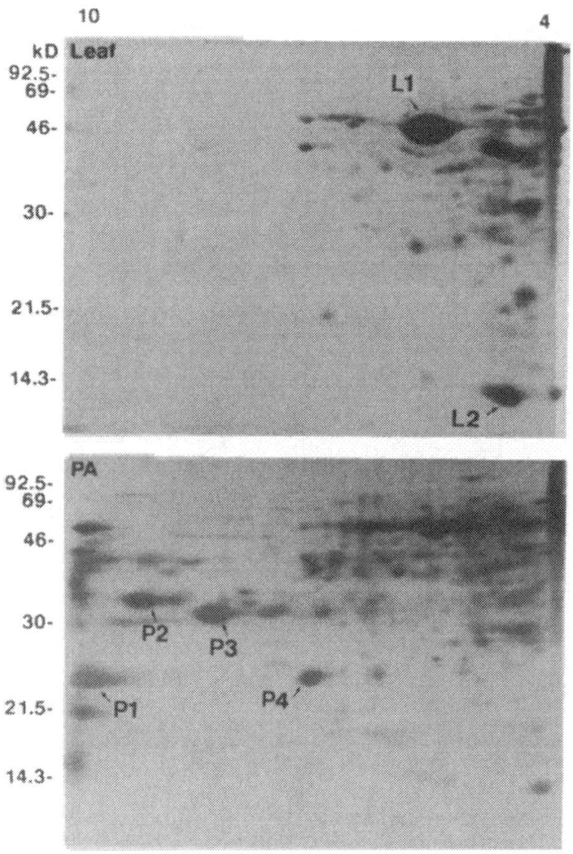

Fig. 2. PVDF membrane blots of proteins separated by 2D-PAGE. Total cellular protein extracts of leaf (*leaf*) and photoautotrophic (*PA*) cells are separated by pre-equilibrium pH gel electrophoresis (horizontal direction) and SDS-PAGE (vertical direction). Proteins were electroblotted onto PVDF membranes and detected by Coomasie Blue. pI values, measured in the first-dimension gel, and molecular weight (*KD*) markers used in the second-dimension gel are indicated at the top and left side of the figure. Proteins selected for sequence analysis are numbered

(A)

```
P1  A T I - E V R N N X P Y T V X A A S T P I G G G R R L D R G Q T X V I N A P
OSM A T I - E V R N N C P Y T V W A A S T P I G G G R R L D R G Q T W V I N A P
P4  S G V F E V H N N X P Y T V W A A A T P V G G G R R L E X G Q S W X F X A P
TPR A T - F D I V N Q C T Y T V W A A A S P - G G G R Q L N S G Q S W S I N V N
```

(B)

```
P2  E Q X G S Q A G G A R X P S G L X X S K F G X X G N T N D Y X G P G N
CHI E Q C G S Q A G G A R C A S G L C C S K F G W C G N T N D Y C G P G N
```

Fig. 3. A Amino acid sequence alignment of protein *P1* and *P4*, osmotin I (*OSM*), thaumatin (*THA*) and tobacco pathogenesis-related protein (*TPR*). **B** Amino acid sequence alignment of protein *P2*, and tobacco-chitinase (*CHI*). Homologous amino acids are enclosed in *boxes*. Gaps (—) are indicated which produce optimal alignment

In the sequencing, P3 showed that the NH_2 terminal was blocked, and did not yield any amino acid sequence. The remaining P1, P2 and P4 spots could be sequenced for 35–38 cycles. The average repetitive sequencing yield, calculated from the recoveries of PTH-alanine and valine of the P1 protein at positions 1, 15, and 5, 13 respectively was more than 90%.

All protein sequences obtained were compared with the information in the Protein Sequence Data Bank. Our search revealed significant homology between the NH_2-terminal sequences of three *Nicotiana* proteins and the corresponding region of previously identified proteins (Fig. 3): (1) the NH_2-terminal sequence of protein P1, containing 37 amino acid residues, is identical to the NH_2-terminal amino acid sequence of osmotin except for three residues which were not identified in our sample. The sequence containing residues of P4 also showed high homology to the NH_2-terminal of osmotin. Osmotin has been shown to accumulate in NaCl-adapted cells [18]. (2) The sequence-containing residue of P2 showed very high homology with the NH_2 terminal of tobacco chitinase [19].

Osmotin is known to be induced by NaCl stress, while chitinase is known to be induced as a pathogenesis-related protein and also induced by the plant hormone ethylene. The proteins, P1-P4, were found not only in PA and PM cells, but also in newly established suspension cells as well as in regenerating adventitious shoots grown in vitro. These results suggested that in vitro cultured cells are subjected to a highly stressed condition, although the factors stressing cultured cells are still obscure.

7 A Possibility that Alternation of Chloroplast or Nuclear Genome Contributes to the Changes in the Gene Expression in Chloroplast

Chloroplasts have their own DNA and replicate semi-autonomously. In recent years, some deletion or methylation of chloroplast (plastid) DNA was reported to cause the albino formation or suppression of chloroplast development. We examined the possibility whether the alternation of gene structure of chloroplast and nuclear DNA affects the gene expression in chloroplast function in our cultured PM/PA cells.

Chloroplast DNA was isolated from PM cells and green leaves by the method of Hermann [5]. The chloroplast DNAs of both cultured cells and leaves were digested with 12 restriction endonucleases, and separated by agarose gel electrophoresis. The restriction fragment patterns of chloroplast DNA were found to be indistinguishable from those of chloroplasts from leaves of the mother variety.

Southern hybridization with a probe for a large subunit of RuBisCo also gave a good agreement among them. Furthermore, the nucleotide sequence of chloroplast *psbA* gene isolated from cultured tobacco cells was completely identical to that of the mother variety.

The structure of nuclear DNA was also analyzed by Southern hybridization with a probe for rDNA after an endonuclease (Eco RI or Hind III) treatment. The results indicated that the restriction pattern of nuclear rDNAs of PM and mesophyll cells was identical, while the amount of rDNA in nuclei of cultured cells was about one-fourth of that of mesophyll cells.

8 Discussion

Our cultured tobacco cells used can grow without any addition of carbon sources. Many analyses, however, showed that the development of chloroplasts in cultured cells, even in PA cells, was depressed in the comparison with that in mesophyll cells. The biggest difference was the Chl content of cells on a fresh weight basis. Cultured green cells with a comparable content of Chl to that of mesophyll cells on a fresh weight basis have not been reported as yet, except for one cell line of soybean [6,8,10,13,17,21]. The difference in Chl content per chloroplast, however, was slight among cells examined (Fig. 1B). Chloroplast number in a cell also had small effects on the low Chl content of our cultured cells on a fresh weight basis. It was clear that the difference in cell volume between mesophyll cells and cultured cells determined the difference in Chl content on a fresh weight basis among them. The changes in Chl content of cells during growth, however, were explained by the changes in chloroplast number per cell.

Measurement of activities of photochemical reactions also revealed some differences between cultured green cells and mesophyll cells. In green leaves, these activities changed simultaneously with the change in Chl content on a fresh weight basis as reported previously in wheat and soybean leaves [2,4]. In cultured cells, the changes in activities of photochemical reactions did not closely follow the changes in Chl content. These results suggested that in green leaves, the Chl biosynthesis and degradation are almost coincident with the photosynthetic function, while in cultured cells, these parameters were not cooperative.

The large cell volume of cultured cells could contribute to the low photosynthetic activities, especially on the high CO_2 compensation point. Less activities of RuBisCo and carbonic anhydrase and high respiration activity have been thought to be the main reason for the high CO_2 compensation point [9,12–14,20]. We propose that the random distribution of chloroplasts in cultured cells also affects the photosynthetic activity since CO_2 has to diffuse a longer way from the plasma membrane to chloroplasts.

Analyses of the thylakoid membrane and stromal proteins showed that chloroplasts of cultured cells had almost a similar composition of polypeptides to that of mesophyll cells except the decrease in some polypeptides, such as the α- and β-subunit of CF1, LHC (I), LHC(II), and polypeptides of the reaction center of PSI and PSII and so on, especially in PM cells. Thus, the decreases in thylakoid polypeptide are more responsible for the lower photochemical activities of cultured cells than those of leaves on a Chl basis.

All analyses mentioned above indicated that the chloroplasts were less developed and chloroplast activity was lower in cultured green cells, even if they are grown under PA conditions. One of the possibilities to explain the reduction in gene expression in cultured cells is that chloroplast or nuclear genes coding chloroplast proteins were mutated or modified. However, our analyses on chloroplast DNA and nuclear genomes suggested that an alternation of gene structure of the chloroplast or nuclear genome rarely occurred in our PA/PM cultured cells.

The copy number of chloroplast DNA was also reported to be comparable with that of mesophyll cells [3]. Interestingly, the amount of rDNA of PA cells, however, was less than that in mesophyll cells. Our PM cells grow rather slowly compared with the cultured cells freshly established from the same variety of tobacco. Slow growth might allow good development of chloroplasts. The relationship of the reduced amount of rDNA and photoautotrophy of this line will be examined in the future.

The above data indicated that some regulatory mechanisms on transcription and/or translation processes are regulating the low gene expression of chloroplasts in cultured cells. Analyses of total cellular protein indicated clearly that cultured plant cells are under stress. However, it is not clear what stress factors are most severe for cultured cells, but it is highly probable that some stress depresses the gene expression and chloroplast development. A better understanding of the regulatory systems controlling the cultured cells is essential for the modification of the gene expression of cultured cells to increase their productivity.

Photoautotrophic cells would not only be a good model system to investigate the gene expression of plant cells in vitro, but it would also be useful for the selection of mutants with a modified function of the chloroplasts. We recently successfully isolated an atrazine-resistant mutant with a mutation in the gene *psbA* [16].

Intact plants could not be regenerated from our resistant cell line, because our photomixotrophic cells used have been maintained more than 10 years after callus induction. However, we expect that this mutated gene will prove useful for plant breeding by the recent advances in protoplast fusion techniques. We believe that cultured cells offer a valuable system for selection of new mutations in the chloroplast genome as well as in the nuclear genome.

Acknowledgements. We are grateful to Dr. T. Eriksson for his kind reviewing of this manuscript. This research was supported in part by a Grant-in-Aid for Scientific Research on Priority Areas from the Ministry of Education, Science and Culture, Japan.

References

1. Bauw G, De Loose M, Inze D, Van Montagu M, Vandekerckhove J (1987) Alternations in the phenotype of plant cells studied by NH_2-terminal amino acid-sequence analysis of proteins electroblotted from two-dimensional gel-separated total extracts. Proc Natl Acad Sci USA 84:4806–4810
2. Camp PJ, Huber SC, Burke JJ, Moreland DE (1982) Biochemical changes that occur during senescence of wheat leaves: basis for the reduction of photosynthesis. Plant Physiol 70:1641–1646
3. Cannon G, Heinhorst S, Siedlecki J, Weissbach A (1985) Chloroplast DNA synthesis in light and dark grown cultured *Nicotiana tabacum* cells as determined by molecular hybridization. Plant Cell Rep 4:41–45
4. Ford DM, Shibles R (1988) Photosynthesis and other traits in relation to chloroplast number during soybean leaf senescence. Plant Physiol 86:108–111
5. Hermann RG (1982) The preparation of circular DNA from plastids. In: Edelman M, Hallick RB, Chua N-H (eds) Methods in chloroplast molecular biology. Elsevier Biomedical Press, Amsterdam, pp 259–280
6. Horn ME, Sherrard JH, Widholm JM (1983) Photoautotrophic growth of soybean cells in suspension culture. I. Establishment of photoautotrophic cultures. Plant Physiol 72:426–429
7. Horn ME, Widholm J (1984) Aspects of photosynthetic plant tissue cultures. In: Collins GB, Petolino JG (eds) Applications of genetic engineering to crop improvement. Nijholf, Junk, Boston, pp 113–161
8. Husemann W, Barz W (1977) Photoautotrophic growth and photosynthesis in cell suspension cultures of *Chenopodium rubrum*. Physiol Plant 40:77–81
9. Husemann W, Plohr A, Barz W (1979) Photosynthetic characteristics of photomixotrophic and photoautotrophic cell suspension cultures of *Chenopodium rubrum*. Protoplasma 100:101–112
10. La Rosa PC, Hasegawa PM, Bressan RA (1984) Photoautotrophic potato cells: transition from heterotrophic to autotrophic growth. Physiol Plant 61:279–286
11. Matsudaira P (1987) Sequence from picomole quantities of proteins electroblotted onto polyvinyliden difluoride membranes. J Biol Chem 262:10035–10038
12. Nato A, Mathieu Y, Brangeon J (1981) Heterotrophic tobacco cell cultures during greening II, Physiological and biochemical aspects. Physiol Plant 53:335–341
13. Roger SMD, Ogren WL, Widholm JM (1987) Photosynthetic characteristics of a photoautotrophic cell suspension culture of soybean. Plant Physiol 84:1451–1456
14. Sato F, Nishida K, Yamada Y (1980) Activities of carboxylation enzymes and products of $^{14}CO_2$ fixation in photoautotrophically cultured cells. Plant Sci Lett 20:91–97
15. Sato F, Takeda S, Yamada Y (1987) A comparison of effects of several herbicides on photoautotrophic, photomixotrophic and heterotrophic cultured tobacco cells and seedlings. Plant Cell Rep 6:401–404
16. Sato F, Shigematsu Y, Yamada Y (1988) Selection of an atrazine-resistant tobacco cell line having a mutant *psbA* gene. Mol Gen Genet (in press)
17. Seeni S, Gnanam A (1982) Growth of photoheterotrophic cells of peanut (*Arachis hypogaea L.*) in still nutrient medium. Plant Physiol 70:815–822
18. Singh NK, Bracker CA, Hasegawa PM, Handa AK, Buckel S, Hermodson MA, Pfankock E, Regnier FE, Bressan RA (1987) Characterization of osmotin: a thaumatin-like protein associated with osmotic adaptation in plant cells. Plant Physiol 85:529–536
19. Shinshi H, Mhonen D, Meins F Jr (1987) Regulation of a plant pathogenesis-related enzyme: inhibition of chitinase and chitinase mRNA accumulation in cultured tobacco tissues by auxin and cytokinin. Proc Natl Acad Sci USA 84:89–93
20. Tsuzuki M, Miyachi S, Sato F, Yamada Y (1981) Photosynthetic characteristics and carbonic anhydrase activity in cells cultured photoautotrophically and mixotrophically and cells isolated from leaves. Plant Cell Physiol 22:51–57
21. Yamada Y, Sato F (1978) The photoautotrophic culture of chlorophyllous cells. Plant Cell Physiol 19:691–699
22. Yamada Y, Sato F, Watanabe K (1982) Photosynthetic carbon metabolism in cultured photoautotrophic cells. In: Fujiwara A (ed) Plant tissue and cell culture. Maruzen, Tokyo, pp 249–250

Photoautotrophic Plant Cell Cultures for Studies on Primary and Secondary Metabolism

W. Hüsemann[1], K. Fischer[1], I. Mittelbach[1], S. Hübner[1], G. Richter[2], and W. Barz[1]

1 Introduction

Photoautotrophic growth of in vitro cultured plant cells in a sugar-free medium but in the presence of CO_2-enriched air (1–2%, v/v) has been achieved for about 25 different plant species with the exception of the Gramineae.

Photoautotrophic plant cell suspension cultures "reduce" a complex plant organism to a single cell level with regard to its growth requirements and its photosynthetic capacities, including chloroplast differentiation. This facilitates the use of photoautotrophic cell cultures to study the physiology, biochemistry and molecular biology of the green plant cell. The following topics are presently studied by our research groups:

Regulation of carboxylation reactions and photosynthetic carbon metabolism; physiological, biochemical and molecular biological events in chloroplast differentiation; the involvement of the chloroplast in secondary product formation; viroid infection, viroid replication and its effect on the photosynthetic potential of the green plant cell; formation of phytoalexins in photoautotrophic in comparison to heterotrophic cells; isolation and characterization of herbicide-resistant mutants; comparative metabolism of xenobiochemicals in heterotrophic and photoautotrophic plant cell cultures.

This chapter mainly describes recent results to characterize the physiological, biochemical and molecular biological reactions involved in the expression and maintenance of photoautotrophism in cell suspension cultures from *Chenopodium rubrum.* The question as to whether the physiological and biochemical properties of photoautotrophically cultured plant cells resemble those of mesophyll cells, can now be affirmed for many respects. This is especially true for the composition of the chlorophylls, carotenoids, lipids and their constituent fatty acids as well as for membrane transport capacities and photosynthetic capacities (Hüsemann 1985, 1988a, b). Yet, there are some remarkable differences between photoautotrophic plant cell cultures and the intact plants. This concerns the demand for high CO_2 concentrations (1–2%; v/v) for the induction and maintenance of photoautotrophic growth, the high amounts of malate formed by the PEP-carboxylase reaction and finally the blue light-dependent chloroplast differentiation.

[1]Lehrstuhl für Biochemie der Pflanzen, Westfälische-Wilhelms-Universität Münster, Hindenburgplatz 55, D-4400 Münster, FRG
[2]Institut für Botanik, Universität Hannover, Herrenhäuser Straße 2, D-3000 Hannover, FRG

2 Chloroplast Differentiation and Photoautotrophic Growth in Cell Cultures from *Chenopodium rubrum* Under Blue and Red Light Irradiation

Photoautotrophic cell suspension cultures from *C. rubrum* have been used to test whether the blue light dependency of chloroplast differentiation is a phenomenon also observed in photoautotrophically cultured plant cells or whether it is restricted to cells growing heterotrophically in a complex medium in the presence of sugar, phytohormones and vitamins (Richter et al. 1984, 1986, 1987).

Cell suspension cultures from *C. rubrum* growing in a simple mineral salt medium in the presence of 2% (v/v) CO_2 were transferred from white light (24 W m^{-2}) to either red light (fluorescent tubes, Philips TLD 36/15; E_{max} 660 nm) or blue light (fluorescent tubes, Philips TLD 36/18; E_{max} 440 nm) of equal energy flux rate (14 W m^{-2}). Exposure to red light of the photoautotrophic cells blocked their chlorophyll synthesis and impaired their growth under photoautotrophic conditions. Chlorophyll content of such red-light grown cells was less than half that observed in cells cultured under blue light.

Cloned chloroplast genes encoding for the large subunit of the ribulose-bisphosphate carboxylase (RuBP-carboxylase) were used to study the effect of the two light qualities on the steady state level of the complementary mRNA sequences in photoautotrophic cells (Fig. 1). For the mRNAs of the small subunit of the RuBP-carboxylase and the light-harvesting chlorophyll a,b-protein (LHCP), the analyses established a similar rate of accumulation, which only occurred to a much

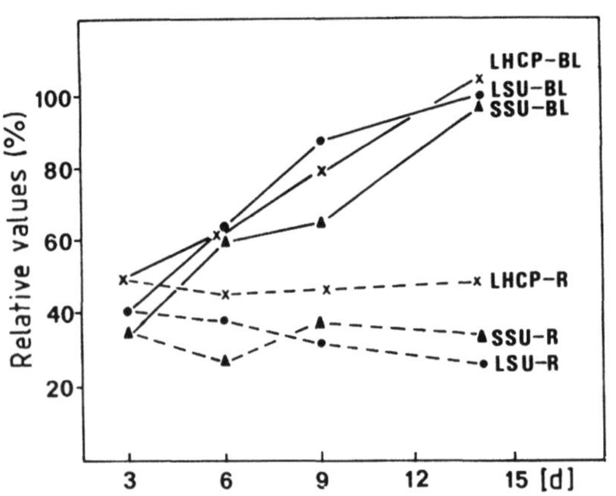

Fig. 1. Determination by dot hybridization of changes in the steady-state concentration of the mRNA of the large subunit (*LSU*) and small subunit (*SSU*) of the RuBP-carboxylase and the light-harvesting chlorophyll a,b protein (*LHCP*) during blue light (*BL*) and red light (*R*) irradiation of photoautotrophically cultured cells from *Chenopodium rubrum*. Total RNA was isolated 3,6,9 and 14 days after the onset of irradiation and analyzed. Total RNA applied to nitrocellulose membrane; hybridization to the ^{32}P-labelled DNA of pSA 204 containing the LSU gene and of clone pNTS 142/1 containing the LHCP. Dots were cut out and their radioactivity was counted; the average ^{32}P-label in hybrids as a function of the RNA applied was calculated. The corresponding value for RNA preparation of fully greened cells (day 15) was set at 100%, and all other values referred to it

lower degree under red-light conditions (Fig. 1). These results indicate that chloroplast differentiation is induced and predominantly maintained by blue light. The mRNA level is under blue light control, but how close is this photoprocess to transcription?

We succeeded in isolating and purifying a transcriptional complex (transcriptional active chromosome; TAC) from chloroplasts of suspension cultured cells from both blue- and red-light irradiated photoautotrophic cell cultures from *C. rubrum* following the procedure of Reiss and Link (1985). The plastid transcripts in the RNA synthesized by the TAC from chloroplasts of either blue- or red-light irradiated photoautotrophic cells were analyzed and quantified by hybridization with cloned chloroplast DNA fragments using gel transfer techniques as well as dot hybridization. The labelled transcripts of both TAC systems generated about the same pattern of radioactive dots but with a much higher intensity for RNA synthesized from TAC from the chloroplasts of blue-light irradiated cells when compared to red-light grown cells. The ^{32}P-label of the hybrids and hence the amounts of gene-specific transcripts synthesized by TAC from blue-light irradiated cells were approximately 40% higher compared to those produced by TAC from red-light grown cells. Possibly, the RNA polymerase activity from blue-light cells is much higher than those from red-light cells. Though chlorophyll formation and steady-state levels of mRNA encoding for prominant chloroplast proteins are extremely reduced in red-light irradiated photoautotrophic cell cultures compared to blue-light treated cells, photoautotrophic growth did not cease completely after a 14-day period of irradiation with red light. Therefore, it seemed reasonable to try to select cell lines capable of unlimited photoautotrophic growth under red-light conditions. We succeeded in establishing cell cultures from *C. rubrum*, which have now been growing photoautotrophically for more than 6 months exclusively under red or blue light of equal photon flux density (27 μEinstein m^{-2} s^{-1}). Our experimental data clearly indicate that light-dependent chloroplast differentiation in cultured plant cells changes according to the nutritional status of the cells. Although the induction of chloroplast differentiation from etioplast/leucoplasts in heterotrophically grown cells is strictly blue light-dependent (Richter et al. 1986), the maintenance of chloroplast structure and photosynthetic function proceeds both under blue and red light (Hübner 1988).

Photoautotrophic growth in cell suspension cultures from *C. rubrum* is drastically reduced by blue-light irradiation compared to red light of equal photon flux density (Table 1). This leads to the question whether growth inhibition is due to an overall reduction of photosynthetic CO_2 assimilation and/or to profound alterations in the formation of individual photosynthetic products. We found that photosynthetic CO_2 assimilation of photoautotrophic cell cultures from *C. rubrum* is drastically reduced by blue light compared to red light when calculated on a chlorophyll basis (Table 2). At the moment we cannot decide whether the low photosynthetic rates under blue light are due to restrictions in photochemical and/or biochemical capacities of the chloroplasts. A detailed study of the formation of ^{14}C-labelled photosynthetic products after $^{14}CO_2$ photosynthesis only revealed some quantitative differences in the photosynthetic carbon incorporation into the main fractions (Table 2). Blue light stimulated ^{14}C-incorporation into amino acids and proteins at the expense of ^{14}C-labelling of organic acids and

Table 1. Some physiological characteristics of photoautotrophic cell cultures from *Chenopodium rubrum* grown under blue and red light[a]

	Blue-light		Red-light	
	Culture Age (d)			
	7	21	7	21
Culture growth:				
(Increase in cell number %)	15	176	95	413
Chlorophyll: $\mu g/10^6$ cells	25.5	36.0	17.6	22.3
Protein: $\mu g/10^6$ cells	306	254	248	205
Enzyme activities				
$\left(\frac{\text{nmol substrate}}{\text{mg protein} \times \text{min}}\right)$				
RuBP-carboxylase	210	340	300	208
PEP-carboxylase	70	49	86	50
CO_2-assimilation	8.2	10.5	17.8	21.0
(μmol CO_2/mg CHL. $\times$ h)				

[a] Blue and red light were of equal photon flux density (27 μEinstein m^{-2} s^{-1}). The cells were photoautotrophically grown in a sugar-free mineral salt medium in the presence of 2% (v/v) CO_2. The data are the mean of four independent experiments.

Table 2. Pattern of ^{14}C-labelled photosynthetic products after $^{14}CO_2$ photosynthesis in photoautotrophic cell cultures from *Chenopodium rubrum* under blue and red light[a]

^{14}C-Label distribution (%)	Blue light	Red light
Lipid/pigment fraction	5.6 ± 0.4	4.7 ± 0.8
Ion-exchange chromatography		
Basic fraction: amino acids	32.5 ± 2.7	26.1 ± 3.0
Acidic fraction:		
Organic acids	9.2 ± 2.5	15.5 ± 3.2
Sugar phosphates	31.7 ± 4.2	30.0 ± 4.1
Neutral fraction: sugars	8.5 ± 2.8	12.7 ± 2.1
Starch fraction	2.7 ± 0.3	4.0 ± 0.8
Protein fraction	7.1 ± 0.4	6.0 ± 0.4
Insoluble residue	3.5 ± 0.8	1.0 ± 0.2

[a] Blue and red light were of equal photon flux density (27 μEinstein m^{-2} s^{-1}). Rapidly dividing cells (day 7–8) were used. $^{14}CO_2$ photosynthesis was performed under standard culture conditions (i.e. pH and composition of the culture medium in which the cells had been grown before, 2% (v/v) CO_2 in gas phase in the presence of 180 kBq $^{14}CO_2$). The data are the mean of five independent experiments. Variations of the average are expressed as the standard error of the mean.

carbohydrates when compared to red light. At the present stage of knowledge, the question cannot be answered whether blue-light induced growth inhibition in photoautotrophic cell cultures from *C. rubrum* is only due to a reduction in the photosynthetic CO_2 assimilation and product formation or whether non-photosynthetic blue light effects on cell metabolism must also be considered (Ruyters 1984).

3 Photosynthetic Studies: pH-Dependent Changes in the Mode of CO_2 Fixation in Photoautotrophic Cell Cultures from Chenopodium rubrum

Photoautotrophic cell suspension cultures from *C. rubrum* assimilate CO_2 predominantly by a $C_1 \rightarrow C_5$ carboxylation sequence using the RuBP-carboxylase and they thus represent a C_3-type of photosynthesis (Hüsemann 6988a). But due to the high in vitro activity of the PEP-carboxylase, high amounts of malate and, to a lesser degree, aspartate are also formed, especially in actively dividing cells during the exponential growth phase. Up to 20% of the total $^{14}CO_2$ radioactivity is incorporated into malate and aspartate after 10 s of $^{14}CO_2$ photosynthesis under standard culture conditions [i.e. pH, use of the culture medium in which the cells had been grown before, 2% (v/v) CO_2 in the gas phase; Herzbeck and Hüsemann 1985].

The rate of malate formation turned out to be clearly pH-dependent. If the pH of the incubation medium is increased from initially 4.5 to 6.0 or 7.0, the rate of ^{14}C-incorporation into malate and aspartate after short-term $^{14}CO_2$ photosynthesis is nearly doubled at the expense of ^{14}C-labelling of sugar phosphates (Table 3).

The intramolecular pattern of ^{14}C-labelling of malate formed during short-term $^{14}CO_2$ photosynthesis was determined after specific decarboxylation of malate at 4-C atom using the malic enzyme (Hatch 1972). Approximately 80% of the total ^{14}C-radioactivity is located in the C-4 atom of malate, indicating its

Table 3. pH Dependency of ^{14}C-labelling of photosynthetic intermediates after short-term photosynthesis (10 s) in photoautotrophic cell suspension cultures from *Chenopodium rubrum*[a]

	pH of the incubation medium		
	4.5	6.0	7.0
^{14}C-Label distribution (%):			
Sugar phosphates	45.3 ± 4.0	29.4 ± 4.0	28.0 ± 4.5
Phosphoglyceric acid	14.2 ± 3.7	11.4 ± 1.7	15.7 ± 3.8
Phosphoenol pyruvate	3.2 ± 1.1	1.6 ± 0.8	1.4 ± 0.3
Citrate	3.9 ± 2.1	1.6 ± 0.8	1.8 ± 0.8
Malate	19.5 ± 3.9	41.7 ± 5.5	41.0 ± 1.4
Aspartate	5.0 ± 1.4	9.0 ± 2.4	7.4 ± 2.4
Glycine, serine, glutamine	3.3 ± 0.9	2.0 ± 0.8	1.4 ± 0.4
Alanine	2.9 ± 0.9	1.8 ± 0.7	2.0 ± 0.8
Sugar	0.6 ± 0.2	–	–
Not identified	2.1 ± 1.0	1.5 ± 0.6	1.3 ± 0.6

[a] Cells from exponential growth phase were used. The pH of the culture medium was adjusted to 4.5, 6.0 in the presence of 50 mM MES-NaOH buffer and to 7.0 using 50 mM HEPES-NaOH buffer. A 10-s $^{14}CO_2$ fixation under white light (110 μEinstein m^{-2} s^{-1}) was started by injecting 2400 nMol $NaH^{14}CO_3$ solution (specific activity: 693 Bq $nmol^{-1}$) directly into the cell suspension. Isolation and quantification of the radioactively labelled photosynthetic intermediates was performed according to Hüsemann et al. (1984). The data are the mean of five independent experiments. Variations of the average are expressed as the standard error of the mean.

synthesis by β-carboxylation of phosphoenolpyruvate mediated by the PEP-carboxylase (Table 4).

Under the appropriate pH conditions, the PEP-carboxylase will contribute substantially to the overall photosynthetic CO_2 assimilation in photoautotrophic cell cultures of *C. rubrum* (Table 3). This means that the expression of the C_4-type of carbon assimilation in the photoautotrophic cell can be regulated by the extracellular pH value.

What are the mechanism(s) by which the extracellular pH exerts its stimulating effect on PEP-carboxylase mediated carboxylation leading to the observed increase in malate formation?

An increase in the pH value of the culture medium results in an increase in the concentration of dissolved inorganic carbon (Table 5), mainly due to higher amounts of dissolved HCO_3^- at pH values above 6.

In algae (Coleman and Colman 1981) and in isolated mesophyll cells (Servaites and Ogren 1977; Espie and Coleman 1982) the highest photosynthetic activities of the cells were obtained at pH 7.0–7.5 of the incubation medium, whereas at lower or higher pH values photosynthesis was reduced. In contrast to algae (Kaplan et al. 1980; Coleman and Colman 1981), the uptake of HCO_3^- into

Table 4. Intramolecular ^{14}C-label distribution of malate formed after short-term (10 s) $^{14}CO_2$ photosynthesis in photoautotrophic cell suspension cultures from *Chenopodium rubrum*[a]

pH of the culture medium	^{14}C-Label distribution in malate (%) C_4	C_{1-3}
4.5	79.8 ± 1.0	20.2 ± 1.0
6.0; 7.0	78.0 ± 0.6	22.0 ± 0.6

[a] Radioactively labelled malate was isolated by TLC. The intramolecular distribution of ^{14}C-label was measured according to Hatch (1972). The data are the averages of five independent experiments.

Table 5. Availability of dissolved inorganic carbon ($CO_2 + HCO_3^-$) in the culture medium at different pH values[a]

CO_2 Content in the gaseous atmosphere	($CO_2 + HCO_3^-$): nmol/ml Culture medium			
	pH of the culture medium			
	4.5	5.5	6.0	7.0
2% (V/V)	773 ± 26	880 ± 21	979 ± 32	2392 ± 140

[a] Volume-calibrated (35-ml) Warburg vessels consisting of two glass flasks connected by a giass tube were used. Flask A contained 3 ml culture medium, flask B contained the appropriate amount of $K_2\,^{14}CO_3$ solution (specific activity: 6850 Bq μ mol^{-1}) to establish a 2% (v/v) CO_2 content in the gaseous atmosphere of the reaction vessel. $^{14}CO_2$ was set free by injecting 1 ml 1 N sulfuric acid. The concentration of the dissolved inorganic carbon in the culture medium equilibrated with 2% CO_2 in the gaseous atmosphere was calculated from the amount of ^{14}C-radioactivity accumulated in the culture medium. The temperature was adjusted to 25 ± 1°C.

cells from land-grown angiosperms is restricted to about 5% of the total inorganic carbon uptake at pH values above 8 (Volokita et al. 1981; Espie and Coleman 1986). Therefore, it should be excluded that the stimulation of PEP-carboxylase mediated malic acid formation in photoautotrophic cell cultures from *C. rubrum* results from an improved supply of the cells with HCO_3^- from the culture medium.

The observed increase in malate formation after changing the extracellular pH from 4.5 to 6.0 or 7.0 might be due to a pH-dependent stimulation of the PEP-carboxylase activity. This implies that changes in the extracellular pH will also alter the intracellular (cytoplasmic) pH. To solve this problem, we applied the ^{31}P-NMR spectroscopy as a non-invasive, non-destructive in-vivo method for measuring the intracellular (cytoplasmic, vacuolar) pH of the plant cells (Roberts et al. 1980). The ^{31}P-NMR studies were performed by Prof. Leibfritz, Dr. Callis, Department of Chemistry, University of Bremen. The cells were irradiated with white light (40 μEinstein m^{-2} s^{-1}) and flushed with CO_2-enriched (2% CO_2; v/v) air during NMR spectroscopy to maintain photoautotrophic conditions. Raising the pH of the incubation medium from initially 4.5 (50 mM MES-NaOH-buffer) to 6.0 (50 mM MES-NaOH buffer), the cytoplasmic pH only changed for 0.1–0.2 pH units from 7.2/7.3 to 7.3/7.4, whereas a drastic increase in the pH value of the vacuole for about 1.2 pH units ranging from pH 5.1/5.3 to 6.2/6.5 could be observed. Though these experimental data have to be regarded as preliminary, they will give a first estimate of the cytoplasmic pH in photoautotrophically cultured cells from *C. rubrum*. There seems to be a strong ion flux from the vacuole to the extracellular medium. For example, photoautotrophic cell suspensions (approximately 3×10^6 cells ml^{-1}) acidified the culture medium for about 0.1 or 0.5 pH units per hour if the incubation medium was adjusted to pH 6.0 in the presence of 50 mM MES-NaOH buffer or by NaOH respectively. Similar results have been reported for mesophyll cells from higher plants (Bown 1982; Marré et al. 1988) and for heterotrophic plant cell cultures (Guern et al. 1986; Mathieu et al. 1986).

Although the cytoplasmic pH in photoautotrophically cultured cells from *C. rubrum* only changed for about 0.1–0.2 pH units when the extracellular pH value was increased from 4.5 to 6.0, this raise in pH might have substantial stimulatory effects on the activity of the PEP-carboxylase (Table 6). We found that the in-vitro activity of the PEP-carboxylase from photoautotrophic cell cultures of *C. rubrum* is strongly pH-dependent. For example, a rise in the pH of the assay medium from 7.2 $\rightarrow$ 7.4 (equivalent to the change in cytoplasmic pH if the extracellular pH is increased from 4.5 to 6.0) will stimulate PEP-carboxylase activity about 1.8-fold. This increase should be sufficient to account for the observed increase in malate formation after short-term $^{14}CO_2$ photosynthesis at pH 6.

The cytosolic origin of the PEP-carboxylase in photoautotrophic cell cultures from *C. rubrum* has been proven by cell fractionation of isolated protoplasts (Table 7). The silicone-oil centrifugation technique (Robinson and Walker 1979; Wirtz et al. 1980) has been applied. Separation of the chloroplasts from other cytoplasmic components was monitored by the activity of marker enzymes specific for chloroplasts (NADP-glyceraldehyde-3-phosphate dehydrogenase; NADP-GAPDH), mitochondria (fumarase), peroxisomes (catalase) and cytosol (pyruvate-kinase). Within 8 s of centrifugation about 80% of the chloroplasts, as indicated by the activity of the NADP-GAPDH, was separated into a pellet fraction, which was

Table 6. pH-Dependency of the in-vitro activity of the PEP-carboxylase isolated from photoautotrophic cell cultures of *Chenopodium rubrum*[a]

Change in the pH of the assay medium	Change in PEP-carboxylase activity	
	nmol subtrate / mg protein x min	Factor of increase
7.2 → 7.4	10.2 → 18.5	1.8
7.4 → 7.6	18.5 → 50.0	2.7
7.6 → 7.8	50.0 → 94.6	1.9
7.8 → 8.0	94.6 → 80.4	0.85

[a] PEP-carboxylase activity was determined as described by Hüsemann et al. (1984). The data are the mean of five independent experiments.

Table 7. Distribution of enzyme activity after fractionation of protoplasts from photoautotrophic cell suspension cultures of *Chenopodium rubrum*[a]

Enzyme	Distribution of enzyme activity (%)		Summed enzyme
	Chloroplast pellet	Extrachloroplastic supernatant	activities in % of control
NADP-GAPDH	80.0	20.0	114.6
Fumarase	34.7	65.3	95.3
Catalase	17.5	82.5	86.7
Pyruvate kinase	5.3	94.7	108.0
PEP-carboxylase	5.0	95.0	98.5
NAD-MDH	29.5	70.5	104.8
NADP-MDH	77.0	23.0	102.5

[a] Control: in-vitro activity of enzymes directly extracted from protoplasts without fractionation. The data are the mean of five independent experiments.

essentially free of soluble cytoplasmic components (5% pyruvate kinase activity), but still contained more than 30% contamination by mitochondria. The unsatisfactory separation of mitochondria from the chloroplasts may partly be due to the small difference in size of these organelles in actively dividing cells (Mittelbach 1988) as well as by an adhesion of chloroplasts and mitochondria to multiorganelle complexes formed during centrifugation (Lörz 1985). Nevertheless, the subcellular fractionation of protoplasts by the silicone oil-centrifugation technique confirms that the PEP-carboxylase is a cytosolic enzyme. It gives a first estimate that the NADP-malic acid dehydrogenase (NADP-MDH) is localized in the chloroplasts and cytoplasm, whereas the NAD-malic acid dehydrogenase (NAD-MDH) is obviously restricted to the mitochondria.

Such an intracellular distribution of enzymes would corroborate their possible involvement in anapleurotic metabolic reactions allowing an additional formation of oxaloacetate and malate as intermediates of the citric acid cycle to support its synthetic function (Herzbeck and Hüsemann 1985).

4 Secondary Product Formation in Photoautotrophic Cell Cultures

Our investigations on the accumulation of secondary plant products in photoautotrophic cell cultures are based on the hypothesis that green cell cultures will preferentially express the biosynthesis of secondary products which are partly or totally produced by chloroplast-localized enzymes (Wink et al. 1980; Barz and Hüsemann 1982; Schoofs et al. 1983). This principle is best demonstrated by comparative measurements of secondary metabolites in heterotrophic, photomixotrophic and photoautotrophic cell cultures derived from one plant species. Such reversion experiments of cell cultures reveal how the mode of growth together with the inherent, specific pattern of cellular metabolism will determine the potential for biosynthesis of secondary products (Barz 1988).

Alternative and reversible formation of anthraquinone and lipoquinones has been observed in cell suspension cultures of *Morinda lucida* (Igbavboa et al. 1985). Anthraquinone pigments are confined to heterotrophic cultures, whereas the lipoquinones (vitamin K_1 plastoquinone, α-tocopherol and ubiquinone) are only found in photoautotrophic cultures. Such investigations allow insight into the regulatory parameters of aromatic compound metabolism and they show the involvement of chloroplasts as the site of lipoquinone biosynthesis.

Heterotrophic and photosynthetically active cell cultures of *Ruta graveolens* significantly differ in their volatile constituents. Heterotrophic cells produce a mixture of C_6–C_8 ketones, acetic acid n-butylester and a series of C_4–C_9 primary and secondary alcohols (Jordan et al. 1986). Green cell cultures are characterized by C_9–C_{13} methylketones and esters of 2-methylbutyric acid and 3-methylbutyric acid esterified with straight chain or branched C_8–C_{11} alcohols (Drawert et al. 1984).

Photoautotrophic cell suspension cultures of *Catharanthus roseus* are characterized by very low levels of indole alkaloids and tryptamine as well as low activities of tryptophan decarboxylase. Upon reversion of such cultures to photomixotrophic and subsequently heterotrophic cultures, the level of these constituents is drastically increased. Inoculation of photoautotrophic cells into sucrose-rich media leads to an almost instantaneous induction of tryptophan decarboxylase and tryptamine accumulation (Röwer 1986; Beimen and Barz, unpublished). These observations indicate that indole alkaloid biosynthesis and expression of relevant biosynthetic enzymes appear to be restricted to heterotrophic cells.

Chenopodium rubrum cell cultures produce the glucose esters of p-coumaric and ferulic acid (Strack et al. 1984; Dräger 1986). Such esters are especially prominent in photomixotrophic cultures because the biosynthesis is regulated both by light induction of phenylalanine ammonia lyase and by sufficiently high levels of sucrose. Under photoautotrophic conditions the aromatic glucose esters are barely detectable in the cells and phenylalanine ammonia lyase shows very low enzyme activity. The induction of both PAL and the accumulation of the cinnamic acid esters can be titrated with sucrose. Below 0.5% of the carbon source PAL activity and the levels of the cinnamic acid esters remain very low, whereas in the range of 0.5–1% sucrose maximum expression has been found. This concentration of approximately 0.5% sucrose has repeatedly been found in our reversion experiments as the regulatory switching point between the photoautotrophic and heterotrophic mode of nutrition.

Our present knowledge of secondary product formation in photosynthetically active plant cell cultures is summarized in the hypothesis that green cell cultures are suitable models if chloroplast-derived products are being investigated and, furthermore, that sufficiently high concentration of sucrose are available to sustain secondary product biosynthesis of a suitable rate. Therefore, future investigations will best be carried out with photomixotrophic cell cultures because they are characterized by functioning chloroplasts and sufficiently high levels of a metabolizable carbon source.

5 Conclusions

The differential sensitivity of cultured cells from *C. rubrum* to blue and red light in chloroplast differentiation obviously depends on the nutritional status (heterotrophic-photoautotrophic) of the cells. Investigations on the mRNA steady-state concentrations and on the transcriptional activities of the cells are continued to quantify this change in light-dependent chloroplast differentiation on the molecular biological level. Furthermore, it remains to be studied, whether this change in the blue/red light sensitivity of the cells is accompanied by a switch from a specific blue-light photoreceptor to phytochrome.

It has been shown that the expression of $C_1 \rightarrow C_3$ carboxylation sequences and thus the mode of CO_2 fixation using the PEP-carboxylase is pH-dependent. Whether this regulatory principle is a common phenomenon of photosynthesizing cells has to be studied with photoautotrophic cell cultures from different plant species.

Finally, in-vitro cultured photoautotrophic plant cells differ from mesophyll cells with respect to their high demand for CO_2. Meanwhile, we succeeded in establishing a cell line capable of sustained photoautotrophic growth under normal air. Whether this change in the photosynthetic potential of the cells is due to an improved CO_2 uptake and intracellular CO_2 concentration and/or to increased carboxylation capacities of the cells is presently being studied.

Acknowledgements. Our investigations were supported by the Deutsche Forschungsgemeinschaft and by the Bundesminister für Forschung und Technologie. Thanks are expressed to Prof. Dr. Leibfritz/R. Callis, Bremen University for 31-P NMR measurements.

References

Barz W (1988) Investigations on plant cell cultures. Bioengineering 4:145–148

Barz W, Hüsemann W (1982) Aspects of photoautotrophic cell suspension cultures. In: Fujiwara A (ed) Plant tissue culture 1982. Mazuren, Tokyo, 245–248

Bown AW (1982) An investigation into the roles of photosynthesis and respiration in H^+ efflux from aerated suspensions of *Asparagus* mesophyll cells. Plant Physiol 70:803–810

Coleman JR, Colman B (1981) Inorganic carbon accumulation and photosynthesis in a bluegreen alga as a function of external pH. Plant Physiol 67:917–921

Dräger B (1986) Untersuchungen zum Sekundärstoffwechsel in Zellkulturen von *Chenopodium rubrum*. D Thesis, Münster Univ

Drawert F, Berger RG, Godelmann R, Collin S, Barz W (1984) Characterization of volatile constituents from photomixotrophic cell suspension cultures of *Ruta graveolens*. Z Naturforsch 39c: 525–530

Espie GS, Colman B (1982) Photosynthesis and inorganic carbon transport in isolated *Asparagus* mesophyll cells. Plant Physiol 70:649–654

Espie GS, Owttrim GW, Colman B (1986) Inorganic carbon uptake during photosynthesis. Plant Physiol 80:870–876

Guern J, Mathieu Y, Pean M, Pasquier C, Beloeil JC, Lallemand JY (1986) Cytoplasmic pH regulation in *Acer pseudoplatanus* cells. I. ^{31}P NMR description of acid-load effects. Plant Physiol 82:840–845

Hatch MD (1972) Synthesis of L-malate-4-^{14}C and determination of label in the C-4 carboxyl of L-malate. Anal Biochem 47:174–183

Herzbeck H, Hüsemann W (1985) Photosynthetic carbon metabolism in photoautotrophic cell suspension cultures of *Chenopodium rubrum*. In: Neumann KH, Barz W, Reinhard E (eds) Primary and secondary metabolism of plant cell cultures. Springer, Berlin Heidelberg New York Tokyo, 15–23

Hübner S (1988) Wachstum und photosynthetische CO_2-Assimilation in photoautotrophen Zellkulturen von *Chenopodium rubrum* unter Rotlicht- und Blaulichtbedingungen. Diplomarbeit (Master's Thesis) Münster Univ

Hüsemann W (1985) Photoautotrophic growth of cells in culture. In: Vasil IK (ed) Cell culture and somatic cell genetics of plants, vol 2: Cell growth, nutrition, cytodifferentiation, and cryopreservation. Academic Press, New York London, pp 213–252

Hüsemann W (1988a) The photosynthetic potential and growth of photoautotrophic cell suspension cultures from *Chenopodium rubrum*: a case study. In: Marin B (ed) Plant vacuoles. NATO ASI Ser A Plenum, New York London, pp 67–78

Hüsemann W (1988b) Physiological and biochemical characteristics of photoautotrophic plant cell cultures. In: Pais (ed) Plant cell biotechnology. NATO ASI Ser A. Springer, Berlin Heidelberg New York Tokyo, pp 179–183

Hüsemann W, Herzbeck H, Robenek H (1984) Photosynthesis and carbon metabolism in photoautotrophic cell suspensions of *Chenopodium rubrum* from different phases of batch growth. Physiol Plant 62:349–355

Igbavboa U, Sieweke HJ, Leistner E, Röwer I, Hüsemann W, Barz W (1985) Alternative formation of anthraquinones and lipoquinones in heterotrophic and photoautotrophic cell suspension cultures of *Morinda lucida* Benth. Planta 166:537–544

Jordan M, Rolfs CH, Barz W, Berger RG, Kohlmannsberger H, Drawert F (1986) Characterization of volatile constituents from heterotrophic cell suspension cultures of *Ruta graveolens*. Z Naturforsch 41c:809–812

Kaplan A, Badger MR, Berry JA (1980) Photosynthesis and intracellular inorganic carbon pool in the bluegreen alga *Anabaena variabilis*: response to external CO_2 concentration. Planta 149:219–226

Lörz H (1984) Enucleation of protoplasts: preparation of cytoplasts and miniprotoplasts. In: Vasil IK (ed) Cell culture and somatic cell genetics of plants, vol 1: Laboratory procedures and their applications. Academic Press, New York London, pp 448–453

Marré E, Beffagna N, Romani G (1988) Potassium transport and regulation of intracellular pH in *Elodea densa* leaves. Bot Acta 101:24–31

Mathieu Y, Guern J, Pean M, Pasquier C, Beloeil JC, Lallemend JY (1986) Cytoplasmic pH regulation in *Acer pseudoplatanus* cells. II. Possible mechanisms involved in pH regulation during acid-load. Plant Physiol 82:846–852

Mittelbach I (1988) Aktivitätsmuster und intrazelluläre Lokalisation von Enzymen des photosynthetischen Kohlenstoffmetabolismus in photoautotrophen Zellsuspensionskulturen von *Chenopodium rubrum*. Diplomarbeit (Master's Thesis) Münster Univ

Reiss Th, Link G (1985) Characterization of transcriptionally active DNA protein complexes from chloroplasts and etioplasts of mustard (*Sinapis alba* L.) Eur J Biochem 148:207–212

Richter G, Hundrieser J, Groß M, Schultz S, Bottländer K, Schneider Ch (1984) Blue light effects in cell cultures. In: Senger H (ed) Blue light effects in biological systems. Springer, Berlin Heidelberg New York, pp 387–396

Richter G, Einspanier R, Hüsemann W, Dudel A, Wessel K (1986) Gene expression in blue light-dependent chloroplast differentiation of cultured plant cells. In: Akoyunoglou G (ed) Regulation of chloroplast differentiation. Liss, New York, pp 549–558

Richter G, Dudel A, Einspanier R, Dannhauer I, Hüsemann W (1987) Blue-light control of mRNA level and transcription during chloroplast differentiation in photomixotrophic and photoautotrophic cell cultures (*Chenopodium rubrum* L.) Planta 172:79–87

Roberts JKM, Ray PM, Wade-Jardetzky N, Jardetzky O (1980) Estimation of cytoplasmic and vacuolar pH in higher plant cells by ^{31}P NMR. Nature (London) 283:870–872

Robinson SP, Walker DA (1979) Rapid separation of the chloroplast and cytoplasmic fractions from intact leaf protoplasts. Arch Biochem Biophys 196:319–323

Röwer I (1986) Etablierung, Wachtsumseigenschaften und Sekundärstoffwechsel photosynthetisch aktiver Zellen aus höheren Pflanzen. D Thesis, Münster Univ

Ruyters G (1984) Effects of blue light on enzymes. In: Senger H (ed) Blue light effects in biological systems. Springer, Berlin Heidelberg New York, pp 283–301

Schoofs G, Teichmann S, Hartmann T, Wink M (1983) Lysine decarboxylase in plants and its integration in quinolizidine alkaloid biosynthesis. Phytochemistry 22:65–69

Servaites JC, Ogren WL (1977) pH dependency of photosynthesis and photorespiration in soybean leaf cells. Plant Physiol 60:693–696

Strack D, Bokern M, Berlin J, Sieg S (1984) Metabolic activity of hydroxy cinnamic acid esters in cell suspension cultures of *Chenopodium rubrum*. Z Naturforsch 39c:902–907

Volokita M, Kaplan A, Reinhold L (1981) Evidence for mediated HCO_3^- transport in isolated pea mesophyll protoplasts. Plant Physiol 67:1119–1123

Wink M, Hartman T, Witte L (1980) Enzymic synthesis of quinolizidine alkaloids in lupin chloroplasts. Z Naturforsch 35c:93–97

Wirtz W, Stitt M, Heldt HW (1980) Enzymic determination of metabolites in the subcellular compartments of spinach protoplasts. Plant Physiol 66:187–193

Part II
Regulation of Secondary Metabolite Biosynthesis

Regulation of Secondary Metabolism in Relation to Growth and Differentiation

A. Komamine[1], M. Sakuta[2], M. Hirose[3], H. Hirano[1], T. Takagi[4], K. Kakegawa[1], and Y. Ozeki[5]

It is well known that most secondary metabolites are accumulated in specific tissues and cells in higher plants, implying that the expression of secondary metabolism is closely correlated with the growth and morphological differentiation of cells. On the other hand, some secondary metabolites are not localized in specific tissues. For example, accumulation of betacyanin is observed in all tissues in roots of red beet. These facts suggest that there are two different patterns of accumulation of secondary metabolites in relation to growth and differentiation in plants.

We investigated anthocyanin accumulation in suspension cultures of *Daucus carota* and *Vitis vinifera* as examples of the former type of secondary metabolites and betacyanin in *Phytolacca americana* suspension cultures as an example of the latter.

As we reported previously in carrot suspension cultures (Ozeki and Komamine 1981), anthocyanin synthesis was induced by the same trigger, depriving auxin from media, as for the induction of embryogenesis. Small cell aggregates (31–81 μm) were selected by sieving through nylon screens at every subculture. Cells in suspension were fractionated by density gradient centrifugation in Ficoll solution. In cells in lighter fractions than 14% Ficoll, anthocyanin synthesis was induced when cells were transferred to an auxin-free medium. Cells in heavier fractions than 14% Ficoll differentiated to embryos by the same trigger. In lighter cells, anthocyanin synthesis occurred when cell division ceased and cell elongation initiated. The readdition of auxin caused division of elongated cells and the disappearance of anthocyanin. These results indicate that anthocyanin synthesis is closely correlated with morphological differentiation and shows a negative correlation with cell division or proliferation.

Another example of the negative correlation of anthocyanin accumulation with cell division was shown in *Vitis vinifera* suspension cultures in which accumulation of anthocyanin occurs during culture in Murashige and Skoog medium containing auxin (2,4-D 2.3×10^{-7} M, kinetin 9.3×10^{-7} M) in the dark without differentiation. This suspension culture was established in Prof. Kodama's laboratory at the University of Tokyo (Yamakawa et al. 1983). In *Vitis* suspension cultures, an increase in the accumulation of anthocyanin was not observed during the lag and log phases but a marked increase occurred in the stationary phase when

[1] Biological Institute, Faculty of Science, Tohoku University, Sendai, Japan
[2] Department of Biology, Faculty of Science, Ochanomizu University, Otsuka, Tokyo, Japan
[3] Research Laboratory for Development, Mitsubishi Oil Company, Kawasaki, Japan
[4] Nishiki Research Laboratory, Kureha Chemical Industry Co., Ltd., Iwaki, Fukushima, Japan
[5] Department of Biology, Faculty of Arts and Sciences, University of Tokyo, Komaba, Tokyo, Japan

Primary and Secondary Metabolism of Plant Cell Cultures, ed. by W.G.W. Kurz

cell division ceased completely. When *Vitis* cells were cultured under conditions which caused the cessation of cell division, such as culturing in lowered phosphate concentrations in media or in the presence of aphidicolin, an inhibitor of DNA synthesis, rapid accumulation of anthocyanin occurred in coincidence with cessation of cell division. An exception to the effects of inhibitors on cell division was that of an inhibitor of protein synthesis, cycloheximide, which inhibited both accumulation of anthocyanin and cell division. However, de novo synthesis of enzymes involved in the anthocyanin synthetic pathway is required for induction of anthocyanin synthesis in carrot suspension cultures (Ozeki and Komamine 1985). In *Vitis* cells, activities of some enzymes involved in the anthocyanin biosynthetic pathway, such as glucose-6-phosphate dehydrogenase, phenylalanine ammonia lyase and chalcone flavanone isomerase, increased prior to or at the time of accumulation of anthocyanin. Cycloheximide may inhibit the synthesis of these enzymes, leading to the inhibition of anthocyanin accumulation.

A different pattern was observed in the accumulation of betacyanin in *Phytolacca americana* suspension cultures which were maintained in Murashige and Skoog medium containing 5×10^{-6} M 2,4-D in the dark. The peak of betacyanin content per cell was shown at the log phase of the growth cycle, suggesting a positive correlation of betacyanin accumulation with cell proliferation (Sakuta et al. 1986). Other evidence for it was provided by experiments using factors regulating cell division (Sakuta et al. 1987a,b).

Increasing the phosphate concentrations in media up to 1.25×10^{-3} M promoted cell division as well as betacyanin accumulation. In the absence of phosphate in media, cell division did not occur and betacyanin accumulation was also stopped. Readdition of phosphate to cells starved with phosphate induced both cell division and betacyanin accumulation. Aphidicolin (5×10^{-5} M), an inhibitor of DNA synthesis, completely inhibited both cell division and betacyanin accumulation. When aphidicolin was removed from cells by washing, partially synchronous cell division was induced and betacyanin accumulation also occurred, associated with cell division. Cycloheximide also inhibited both cell division and betacyanin accumulation.

Addition of tyrosine, a precursor of betacyanin, promoted accumulation of betacyanin in proliferating *Phytolacca* cells. It was confirmed by tracer experiments and by measuring enzyme activities and pool sizes of intermediates that the supply of tyrosine was a key regulation step of betacyanin accumulation. However, even when tyrosine was added to media, inhibition of betacyanin accumulation by aphidicolin was not recovered. Thus, a physiological situation, in which proliferation can occur, is considered to be essential for betacyanin synthesis. In contrast to betacyanin, anthocyanin synthesis occurs when cell division ceases and/or morphological differentiation is induced.

In conclusion, patterns of secondary metabolite accumulation in relation to proliferation and differentiation fall into two categories: (1) accumulation during the stationary phase or associated with differentiation, such as anthocyanin. A negative correlation of accumulation is observed with proliferation in this case. Patterns of most secondary metabolite accumulations fall into this category. (2) Accumulation during the logarithmic phase, such as betacyanin accumulation in

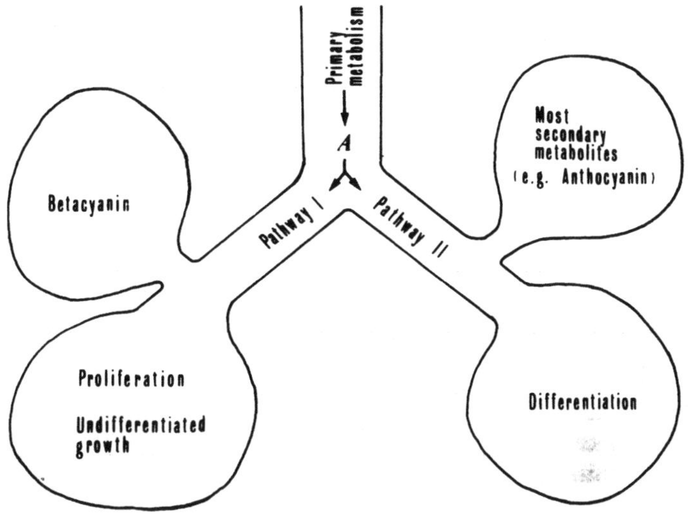

Fig. 1. Possible scheme of primary and secondary metabolic flows in relation to proliferation and differentiation (Sakuta and Komamine 1987)

suspension cultures of *Phytolacca americana*. A positive correlation of accumulation with proliferation is observed.

From the results obtained, a working hypothesis is proposed as shown in Fig. 1 (Sakuta and Komamine 1987). The primary metabolic flow branches at Point A into two pathways: one is the pathway to proliferation or undifferentiated growth (Pathway I), and the other is the pathway (Pathway II) to differentiation. It is presumed that switching of the metabolic flow between Pathways I and II occurs at Point A. Some secondary metabolites such as betacyanin accumulate when primary metabolism flows to Pathway I in sufficient amounts. On the other hand, most secondary metabolites are synthesized and accumulated when metabolic flows are switched to Pathway II. This hypothesis can be applied at least to betacyanin in suspension cultures of *Phytolacca americana* and anthocyanin in suspension cultures of *Daucus carota* and *Vitis vinifera*. Further investigation is necessary, however, to reveal whether or not it can be generalized to other secondary metabolism in cell cultures.

References

Ozeki Y, Komamine A (1981) Induction of anthocyanin synthesis in relation to embryogenesis in a carrot suspension culture; correlation of metabolic differentiation with morphological differentiation. Physiol Plant 53:570–577

Ozeki Y, Komamine A (1985) Changes in activities of enzymes involved in general phenylpropanoid metabolism during the induction and reduction of anthocyanin synthesis in a carrot suspension culture as regulated by 2,4-D. Plant Cell Physiol 26:903–911

Ozeki Y, Komamine A (1986) Effects of growth regulators on the induction of anthocyanin synthesis in carrot suspension cultures. Plant Cell Physiol 27:1361–1368

Sakuta M, Takagi T, Komamine A (1986) Growth related accumulation of betacyanin in suspension cultures of *Phytolacca americana*. L J Plant Physiol 125:337–343

Sakuta M, Komamine A (1987) Cell growth and accumulation of secondary metabolites. In: Vasil KI (ed) Cell culture and somatic cell genetics of plants, vol 4. Academic Press. San Diego, New York, Berkeley, Boston, London, Sydney, Tokyo, Toronto pp 97–114

Sakuta M, Takagi T, Komamine A (1987a) Effects of sucrose on betacyanin accumulation and growth in suspension cultures of *Phytolacca americana*. Physiol Plant 71:455–458

Sakuta M, Takagi T, Komamine A (1987b) Effects of nitrogen source on betacyanin accumulation and growth in suspension cultures of *Phytolacca americana*. Physiol Plant 71:459–463

Yamakawa T, Kato S, Ishida K, Kodama T, Minoda Y (1983) Production of anthocyanins by *Vitis* cells in suspension culture. Agric Biol Chem 47:2185–2191

On the Nature of the Variation in Yields of Phytochemicals Observed Within Plant Cell Cultures

D.L. Vogelein and D.K. Dougall[1]

1 Introduction

Great differences in the yields of phytochemicals within a population of clones obtained from plant cell cultures have been reported by a number of investigators and these observations have been reviewed by Deus-Neuman and Zenk (1984), Dougall (1987), and Yamada and Hashimoto (1984). These differences have been exploited to increase yields in plant cell cultures. Only a few of the studies on variability in plant cell cultures have been done in such a way that they can distinguish between variation in the parental culture due to heterogeneity in the original explant and variation in phytochemical yield arising from single cells within the culture. These studies (Dougall et al. 1980; Matsumoto et al. 1980; Deus-Neuman and Zenk 1984; Ellis 1985) show that clones give rise to subclones displaying a wide range of yields. Using wild carrot cultures and their production of anthocyanin, we have examined some aspects of clonal variation and begun to characterize the differences between high and low yielding clones.

2 Materials and Methods

All clones were derived from subclones WC-63-1-9-1, described by Dougall et al. (1980), which had been cloned three times, each time taking the highest yielding clone for the next step. Cloning of this line was performed as described by Dougall et al. (1980). In this procedure 6 weeks are required for development of colonies to a size suitable to initiate cultures. The colonies are individually transferred to tubes and grown in 2.5 ml of stock culture medium for 3 weeks and then the 2.5 ml culture is transferred into 25 ml of medium for improved anthocyanin production (WCM-4, WCM-Imp). After 12 days of growth, the anthocyanin/ml of culture is measured using duplicate samples from each culture.

To test the responses of clones to naringenin, dihydroquercetin, or sinapic acid, these compounds were provided to the cultures at days 4 and 8 in DMSO to give concentrations of 1% v/v for DMSO, 0.1 mM naringenin, 0.25 mM dihydroquercetin, and 0.05 mM sinapic acid. These concentrations were established as those giving maximum stimulation of anthocyanin accumulation with no growth inhibition of the parental culture WC-63. DMSO was used because naringenin is

[1]Department of Botany, The University of Tennessee, Knoxville, TN 37996-1100, USA

Primary and Secondary Metabolism of Plant Cell Cultures, ed. by W.G.W. Kurz

very insoluble in water and treatments with DMSO alone were used as the control for treatments with intermediates. The effects of these compounds on anthocyanin accumulation and growth was determined at 12 days of culture. Anthocyanin/ml culture is expressed as OD_{530nm} after extraction of the tissue in culture samples with an equal volume of 95% MeOH containing 1% v/v conc. HCl as described by Dougall and Weyrauch (1980). Dry weight is given either as total dry wt. obtained from independent samples or as extracted dry wt. which is the weight of the residue after measurement of anthocyanin content of samples.

3 Results

Cloning Experiment. When high yielding subclones were selected after cloning the culture WC63-1-9-1 and recloned several times, each time selecting the highest yielding subclone from approximately 30 clones, the mean anthocyanin content of each population of clones remained relatively constant within each series. However, some of the population means were almost double others. The anthocyanin in the highest yielding clone in each set varied quite substantially. This can be explained as random selection from populations which have a continuous distribution of yields. At each cloning the standard deviation of the population is one-half to two-thirds of the mean. Individual clones in a set have anthocyanin contents of 10% or less of the mean value, showing the capacity of individual cells in the population to decrease the level of anthocyanin they accumulate.

When low yielding subclones were selected initially and serially recloned with selection for the highest yielding clones at each step, the population means at the second and third serial cloning were very similar in three of four sets. In the fourth set the mean doubled at the third cloning. The means in these cases where one selection for low yield had been interposed were substantially lower in two of four cases than the means of the sets where selection for high yield had only been applied. The other two cases gave population means which were as high as those obtained with selection for high yield at every step. In each case the standard deviations were 50% or more of the means and the populations contained many individuals whose yield of anthocyanin exceeded that measured in the initial clone from which they had been derived. In two cases the increase was almost 50-fold after two or three reclonings with selection for high yield at each cloning. These data show that a clone selected once for low yield can recover the ability to accumulate anthocyanin. We have seen in the selections for high yield only, clones which demonstrate the decrease of capacity to accumulate anthocyanin. These two selection schemes show clearly that cells in the carrot cultures can alter the amount of anthocyanin they accumulate in both directions but they give no indication of the rate or frequency at which this occurs.

When low yielding clones were serially recloned and the lowest yielding subclone selected from each population, then the population means declined at each of the three serial clonings. The standard deviations were between 30 and 100% of the means. The range of yields in the populations became relatively narrow so that in three sets, all clones were in the range of 0.002 to 0.076 OD units and in the fourth set in the range of 0.01 to 0.33 OD units after the third cloning. In the two cases where we have selected the highest yielding clone from these sets and recloned them, the mean and standard deviations of those populations do not differ from the previous population. This suggests that the effects of three serial clonings with selection of the lowest yielding clone at each step is not readily reversible by subsequent selection and recloning of the highest yielding clones in these populations. This then appears to be a second phenomenon associated with selection for yields and cloning of these populations, i.e., repeated selection for low yields at each cloning appears to irreversibly reduce the ability of the cells to accumulate anthocyanin.

Abilities of Subclones to Utilize Intermediates in the Anthocyanidin Biosynthetic Pathway. The wide differences found in the abilities of subclones to accumulate anthocyanin led us to attempt to determine whether the low yielding subclones were deficient in enzymes or substrates for biosynthesis of anthocyanidins. To this end we provided known intermediates in anthocyanidin biosynthesis to subclones and asked whether the intermediates increased anthocyanin accumulation. If that occurred, it would suggest that anthocyanin accumulation in such clones was substrate-limited. If no stimulation occurred, then anthocyanin accumulation was presumably enzyme-limited at a point beyond the intermediate provided. Both responses were obtained as shown in Table 1. Both naringenin and dihydroquercetin increased the anthocyanin accumulation by some subclones but not by others. DMSO also increased anthocyanin accumulation in some of these cultures. Of 43 clones examined this way, 30 gave clear increases in anthocyanin with naringenin and dihydroquercetin and 23 of those gave increased anthocyanin with DMSO alone. The remaining 13 gave no increase in anthocyanin with DMSO alone or with naringenin or dihydroquercetin and 11 of the 13 gave anthocyanin levels in untreated cultures of less than 0.1.

Table 1. Effects of naringenin and dihydroquercetin on anthocyanin accumulation and growth of wild carrot subclones

Subclone[a]			Control	DMSO[b]	+Nar[b]	+DHQ[b]
1B	HH	OD	2.64 ± 0.08	4.94 ± 0.08	5.91 ± 0.26	7.00 ± 0.19
		DW	3.2 ± 0.10	2.8 ± 0.0	2.4 ± 0.1	2.5 ± 0.1
1B	LLL	OD	0.13 ± 0.01	0.35 ± 0.006	0.53 ± 0.01	1.02 ± 0.01
		DW	4.0 ± 0.1	4.0 ± 0.2	4.1 ± 0.0	3.7 ± 0.3
1A	LLL	OD	0.029 ± 0.004	0.026 ± 0.002	0.027 ± 0.002	0.027 ± 0.003
		DW	5.4 ± 0.3	4.1 ± 0.1	4.3 ± 0.1	4.0 ± 0.1
2	LLL	OD	0.053 ± 0.01	0.053 ± 0.003	0.053 ± 0.003	0.072 ± 0.002
		DW	2.8 ± 0.0	2.5 ± 0.1	2.7 ± 0.1	2.7 ± 0.2
3	LLL	OD	0.018 ± 0.004	0.016 ± 0.002	0.018 ± 0.002	0.026 ± 0
		DW	3.8 ± 0.1	3.7 ± 0.1	4.1 ± 0.2	3.4 ± 0

[a] The letters H and L represent clonings in which the highest (H) or the lowest (L) yielding clone of a set were selected.

[b] Additions were made on days 4 and 8 to give concentrations of 1% DMSO, 0.1 mM naringenin (Nar), and 0.25 mM dihydroquercetin (DHQ). Data are mean anthocyanin/ml (OD_{530}) and mean total dry wt. ($mg\ ml^{-1}$) ± SD of triplicate samples at day 12.

From this data it is very clear that two types of clones are obtained. Those which accumulated little or no anthocyanin, i.e., which have anthocyanin/ml culture less than OD=0.1, and in which anthocyanin accumulation is not increased by narringenin, dihydroquercetin, or DMSO, are one type. The second type gives an increase of anthocyanin with the two intermediates and in most cases also with DMSO. It is to be noted that while six of these clones gave OD_{530nm} of 6–7 in the presence of DMSO and dihydroquercetin, the remainder gave much lower OD's varying all the way down to 0.25. Thus, there appear to be quantitative differences between clones of this second type. We have no information about the nature of these quantitative differences.

Stimulation of Anthocyanin Accumulation by Sinapic Acid. The anthocyanin accumulated by carrot cell cultures has been identified as cyanidin-3-(sinapoyl-xylosyl-glucosyl-galactoside) (Harborne et al. 1983; Hopp and Seitz 1987). In addition, Hopp and Seitz (1987) have provided evidence showing that the isolated vacuoles from their carrot cell cultures accumulate the sinapoylated anthocyanin but not the anthocyanin with the sinapoyl residue removed. We therefore tested the possibility that the subclones which did not accumulate anthocyanin were deficient in the ability to synthesize sinapic acid and thus unable to synthesize the sinapoylated anthocyanin and accumulate it in the vacuole. Data representative of the 13 subclones tested are shown in Table 2. Again, in clones which accumulated readily detectable amounts of anthocyanin, sinapic acid further increased the accumulation of anthocyanin and no new anthocyanidin was detectable in the sinapic acid-treated cultures. This is consistent with the data of Hopp and Seitz (1987). It also suggests that anthocyanin accumulation is limited by this substrate as well as others. For those subclones which accumulated little or no anthocyanin, sinapic acid did not increase their accumulation with one exception. In this subclone there was a threefold increase in anthocyanin accumulated but this stimulation did not dramatically increase the anthocyanin accumulation and it remained well below that of subclones selected for high levels of accumulation.

Table 2. Effects of sinapic acid on anthocyanin accumulation by subclones of wild carrot

Subclone[a]		Untreated	DMSO[b] control	+ Sinapic acid[b]
3	HHHH	0.600	0.797	1.038
1A	HH	0.343	0.389	0.537
1B	LLLL	0.007	0.015	0.042
1A	LL	0.015	0.012	0.012
2	LLLL	0.011	0.015	0.014
3	LLL	0.004	0.004	0.005

[a] The letters H and L represent clonings in which the highest (H) or the lowest (L) yielding clone of a set were selected.

[b] Additions were made on days 4 and 8 to give concentrations of 1% v/v DMSO and 0.05 mM sinapic acid. Data expressed as OD · mg^{-1} · ml of triplicate samples taken on day 12.

4 Discussion

The studies of the subclones described show that selection for high yield maintains the level of anthocyanin accumulated. However, the populations obtained do not all have the same quantitative level of anthocyanin. The inclusion of one cloning in which the lowest yielding subclone is taken and serially recloned with selection for highest yield leads to decreased population yields in some but not all cases. Further, the data show that the cells can increase and decrease the levels of anthocyanin they accumulate. This capacity is the property of single cells.

Continued selection for low yields with serial cloning leads to loss of anthocyanin accumulation which does not appear to be reversible. This behavior represents a second phenomenon which is qualitatively different from the reversibility demonstrated in serial clonings with selection for high yield.

Biosynthetic intermediates provided in the culture medium increase anthocyanin accumulation only in subclones which always accumulate significant amounts of anthocyanin: This suggests that the anthocyanin accumulation is substrate-limited and not enzyme-limited. The inability of the biosynthetic intermediates to stimulate anthocyanin accumulation in nonaccumulating subclones cannot be clearly interpreted at this time.

These data show clearly that two different qualitative phenomena lead to different levels of anthocyanin accumulation in cells in these carrot cultures. In addition, there are quantitative differences in the extent to which anthocyanin is accumulated in different populations obtained by serial recloning. There may be additional ways in which the level of accumulation can change. The demonstration of these phenomena indicates that we have only just begun to outline the nature of variation in phytochemical yield of cells within a culture.

References

Deus-Neuman B, Zenk MH (1984) Instability of indole alkaloid production in *Catharanthus roseus* cell suspension cultures. Planta Med 50:427–431

Dougall DK (1987) Cell cloning and the selection of high yielding strains. In: Vasil IK, Constabel F (eds) Cell culture and somatic cell genetics of plants, vol 4: Cell culture in phytochemistry. Academic Press, Orlando New York, pp 117–123

Dougall DK, Weyrauch KW (1980) Growth and anthocyanin production by carrot suspension cultures grown under chemostat conditions with phosphate as the limiting nutrient. Biotechnol Bioeng 22:337–352

Dougall DK, Johnson JM, Whitten GH (1980) A clonal analysis of anthocyanin accumulation by cell cultures of wild carrot. Planta 149:292–297

Ellis BE (1985) Characterization of clonal cultures of *Anchusa officinalis* derived from single cells of known productivity. J Plant Physiol 119:149–158

Harborne JB, Mayer AM, Bar-Nun N (1983) Identification of the major anthocyanin of carrot cells in tissue culture as cyanidin 3-(sinapoylxyloslglucosylgalactoside). Z Naturforsch 38c:1055–1056

Hopp W, Seitz HU (1987) The uptake of acylated anthocyanin into isolated vacuoles from a cell suspension culture of *Daucus carota*. Planta 170:74–85

Matsumoto T, Ikeda T, Kanno N, Kisaki T, Noguchi M (1980) Selection of high ubiquinone 10-producing strain of tobacco cultured cells by cell cloning technique. Agric Biol Chem 44:967–969

Matsumoto T, Kanno N, Ikeda T, Obi Y, Kisaki T, Noguchi M (1981) Selection of cultured tobacco cell strains producing high levels of ubiquinone 10 by a cell cloning technique. Agric Biol Chem 45:1627–1633

Yamada Y, Hashimoto T (1984) Secondary products in tissue culture. In: Collins GB, Petolino JG (eds) Applications of genetic engineering to crop improvement.Nijhoff/Junk, Boston, pp 561–604

Regulation of Secondary Metabolism in Transformed Root Cultures

M.J.C. Rhodes, R.J. Robins, E. Lindsay, H. Aird, J. Payne, A.J. Parr, and N.J. Walton[1]

1 Introduction

Root cultures formed following transformation by *Agrobacterium rhizogenes* offer a stable and reproducible system for the study of secondary metabolism (see review Hamill et al. 1987). The production of the tropane alkaloids, hyoscyamine (atropine) and hyoscine (scopolamine) have been studied by a number of different groups in a range of related solanaceous species (Flores and Filner 1985, Kamada et al. 1986, Mano et al. 1986, Payne et al. 1987). Since transformation involves the expression of plasmid DNA derived from *A. rhizogenes* in the plant genome, transformation can be used to achieve the expression of heterologous DNA following its co-integration into the genome of transformed roots. This offers the prospect ultimately of manipulating the activity of secondary product pathways in culture either by introducing structural genes coding for enzymes acting at critical points or by manipulating regulatory genes controlling their overall operation. We have concentrated on cultures developed from a range of species of *Datura*, focussing particularly upon *D. stramonium*. This chapter will describe physiological and biochemical studies aimed at characterizing tropane alkaloid production in this material and at identifying factors regulating the flux early in the pathway in *Datura* and *Nicotiana*.

2 General Characteristics of Tropane Alkaloid Production in Transformed Root Cultures of *Datura*

Plants of *Datura stramonium* and related *Datura* species were transformed with *Agrobacterium rhizogenes* strain LBA 9402 and the derived transformed root cultures freed of bacteria and established in axenic liquid cultures as previously described (Hamill et al. 1986). The transformed nature of these cultures was confirmed by standard Southern blotting techniques using probes for either T_L or T_R-DNA (Payne et al. 1987). Our original transformed culture of *Datura stramonium*, clone DS1, is maintained by subculture at 14-day intervals. Typically, 0.2 g of root inoculum is introduced into 50 ml of a medium containing Gamborg's B5 salts, and 3% sucrose but lacking hormonal additives (B50 medium). Over 20 days (Fig. 1a) this inoculum grows to about 9 g. This culture produces hyoscyamine

[1]Plant Cell Biotechnology Group, Department of Genetics and Microbiology, AFRC Institute of Food Research, Norwich Laboratory, Colney Lane, Norwich, UK

Primary and Secondary Metabolism of Plant Cell Cultures, ed. by W.G.W. Kurz

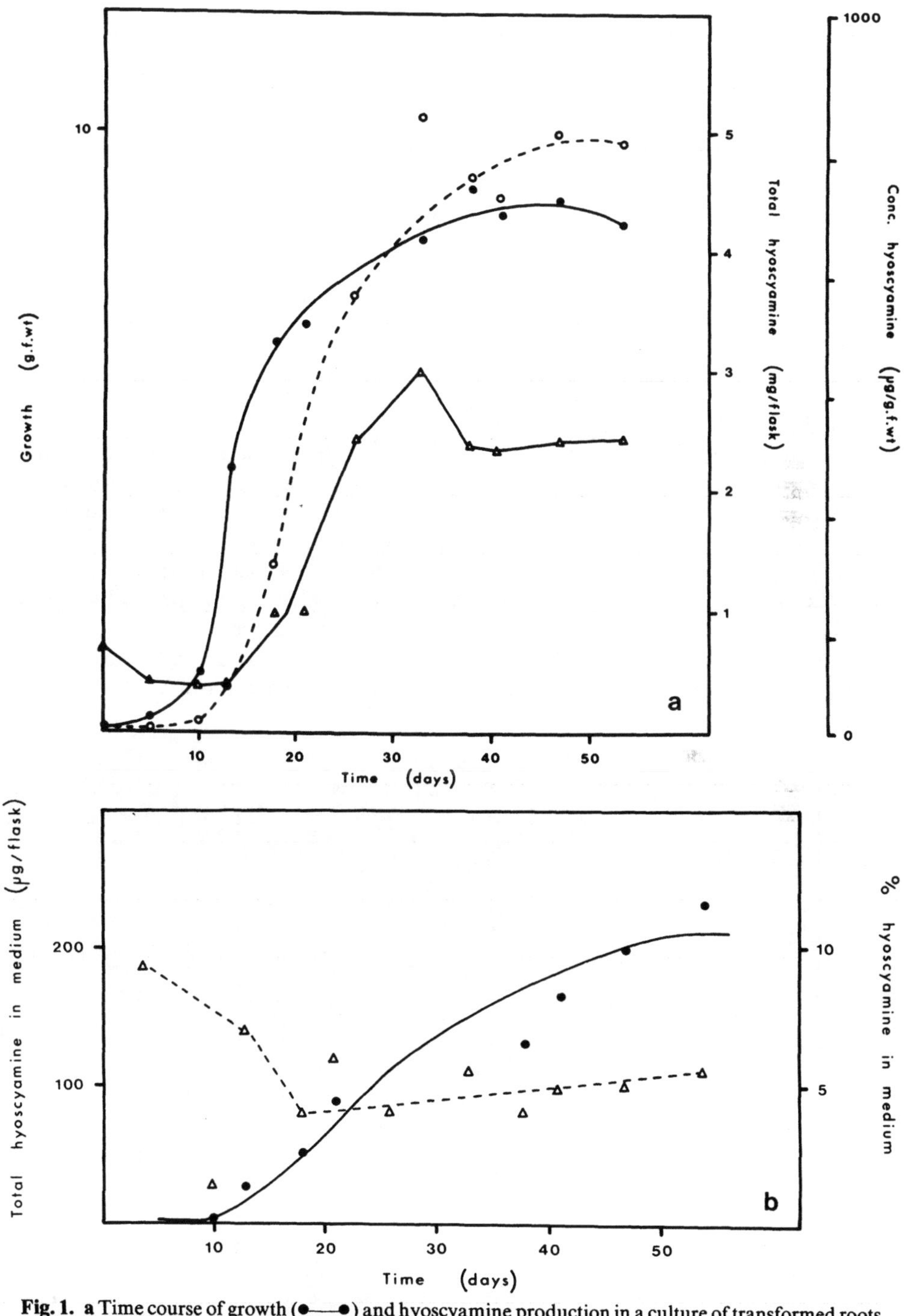

Fig. 1. **a** Time course of growth (●——●) and hyoscyamine production in a culture of transformed roots of *D. stramonium* clone DS1. The inoculum (~ 0.2 g fresh wt), derived from a stock culture subcultured at 14-day intervals, was grown in B50 medium. Hyoscyamine production, expressed as total production (○---○) or as root concentration (△---△) was measured as previously described (Payne et al. 1987). **b** Accumulation of hyoscyamine in the medium during the growth cycle of *D. stramonium* clone DS1. Total hyoscyamine (●——●) and the percentage of the total hyoscyamine present in the medium (△---△) are shown

as the major tropane alkaloid (> 80%) and small amounts of other tropanes including apoatropine, and hyoscine (scopolamine). Alkaloid accumulation (Fig. 1a) lags behind growth in the fashion typical of non-growth related product accumulation in cell suspension cultures. The maximum daily growth rate occurs at about day 11 while the maximum increment in alkaloid content occurs between days 20–22. Three phases of alkaloid accumulation can be discerned, phase 1 up to about day 12 (about the middle part of the fast growth phase) in which the alkaloid concentration is steady or falls slightly, phase 2 in which the alkaloid concentration increases six-fold and finally phase 3 at about the onset of the stationary phase in which the alkaloid concentration remains at this high level. The bulk of the alkaloid produced is retained within the roots. The amount of hyoscyamine in the medium increases (Fig. 1b) as the total hyoscyamine in the system increases but remains at a roughly constant proportion of the total (~ 5%).

If roots from the stationary phase (35-days-old) are subcultured, the same basic pattern is observed (Fig. 2). The growth is slower than with a 14-day-old inoculum, with evidence of a longer lag phase before growth starts, and the maximum biomass level is not reached until day 30. Alkaloid accumulation shows a distinct pattern in phase 1 where there is a steep decrease in alkaloid content from a value typical of phase 3 to a minimum value equivalent to that normally found in phase 1 of 14-day subcultured cells. This is followed by a rise, equivalent to phase 2, back to the level typical of phase 3. Overall growth and total alkaloid accumulation are slightly lower when roots are subcultured at day 14.

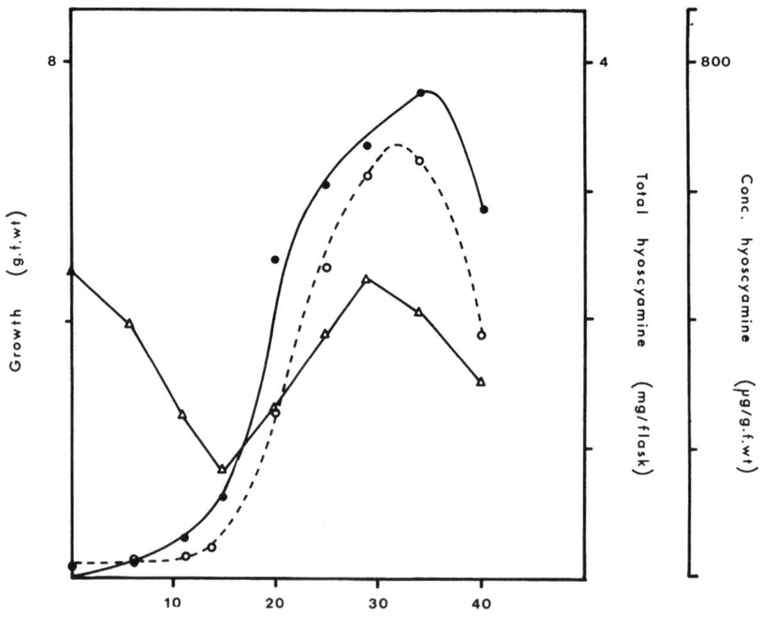

Fig. 2. Time course of growth ●—●) and hyoscyamine production (total production ○---○, root concentration △—△) in transformed roots of *D. stramonium* clone DS1 in which 35-day-old roots rather than 14-day-old roots were used as inoculum

The transition between phases 1 and 2 is dependent on the nutritional status of the culture. If the B50 medium is replenished at 12 days from subculture, there is a 30–40% increase in the final root density over a subsequent 24-day period compared with cultures left in the original medium (Fig. 3a). However, changing the medium suppresses the onset of phase 2 and the associated increase in alkaloid content by 76% (Fig. 3a). This suppression is not observed if the culture is transferred to a medium lacking nitrate (Fig. 3b). In this condition, the concentration achieved at the end of phase 2 is somewhat higher than in the untreated controls. The change into full medium leads to a 50% increase in final root density over the untreated culture while the change into nitrate-free medium leads to a smaller increase (~10–20%). However, the rates of growth are comparable in roots moved to either full or nitrate-free media. There is thus no evidence that suppression of the increase in alkaloid content per unit mass of root material is related to an inhibition of growth. This suppression could be due to an extended phase 1 induced by replenishment of the full medium. Phase 1 normally lasts for about 10 days and in the experiment illustrated in Fig. 3a suppression was followed over 25 days subsequent to changing the medium. In view of the effect of absence of nitrate from the medium, this explanation is unlikely. As a detailed description of the distribution of alkaloid synthetic capacity and accumulation is not available it is not possible to interpret fully the present data. Nevertheless, it does demonstrate that in transformed root cultures the production of a secondary product can be manipulated independently of growth in culture by variation in the composition of the medium and that alkaloid production is not merely an inherent invariant feature of the growth of the organized root structures.

3 Release of Secondary Products by Transformed Root Cultures

Table 1 shows the formation of hyoscyamine over three growth cycles in eight clones of *D. stramonium* derived from plants obtained from a range of sources. The concentration of alkaloid at day 35 varied from 0.3–1.0 mg hyoscyamine per gram fresh weight and for most cultures less than 10% of the production was found in the medium. However, in cultures D2/1 and D2/8 very substantial amounts were, on occasions, found in the medium. These were further investigated as potential high releasing clones. In D2/1 up to 45% was released in the first cycle but less than 5% in cycles 2 and 3. Clone D2/8 released hyoscyamine consistently over the three cycles. Subsequently, it was found that many cultures could give high release of hyoscyamine on occasions. Figure 4 shows the kinetics of hyoscyamine accumulation by one such culture, D15/5. The typical patterns of growth and three phases of intracellular alkaloid concentration are observed and up to day 30 a typical low and steady level of extracellular alkaloid is observed (~5% of the total in the system) but, between days 33 and 40, there is a major increase in hyoscyamine in the medium, up to 30% of the total. This increase occurs while the concentrations of alkaloid in the roots are decreasing. Since the total amount of alkaloid in the flask is also decreasing, it is possible that some of the alkaloid released is degraded. The timing of this late rise in released alkaloid is variable from cycle to cycle and this

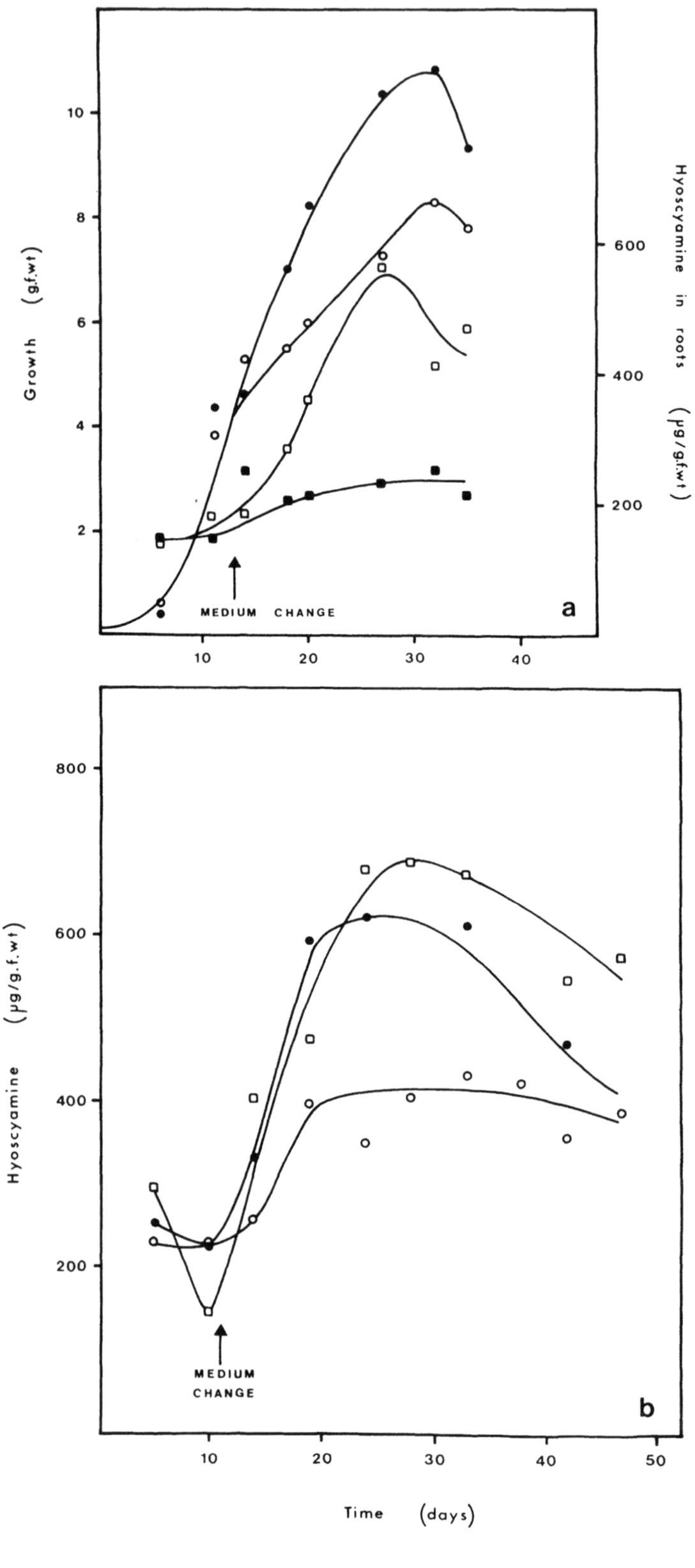

Growth (g.f.wt)
Hyoscyamine in roots (µg/g.f.wt)
MEDIUM CHANGE
a
Hyoscyamine (µg/g.f.wt)
MEDIUM CHANGE
b
Time (days)

Table 1. Production and release of hyoscyamine by transformed root clones of *Datura stramonium*

Hyoscyamine production Cycle No.	1			2			3		
Clone	(mg/ flask)	(mg/g fresh wt)	Percent in medium	(mg/ flask)	(mg/g fresh wt)	Percent in medium	(mg/ flask)	(mg/g fresh wt)	medium)
D3/1	5.9	0.50	5	5.5	0.55	3	5.9	0.65	3
D10/1	8.1	0.78	3	5.9	0.57	2	5.9	0.45	9
D10/9	6.6	0.74	1	6.1	0.86	2	8.0	0.67	6
D10/10	9.2	0.34	2	7.7	0.68	2	7.4	0.89	3
D11/23	8.0	0.67	6	6.8	0.79	3	7.0	0.55	10
D15/11	5.9	0.54	7	7.9	0.62	7	7.0	0.61	6
D2/1	4.8	0.29	45	5.1	0.71	3	7.6	1.02	4
D2/8	6.4	0.47	42	4.2	0.39	28	3.9	0.32	41

explains the variation over several cycles shown in Table 1, which are based on analyses done at a single time point (35 days). It is highly likely that the major rise in the level of extracellular alkaloid is different from the steady low level occurring during the first 30 days of the cycle and is probably related to the onset of senescence in roots maintained for extended periods in stationary phase. It is also clear that the putative high releasing strains (i.e. D2/8, D15/2) do not have the capacity for higher rates of release than other strains under normal conditions. To date no lines of *D. stramonium* with release properties which could be exploited biotechnologically in continuous fermentations have been isolated. This contrasts with the situation in transformed roots of *Nicotiana rustica* where high rates of release were observed in fermenters (Rhodes et al. 1986).

4 Alkaloid Production in Transformed Cultures of Other *Datura Species*

Hyoscyamine is the predominant tropane alkaloid produced in transformed root cultures of *D. stramonium*; only trace amounts of hyoscine are found. This contrasts to the composition of petiole material of the same *D. stramonium* plants from which the cultures were derived. Here, the major alkaloid is hyoscine and the hyoscine:hyoscyamine ratio is between 1:1 and 3:1 (Parr et al. 1988). There is good

←

Fig. 3. a Effect of medium change on the growth (○, ●) and hyoscyamine concentration (□, ■) in transformed roots of *D. stramonium* clone DS1. Roots from a 14-day-old culture were used to inoculate 50 ml of B50 medium. At day 11, the medium was replaced by a further 50 ml of B50 (*filled symbols*) while the control roots (*open symbols*) were left in the original batch of medium. **b** Effect of medium change on the hyoscyamine content of *D. stramonium* clone DS1. At day 11 roots initially in a B50 medium were either untreated (●—●), passaged onto fresh B50 medium (○—○) or passaged into fresh medium lacking nitrate (□—□)

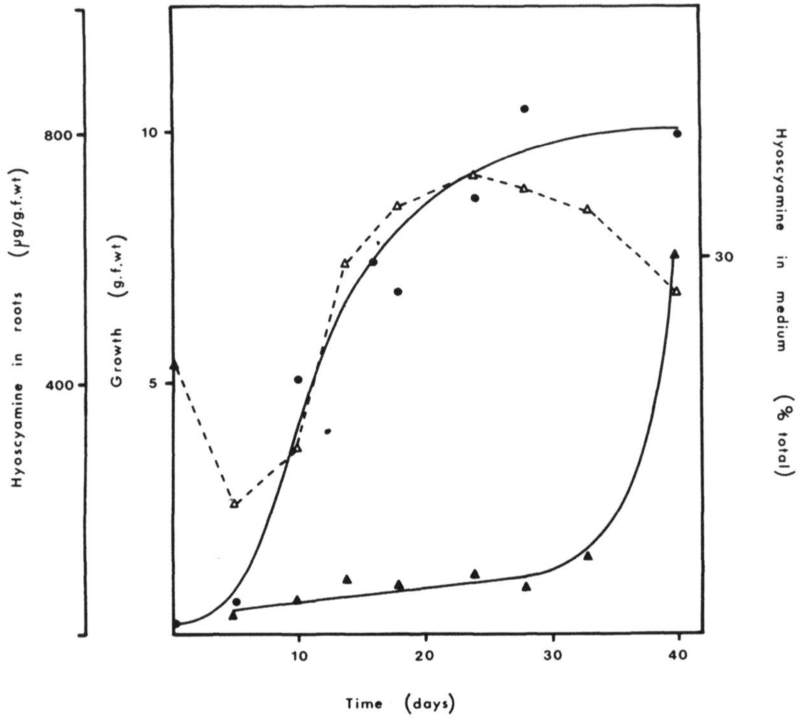

Fig. 4. Time course of growth (●——●), hyoscyamine concentration in the roots (△---△) and the hyoscyamine released into the medium (▲——▲) in a culture of *D. stramonium* clone D15/5 grown on B50 medium

evidence that roots are the exclusive source of hyoscyamine in *Datura* plants and that in some species epoxidation of hyoscyamine to hyoscine mainly occurs following transport to the aerial parts of the plant (Hashimoto et al. 1987). In other species, this epoxidation capacity may be both in the roots and the aerial parts. In our work, some transformed root cultures of *Hyoscyamus muticus* produce twice as much hyoscine as hyoscyamine. We have studied hyoscine production in transformed cultures derived from a *Datura* Fl hybrid (hybrid B5) arising from a cross between *D. candida* and *D. aurea* (El Dabbas and Evans 1982) which produces a high level of total tropane alkaloid with a hyoscine:hyoscyamine ratio of just under 4:1. Transformed root cultures of the material derived using *A. rhizogenes* strain LBA 9402 produce a very complex spectrum of alkaloids amounting to about 3% alkaloid on a dry weight basis. Hyoscine and hyoscyamine account for ~50% of the alkaloid products. Figure 5 shows that the culture derived from hybrid B5 grows rather more slowly than DS1 and that hyoscine production follows growth more closely than in cultures of *D. stramonium.* The ratio of hyoscine:hyoscyamine rises from about 1 at day 5 to over 3 at day 19. The maximal alkaloid levels are typically about 200 µg hyoscyamine and 500 µg hyoscine per gram fresh weight. However, the diversity of accumulated alkaloid, related to its hybrid character, makes such material less attractive for biotechnological exploitation.

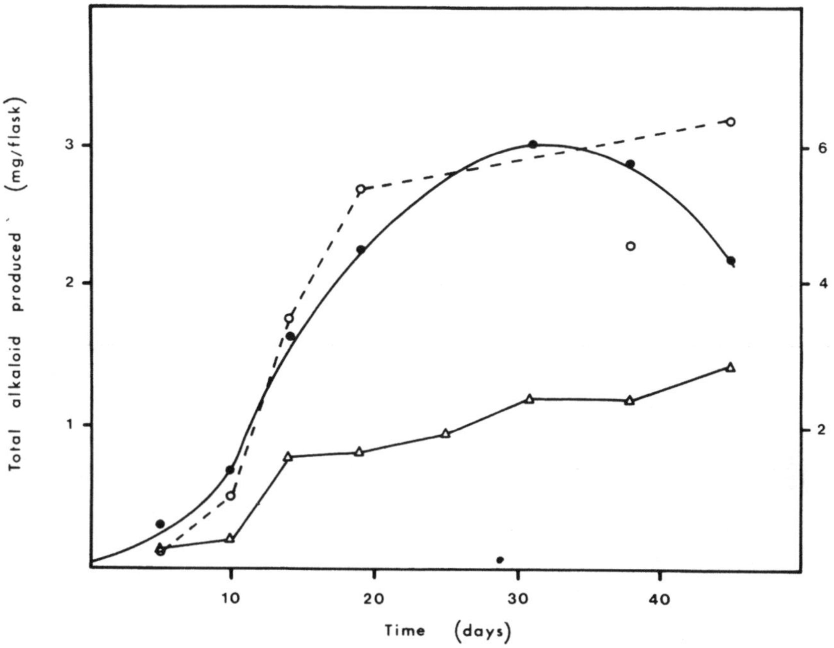

Fig. 5. Time course of growth (●—●), hyoscine (○---○) and hyoscyamine (△—△) production in a transformed root culture of the *Datura* hybrid B5 grown on B50 medium

An interesting species from this viewpoint is *D. wrightii*, which yields transformed root cultures producing an alkaloid spectrum with only two major alkaloid components. These are hyoscyamine and a component present at two to three times the concentration of hyoscyamine which runs very close to hyoscine in the standard HPLC separation (Payne et al. 1987) but which is distinct from it. The production of hyoscyamine in this species (see Fig. 6), as distinct from *D. stramonium*, is approximately growth related and maximal alkaloid accumulation is achieved in 15 days as opposed to 45 days in *D. stramonium* (Fig. 1a). The alkaloid concentration shows a small rise (~ two-fold) early in the culture and rapidly achieves a steady value.

5 Regulation of Alkaloid Biosynthesis in *Datura* and *Nicotiana*

The tropane alkaloids in *Datura* and related pyridine alkaloids in *Nicotiana* are derived from putrescine and very probably share a common pathway to the formation of the N-methyl pyrrolinium ion. The biochemical origin of N-methyl putrescine is a matter of dispute. Three routes (see scheme 1) have been proposed: (1) from ornithine via putrescine under the influence of ornithine decarboxylase (ODC); (2) from ornithine via N-methylornithine; (3) from arginine via agmatine and putrescine following decarboxylation with arginine decarboxylase (ADC). Current research suggests that the routes via ODC and ADC may both be

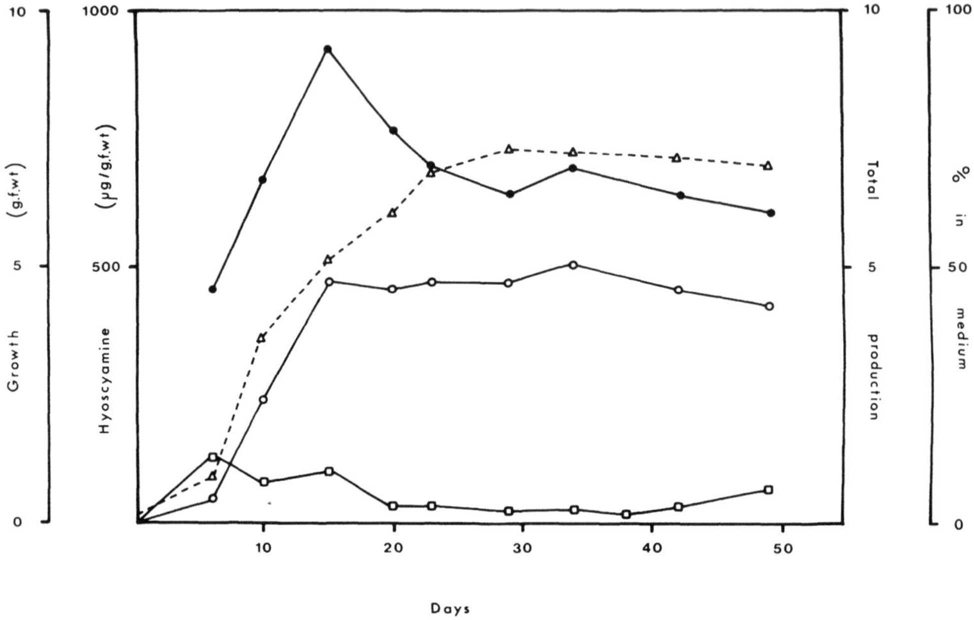

Fig. 6. Time course of growth (△---△), total hyoscyamine production (○—○) hyoscyamine concentration in the roots (●—●) and hyoscyamine in the medium (□—□) in a transformed root culture of *Datura wrightii* grown on B50 medium

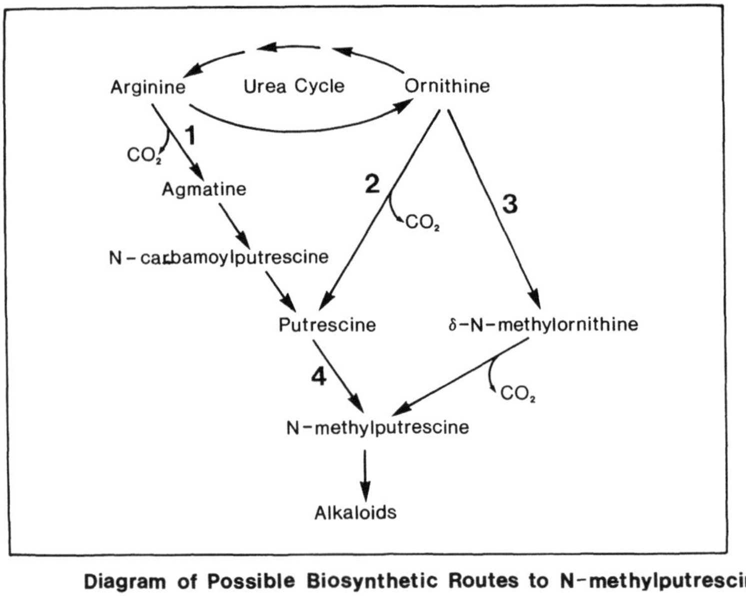

Diagram of Possible Biosynthetic Routes to N-methylputrescine

Scheme 1. The various possible pathways by which N-methylputrescine may be derived from ornithine or arginine. Enzymes involved are: (1) arginine decarboxylase; (2) ornithine decarboxylase; (3) orninthine δ-N-methylase (putative); (4) SAM:putrescine methyl transferase important but that via ADC may be quantitatively the more important, at least in *Nicotiana* (Tiburcio and Galston 1986). Following the production of N-methylpyrrolinium the pathways diverge; in *Nicotiana* condensation with nicotinic acid yields nicotine while in *Datura* a separate condensation with an acetoacetate unit leads to hygrine formation and subsequently to tropine which condenses with tropic acid to form hyoscyamine. Feeding intermediates to transformed cultures of *Datura stramonium* (Table 2) has only a relatively small effect on overall alkaloid production. Putrescine and ornithine (data not shown) have no significant effect on hyoscyamine formation while low levels of N-methyl putrescine (3 mM) increase the concentrations and total yield of hyoscyamine by 30–40%. However, further amounts of N-methyl putrescine (up to 6 mM) increase the alkaloid content of the roots further but do not further increase total yield since they inhibit growth to a small extent.

As an approach to manipulating alkaloid production in transformed cultures of *D. stramonium* and *N. rustica* by biochemical and genetic means we have studied the steps catalyzed by ODC, ADC, putrescine N-methyltransferase (PMT) and N-methylputrescine oxidase (MPO). Figure 7 shows changes in the levels of activity of these four enzymes during culture of *D. stramonium* D15/5. Particular care has been taken in making measurements of ODC, since this is a tightly regulated enzyme, being subject to control by its product, putrescine, via feedback inhibition, repression and stimulation of a proteinaceous enzyme inhibitor (Karomilas et al. 1985). Furthermore, the problems associated with the assay of the enzyme have been avoided (Smith and Marshall 1988). At all stages the enzymes of putrescine synthesis (ODC and ADC) are present at lower levels than the enzymes of putrescine utilization (PMT and MPO). Three of the four enzymes show rapid increases in activity to a maximum at day 10 just before the period of the most rapid phase of alkaloid accumulation (days 10–14). PMT and MPO show increases in activity of three- and two-fold respectively during the period between 3 and 10 days. A much smaller increase in ODC was observed during the same period while ADC showed no significant change in activity. In all cases there is a

Table 2. Effect of precursor feeding on accumulation of hyoscyamine in transformed root cultures of *D. stramonium*[a]

	Expt. 1		Expt. 2	
	Hyoscyamine production			
	(μg/g fresh wt)	(mg/day/flask)	(μg/g fresh wt)	(mg/day/flask)
Control	290 ± 17	0.18 ± 0.02	352 ± 37	0.27 ± 0.04
Putrescine				
3 mM	321 ± 20	0.21 ± 0.02	–	
6 mM	336 ± 60	0.20 ± 0.08	378 ± 47	0.28 ± 0.04
N-Methyl Putrescine				
3 mM	–	–	470 ± 4	0.38 ± 0.04
6 mM	–	–	530 ± 29	0.38 ± 0.04

[a] The roots were cultured in B50 medium or B50 medium containing the stated amine levels for 14 days (Expt. 1) or 16 days (Expt. 2) and analyzed for hyoscyamine production in the standard manner (Payne et al. 1987).

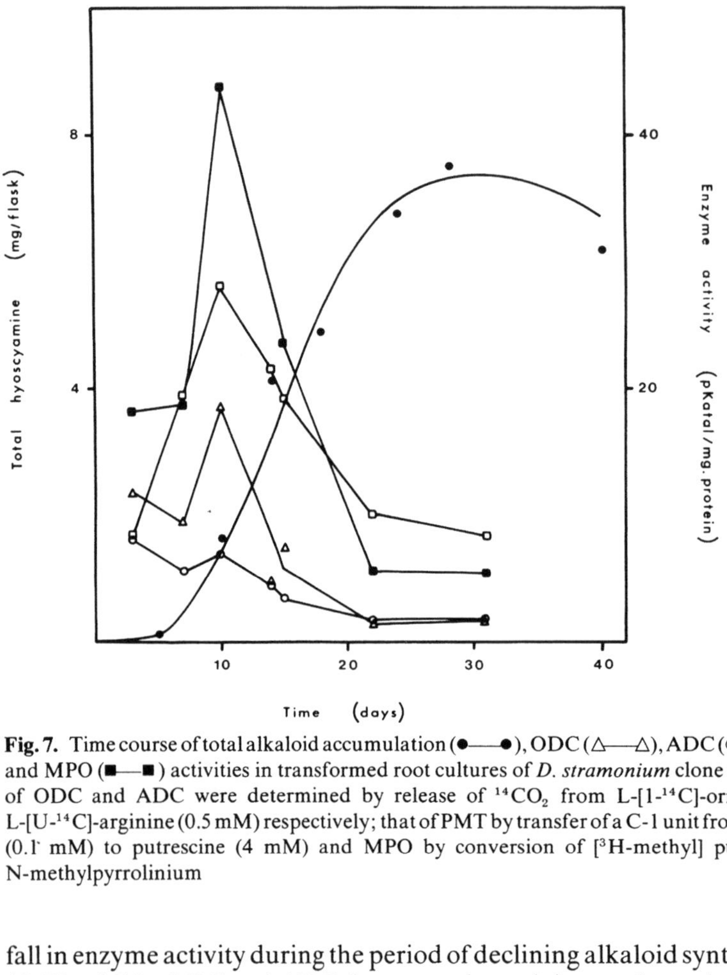

Fig. 7. Time course of total alkaloid accumulation (●——●), ODC (△——△), ADC (○——○), PMT (□——□) and MPO (■——■) activities in transformed root cultures of *D. stramonium* clone D15/5. The activities of ODC and ADC were determined by release of $^{14}CO_2$ from L-[1-^{14}C]-ornithine (2 mM) and L-[U-^{14}C]-arginine (0.5 mM) respectively; that of PMT by transfer of a C-1 unit from [^{14}C-methyl]-SAM (0.1' mM) to putrescine (4 mM) and MPO by conversion of [^{3}H-methyl] putrescine (5 mM) to N-methylpyrrolinium

fall in enzyme activity during the period of declining alkaloid synthesis beyond day 15. The fall in ODC and ADC depresses the activity to a lower level relative to the initial activity than in PMT and MPO. If the potential flux through each of the four enzymes is compared with the mean flux into alkaloids over four different time intervals (Table 3), it is clear that the maximum flux capacity of the enzymes is greater in all cases than the rate of alkaloid formation. However, the excess flux is below six-fold for each enzyme during the phase of maximal accumulation of alkaloid. Since the flux through the enzymes is measured at unphysiological, near-saturating levels of substrate, it is possible that the levels of available enzyme activity could limit the rate of alkaloid biosynthesis. The lowest excess fluxes were found with the two decarboxylases. If either were exclusively committed to tropane formation, the excess would fall to below two-fold. Since these enzymes are involved in a number of pathways other than that to form tropanes (e.g. polyamine and γ-aminobutyric acid formation) a case that the ability to form putrescine may be limiting can be made. The first enzyme solely committed to tropane formation in *Datura*, PMT, is present at only six- to seven-fold excess and it is possible that its level of activity may be limiting under certain conditions.

Table 3. Estimates of the synthetic capacities of individual enzymes of tropane alkaloid biosynthesis and the flux into alkaloids in a transformed root culture of *D. stramonium* D15/5

Time interval (days)	Total alkaloid accumulation over the time interval (μmol/flask) (A)	Total synthetic capacity over time interval of biosynthetic enzymes[a] (μmol/flask) ODC	ADC	ODC+ ADC (B)	Excess B over A (fold)	PMT (C)	Excess C over A (fold)	MPO (D)	Excess D over A (fold)
0– 7	1.4	3.0	2.5	5.5	3.9	9.2	6.6	6.6	4.7
7–14	11.2	14.8	14.7	29.5	2.6	75	6.7	44	3.9
14–22	8.8	22.2	23.5	45.7	5.2	12.3	14.0	69	7.8
22–31	1.4	12.5	12.6	25.1	18	90	64	54	38

[a] For conditions of assay see legend to Fig. 7.

In cell suspensions of *Nicotiana*, there is evidence that the level of PMT activity, and to a lesser extent, that of MPO may be significant factors limiting the rate of alkaloid biosynthesis (Feth et al. 1986). Flores (1986) observed that the ability of transformed root cultures of *Hyoscyamus* to synthesize alkaloids was suppressed in media which induced callus formation. Aird et al. (1988) devised a medium NK (Gamborg's B5 salts, 3% sucrose, NAA 2.0 mg l^{-1} and kinetin 0.2 mg l^{-1}) which rapidly and reproducibly promoted disorganization of the transformed root cultures of *N. rustica* to produce transformed cell suspension cultures and this was associated with the rapid loss of the ability to produce nicotine. This medium was used to investigate the role of PMT in the control of nicotine biosynthesis.

Changes in the activity of the four enzymes (ADC, ODC, PMT and MPO) were followed over an 8-day period in tissue maintained in a medium sustaining root growth, (B50) or in one inducing callus formation (NK) (Fig. 8). Over this period growth measured by dry weight increase, occurs in both media; growth being slightly faster in the callus-inducing medium. Over the first 6 days the nicotine content of roots in the NK medium falls after a lag of 1 day by a factor of 4.5, while there is an increase in dry matter of four-fold, also after a 1-day lag. The total amount of nicotine in such material does not change significantly. Thus, the fall in nicotine content can be explained entirely by the dilution of the nicotine content of the original root inoculum by newly produced cells lacking the capacity for nicotine synthesis without invoking a mechanism involving an enhanced rate of nicotine degradation. Thus, the hormonal treatment directly or indirectly leads to the cessation of nicotine synthesis. This cessation correlates with a very rapid loss of PMT ($>$ ten-fold) for roots in NK medium while ODC activity remains at or slightly above the initial level. The activities of ADC and MPO (data not shown) show a similar pattern of change to that of ODC. The pattern of enzyme changes in *Nicotiana* roots maintained in B50 medium is similar to the pattern observed in *Datura*. This observation strengthens the argument that the level of PMT plays an important role in the formation of alkaloids.

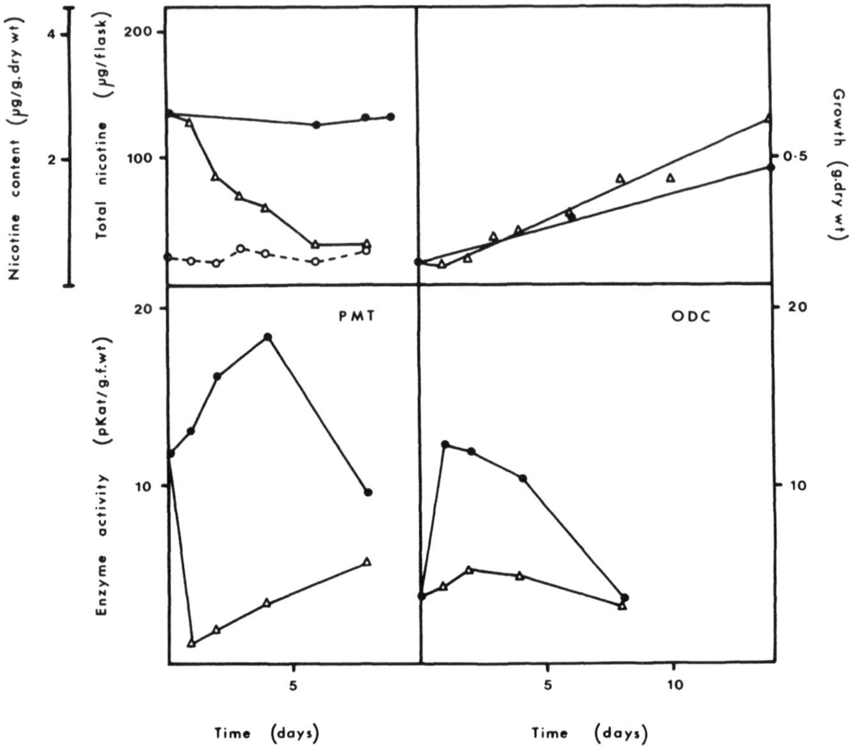

Fig. 8. Change in nicotine concentration (●, △) the total level of nicotine (○) and the activities of ODC and PMT in transformed root cultures of *Nicotiana* cultured either in B50 medium (●—●) or in NK medium (△—△); 1.0 g of roots were incubated in 50 ml of medium in 250-ml flasks and shaken at 90 rpm for the periods specified at 25 °C. Samples were taken for analysis of nicotine levels, the enzyme activities and growth (△, ●) by the techniques previously described

6 Conclusions

Cultures of a range of *Datura* species are capable of reproducible growth at high rates in transformed root culture. By suitable selection of starting material it is possible to generate cultures producing a variety of alkaloid products, some cultures essentially yielding single compounds, others complex mixtures. Thus, this material provides a high degree of flexibility in alkaloid production in initial cultures which can be exploited in subsequent empirical or genetic programs of strain improvement. The possibility of exploiting such cultures in continuous fermentation processes with recovery of products from culture medium is dependent on the development of improved strains which release more product during growth. Such processes will be facilitated by the use of external adsorbents which stimulate the level of release (Rhodes et al. 1986). The transfer of DNA, inherent in the processes leading to transformed root formation, offers considerable potential for the genetic manipulation of transformed roots to stimulate accumulation of specific products. Binary vector systems have been developed for the purpose of inserting de-regulated or foreign genes into transformed roots (Hamill et al. 1988). Such an approach is hampered by a lack of detailed knowledge of the enzymology of secondary pathways, of the genes coding for these enzymes and how the action of such genes is regulated and co-ordinated. In approaching the genetic manipulation of such systems it is essential to target the long and involved work to isolate and characterize genes to those coding for enzymes whose expression or activity limits the flux through the particular pathway. Our work has concentrated on the early part of the tropane and pyridine alkaloid pathway up to the formation of the common intermediate N-methyl pyrrolinium. The evidence from *Datura* suggests that the decarboxylases are present at relatively low levels which, considering the diversity of products formed and the low in vivo levels of substrates, barely account for the observed levels of akaloid accumulation. The manipulation of the genome to effect overexpression of ODC and a concomitant stimulation of the in vivo levels of putrescine might thus be expected to stimulate flux through the pathway. The narrow range of the maximum catalytic capacities measured for the four enzymes might be taken to indicate that the net flux through the pathway is unlikely to be simply controlled only at a single step. However, as the conditions (e.g. of pH, substrate and co-factor concentrations) within the compartment in which the enzymes operate have not been determined, the values presented indicate the potential capacities of the individual enzymes, rather than estimates of their in vivo activities.

Evidence from the precursor feeding experiments suggests that the supply of N-methylputrescine, but not of putrescine, may limit alkaloid formation. Indeed, the large stimulation of activity of the two enzymes of putrescine utilization (PMT and MPO) compared with those of putrescine formation in *Datura* immediately before the period of rapid alkaloid accumulation is consistent with limitation in the pathway beyond that of putrescine formation. This pattern observed in the induction of alkaloid biosynthesis during the growth cycles in transformed roots of *Datura* is rather similar to that observed in cell suspensions of *Nicotiana* transferred from a growth medium to one inducing alkaloid formation (Feth et al. 1986). Not only in both cases is the proportionate increase in PMT activity greater than that in PMO, but also ODC and ADC are affected to a much smaller extent, if at all. These data suggest that the level of PMT activity in both organized and disorganized cultures can be an important determinant of alkaloid biosynthesis. This argument is strengthened by the rapid loss of PMT activity, but not of PMO, ADC and ODC, accompanying the inhibition of alkaloid biosynthesis when transformed roots of *N. rustica* are transferred to callus-inducing media. The fact that PMT activity falls quickly (~ ten-fold in 24 h) suggests that the turnover of this enzyme is rapid and that its level in the tissue is very sensitive to changes in the rate of synthesis of the enzyme protein. The present evidence suggests that the activity of PMT, the first committed enzyme in the pathway of alkaloid biosynthesis, is more sensitive to factors that stimulate or repress nicotine biosynthesis than other enzymes acting at the early stages of the pathway and that, even if this step may not be rate-limiting in the catalytic sense, it is at this step that control of expression of the pathway is determined in both *Nicotiana* and *Datura*.

References

Aird ELH, Hamill JD, Robins RJ, Rhodes MJC (1988) Chromosome stability in transformed root cultures and the properties of variant lines of *Nicotiana rustica* hairy root. In: Robins RJ, Rhodes MJC (eds) Manipulating secondary metabolism in cultures. Cambridge University Press, Cambridge, UK, pp 137–144

El-Dabbas SW, Evans WC (1982) Alkaloids of the genus *Datura*, Section Brugmansia X, Alkaloid content of *Datura* hybrids. Planta Medica 44:184–185

Feth F, Wagner R, Wagner KG (1986) Regulation in tobacco callus of enzyme activities of the nicotine pathway. 1. The route of ornithine to methylpyrroline. Planta 168:402–407

Flores HE, Filner P (1985) Metabolic relationships of putrescine, GABA and alkaloids in cell and root cultures of Solanaceae. In: Neuman K-H, Barz W, Reinhard E (eds) Primary and secondary metabolism of plant cell cultures. Springer, Berlin Heidelberg New York pp 174–185

Flores HE (1986) Use of plant cell and organ culture in the production of biological chemicals. In: Lebaron H, Mumma RO, Honeycutt RC, Dussing JH (eds) Application of biotechnology to agricultural chemistry. Amer Chem Soc Symp Series 190 (in press)

Hamill JD, Parr AJ, Robins RJ, Rhodes MJC (1986) Secondary product formation by cultures of *Beta vulgaris* and *Nicotiana rustica* transformed with *Agrobacterium rhizogenes*. Plant Cell Reports 5:111–114

Hamill JD, Parr AJ, Rhodes MJC, Robins RJ, Walton NJ (1987) New routes to plant secondary products. Biotechnology 5:800–804

Hamill JD, Evans DM, Robins RJ, Rhodes MJC (1988) Foreign gene insertion into transformed roots with binary vectors and *Agrobacterium rhizogenes* – potential for genetic manipulation of secondary metabolism. In: Robins RJ, Rhodes MJC (eds) Manipulating secondary metabolism in cultures. Cambridge University Press, Cambridge, UK, pp 145–153

Hashimoto T, Kohno J, Yamada Y (1987) Epoxidation in vivo of hyoscyamine to scopolamine does not involve a dehydration step. Plant Physiol, 84:144–147

Kamada H, Okamura N, Satake M, Harada H, Shimomura K (1986) Alkaloid production by hairy root cultures in *Atropa belladonna*. Plant Cell Reports 5:239–242

Karomilas AE, Pangiotidis CA, Kyriakidis DA (1985) Regulation and function of ornithine and arginine decarboxylase in germinating barley seeds. In: Recent Progress in Polyamine Research 33:501–508

Mano Y, Nabeshima S, Matsui C, Ohkawa H (1986) Production of tropane alkaloids in hairy root cultures of *Scopolia japonica*. Agric Biol Chem 50:2715–2722

Parr AJ, Hamill JD, Payne J, Rhodes MJC, Robins RJ, Walton NJ (1988) Exploiting variation in tropane alkaloid production by solanaceous plants. In: Robins RJ, Rhodes MJC (eds) Manipulating secondary metabolism in culture. Cambridge University Press, pp 115–122

Payne J, Hamill JD, Robins RJ, Rhodes MJC (1987) Production of hyoscyamine by "hairy root" cultures of *Datura stramonium*. Planta Medica 53:474–478

Rhodes MJC, Hilton M, Parr AJ, Hamill JD, Robins RJ (1986) Nicotine production by "hairy root" cultures of *Nicotiana rustica*: fermentation and product recovery. Biotechnol Lett 8:415–420

Smith TA, Marshall JHA (1988) The oxidative decarboxylation of ornithine by extracts of higher plants. Phytochemistry 27:703–710

Tiburcio AF, Galston AW (1986) Arginine decarboxylase as the source of putrescine for tobacco alkaloids. Phytochemistry 25:107–110

The Regulation of Accumulation of Lower Isoprenoids in Plant Cell Cultures

B.V. Charlwood[1], C. Moustou[1], J.T. Brown[1,3], P.K. Hegarty[1,4], and K.A. Charlwood[2]

In comparison with other classes of natural products, few studies have been carried out concerning the accumulation of monoterpenes in plant cells in culture. The subject has been reviewed (Charlwood et al. 1986) and a compilation of data is available (Koch-Heitzmann and Schultze 1988).

Oil accumulation in a variety of culture systems derived from some 25 genera of essential oil-containing plants has been reported. Monoterpene accumulation occurs most commonly within callus tissue, although significant accumulation in suspension cultures from eight species (*Apium graveolens, Mentha piperita, Ocimum basilicum, Pelargonium fragrans, Perilla frutescens, Ruta graveolens, Thuja occidentalis* and *Vitis vinifera*) has been claimed. Suspensions of *O. basilicum* also accumulate monoterpenes in the form of their glycoside derivatives. Unfortunately, it is very difficult to deduce from the literature the exact degree of cell specialization present in these accumulating systems. However, it may be concluded that, typically, an element of organization must be present in order for *significant* accumulation of the lower isoprenoids to be observed.

A somewhat confused picture exists concerning the actual sites of synthesis of monoterpenes both in vivo and in vitro. In many herbaceous plants synthesis is claimed to occur in epidermal glandular structures, but the producing cells may not necessarily be associated with the trichome itself. In *P. frutescens* essential oil is synthesized in large resin-containing glands on the lower leaf surface, whilst the young epithelial cells of resin ducts are the probable sites of synthesis of monoterpene hydrocarbons in pine.

Studies on species of *Pinus, Citrus* and *Phellodendron* have indicated that within the producing cell, monoterpenes are synthesized on the plastid membrane and transported to the storage site via the endoplasmic reticulum (see Heinrich and Schultze, 1985). It has been clearly demonstrated that the presence of functional chloroplasts is not necessary for monoterpene synthesis (Banthorpe and Wirz-Justice 1972). Indeed, there is a close correlation between the presence of leucoplasts in secretory cells and the accumulation of monoterpenes, but not sesquiterpenes, in 45 species of oil-accumulating higher plants (Cheniclet and Carde

[1]Plant Cell and Molecular Sciences Group, King's College London, Campden Hill Road, London W8 7AH, UK
[2]School of Life Sciences, University of London Goldsmith's College, New Cross, London SE14 6NW, UK
[3]Present address: Chemistry Department, University College London, UK
[4]Present address: Bass Ltd, Burton-on-Trent, Staffordshire, UK

Primary and Secondary Metabolism of Plant Cell Cultures, ed. by W.G.W. Kurz

1985). Intact leucoplasts, isolated from *Citrofortunella mitis,* were able to convert exogenous IPP (isopentenyl pyrophosphate) into the monoterpene hydrocarbons limonene and α- and β-pinene (Gleizes et al. 1987), however, the maximal formation of monoterpenes required an exogenous supply of DMAPP (dimethylallyl pyrophosphate). This finding supports the hypothesis of a metabolic pool of DMAPP, the existence of which was originally suggested to account for the asymmetric labelling pattern established in monoterpenes biosynthesized from [2–^{14}C]-MVA (mevalonic acid). In contrast, isolated chromoplasts from daffodil flowers converted IPP into monoterpene hydrocarbons, linalol and carotenoids in the absence of exogenous DMAPP (Mettal et al. 1988). The demonstration that no sesquiterpenes were synthesized in either isolated leucoplasts or chromoplasts supports the view that the sites of mono- and sesquiterpene synthesis are separate, the latter probably being produced in the endoplasmic reticulum.

The sites of storage of essential oils in whole plants have been well characterized. These are mainly associated with specialized secretory structures such as glandular hairs as in the genus *Pelargonium* or resin ducts as in *Thuja* and *Pinus*. In roots of *Centranthus* species, however, the iridoid monoterpenes (the valepotriates) are stored in lipid vesicles associated with the vacuolar membrane of the parenchyma cells.

It is generally accepted that the biosynthesis of lower isoprenoids is primarily controlled at the mevalonic acid level with only the fine tuning taking place during the later steps of functional elaboration. Unfortunately, the localization of the enzymes involved in the isoprenoid pathway is not well understood. The putative regulatory enzyme HMG-CoA (3-hydroxy-3-methylglutaryl-coenzyme A) reductase has been characterized in chloroplasts, chromoplasts, mitochondria, endoplasmic reticulum and cytoplasm of a number of higher plants (Bach 1987), and segregation and regulation of the different routes of isoprenoid synthesis is probably occasioned by intracellular compartmentalization. It is not clear whether leucoplasts contain their own HMG-CoA reductase. The leucoplastidic membrane is, however, fully permeable to both IPP and DMAPP and regulation of monoterpene synthesis in this organelle may be mediated via the availability of exogenous C5-substrate. Even if leucoplasts are able to synthesize MVA, control of monoterpene synthesis may still be regulated by the size of the exogenous DMAPP pool. Recent histological techniques devised to facilitate the localization of enzymes of the early part of the isoprenoid pathway (Curry 1987) may prove of value in determining the source of the DMAPP pool.

It has often been claimed that a degree of cytodifferentiation, or even morphological differentiation, is a prerequisite for both the synthesis and accumulation of lower isoprenoids. Typical levels of essential oil accumulated by long-term callus material raised from variants of *Pelargonium* and grown under continuous, subdued light (3 $\mu E\ m^{-2}\ s^{-1}$) on MS (Murashige and Skoog) medium, containing 5 mg l^{-1} BAP and 1 mg l^{-1} NAA (6-benzylaminopurine and 1-naphthaleneacetic acid respectively), are shown in Table 1. Different callus lines from the same variant consistently accumulated widely differing amounts of oil. There is clearly no correlation between oil accumulation and morphology of the callus indicating that the regulation of these two characters is not linked.

It is not axiomatic that trace or non-accumulating cultures have lost the enzymatic capability to synthesize essential oils. Cell-free systems derived from

Table 1. Accumulation of essential oil in callus of *Pelargonium* variants

Variant	Cell line	Accumulation of oil [a] (%)	Callus morphology
P. graveolens L'Her	(i)	0.001	White, friable
	(ii)	0.033	White, friable
P. tomentosum Jacq.	(i)	0.001	Green, friable/hard
	(ii)	0.01	Green, friable/hard
	(iii)	0.67	Green, friable/hard
P. fragrans Willd.		0.01	White/purple, friable
P. quercifolium L'Her	(i)	0.35	Purple/orange/green, friable
	(ii)	1.42	Purple/orange/green, friable
P. australe Willd.		2.18	White, friable
P. 'Miss Australia'		5.0	White/pink, friable

[a] Oil accumulation expressed as percent of the oil accumulated by the parent plant.

non-accumulating callus cultures of *Rosmarinus officinalis, Lavandula angustifolia* and *Tanacetum vulgare* exhibited MVA-kinase, MVAP-kinase, MVAPP-decarboxylase and IPP-DMAPP isomerase activities comparable with those of the parent plant (Banthorpe et al. 1986), whilst prenyltransferase activity was some 300-fold greater than in the intact plant. On the other hand, a suspension culture derived from an accumulating callus of *Pinus radiata* neither accumulated, nor possessed the ability to synthesize, monoterpenes. Similarly, suspension cultures (consisting of single cells and small aggregates of up to 15 cells) derived from *Pelargonium fragrans* callus and maintained under continuous, subdued light (growth regime indicated above) accumulated either very low or non-detectable amounts of monoterpenes (Brown and Charlwood 1986a). However, exogenous [2-^{14}C]-MVA was readily absorbed by these cultures and within 24 h approximately 22% of the substrate was incorporated into phytosterols and some 0.6% was trapped in hexane-soluble products. Chromatographic analysis of the hexane extract showed that tracer was almost equally distributed between the mono- and sesquiterpene components normally associated with the oil of the plant (Fig. 1A). The theoretical rate of conversion of MVA into these lower isoprenoids was 0.58 mmol day^{-1} g^{-1} fresh weight of cells. Under similar experimental conditions, conversion of [2-^{14}C]-acetate into mono- and sesquiterpenes (0.24% incorporation) occurred at the rate of 30 nmol day^{-1} g^{-1} fresh weight of cells. This much lower rate of incorporation from acetate is as expected for a multifunctional substrate. It is not clear whether de novo synthesis of monoterpenes was limited to a small number of active producer cells, or occurred at a low level in all cells within the culture.

The quality of light and length of photoperiod are known to play an important regulatory role in the accumulation of monoterpenes in intact plants. The qualitative and quantitative accumulation of α- and β-pinene by callus cultures of *Pinus radiata* was markedly influenced by the light regime (Banthorpe and Njar 1984). In the absence of light the cultures formed toluene and acetone at the expense of the monoterpene hydrocarbons. In *Pelargonium fragrans* suspension cultures, a stimulation in the accumulation of essential oil of some 200-fold (Table 2) could be

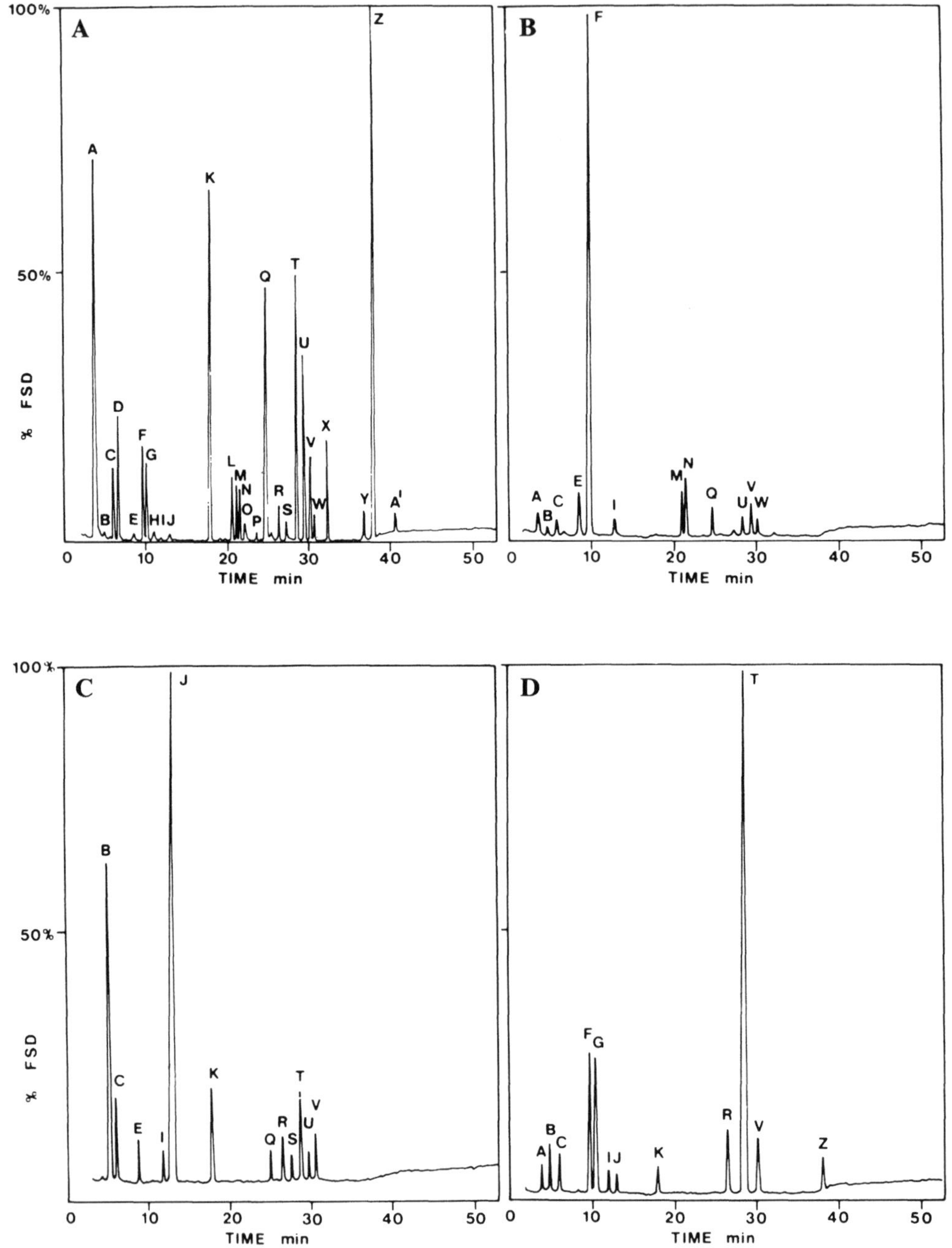

Fig. 1A-H. Oil accumulated by *Pelargonium fragrans*: **A** intact plant; **B** suspension culture (18:6 photoperiod); **C** suspension treated with 10 mM acetate; **D** suspension treated with 10 mM aspartate; **E** aggregate culture; **F** immobilized culture; **G** shoot-proliferation culture; **H** suspension treated with colchicine. *A* α-pinene; *B* camphene; *C* β-pinene; *D* sabinene; *E* myrcene; *F* limonene; *G* 1,8-cineole; *I* γ-terpinene; *J* *p*-cymene; *K* fenchone; *L,M* sesquiterpenes; *N* isomenthone; *O* camphor/sesquiterpene; *P* linalol; *Q* farnesene; *T* sesquiterpene; *V* carvone/cadinene; *Z* methyl eugenol

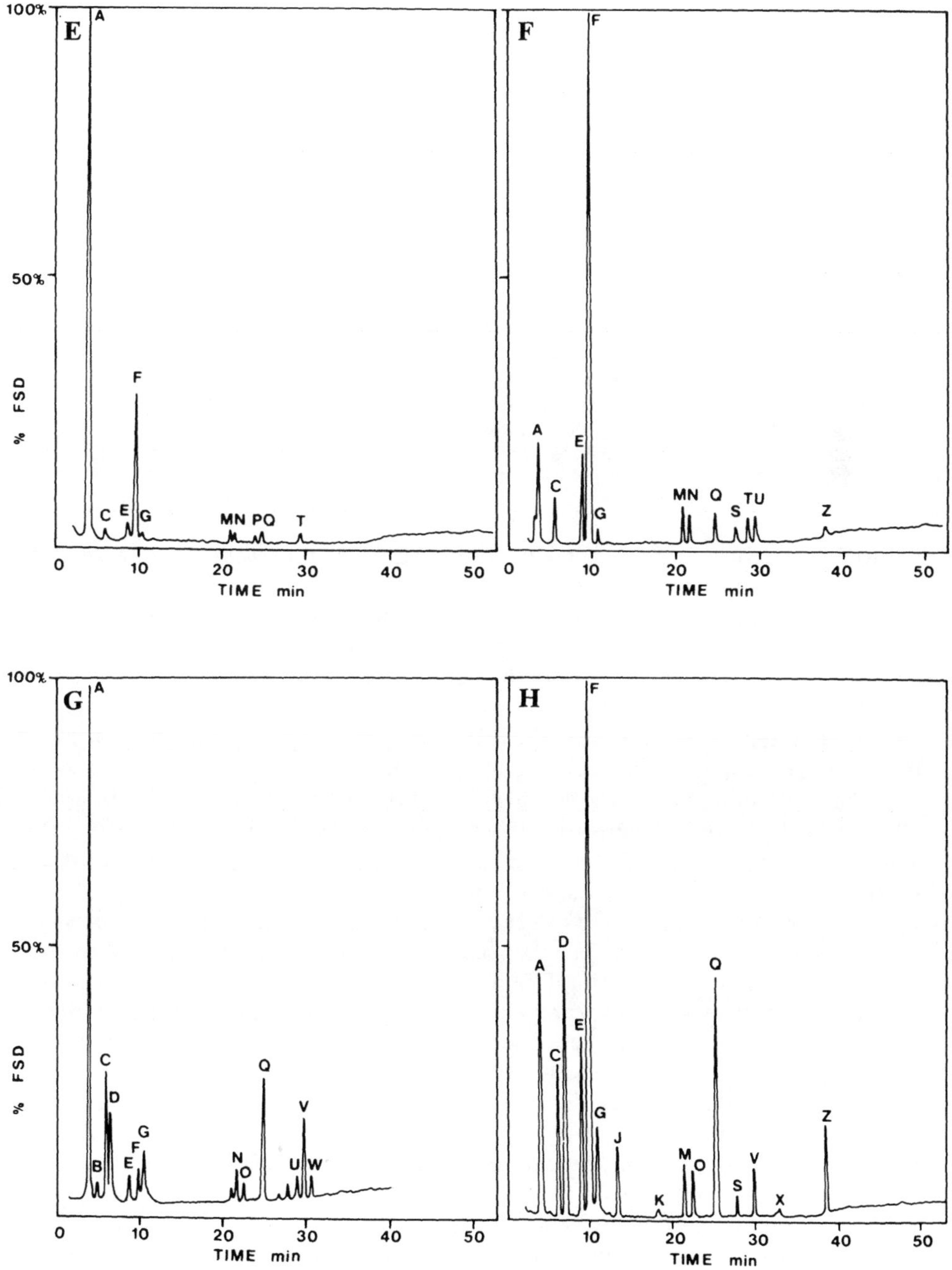

E
F
G
H
100%
50%
% FSD
TIME min
A
B
C
D
E
F
G
J
K
MN
M
N
O
PQ
Q
S
TU
T
U
V
W
X
Z
0
10
20
30
40
50

Fig. 1E-H

Table 2. Accumulation of essential by *Pelargonium frangrans* cultures

Culture type	Growth conditions	Accumulation of oil ($\mu g\ g^{-1}$ fresh weight)
Undifferentiated callus	5 mg $^{-1}$ BAP: 1 mg l^{-1} NAA Continuous, subdued light	0.088
Callus culture (shoot differentiation)	0.5 mg l^{-1} BAP: 0.25 mg l^{-1} NAA Continuous, subdued light	19.3
Suspension culture	5 mg l^{-1} BAP: 1 mg l^{-1} NAA Continuous, subdued light	0.132
Suspension culture	5 mg l^{-1} BAP: 1 mg l^{-1} NAA 16 h light/8 h dark	11.5
Suspension culture	5 mg l^{-1} BAP: 1 mg l^{-1} NAA 18 h light/6 h dark	21.5
Aggregate culture	0.5 mg l^{-1} BAP: 0.25 mg l^{-1} NAA 16 h light/8 h dark	35.0
Immobilized culture	5 mg l^{-1} BAP: 1 mg l^{-1} NAA 16 h light/8 h dark	110

induced, in the short term, by transferring light-grown cultures into the dark or into a long-light photoperiod (Brown and Charlwood 1986b). In each case a large increase in the proportion of monoterpene hydrocarbons (particularly limonene; Fig. 1B) was observed. This is not unexpected since limonene plays a key role as the progenitor of the oxygenated *p*-menthane monoterpenes. Stimulation of monoterpene formation was accompanied by a shift of site of accumulation of the oil from predominantly (90%) within the cell to up to 30% within the medium, but no morphological change in the cell line was apparent.

The regulation of monoterpene accumulation by light may be mediated by its effect on HMG-CoA reductase (Stermer and Bostock 1987). HMG-CoA reductase activity was measured in *P. fragrans* suspension cultures and was shown to be present in the plastid, mitochondria and microsomal fractions but not in the cytoplasm. Microsomal HMG-CoA reductase activity remained unchanged when light-grown cells were transferred to dark conditions, but the activity of the plastidic enzyme fell by almost 70% during this treatment. The overall activity within the culture was, however, very low (60–80 pmol min^{-1} mg^{-1} protein), indeed some 20-fold lower than the level determined in a monoterpene-accumulating culture of *Nepeta cataria* (Arebalo and Mitchell 1984).

It has also been suggested (Banthorpe et al. 1977) that the action of change in the light regime might reduce levels of endogenous degrading enzymes. Epoxidases responsible for monoterpene catabolism, for example, exhibit a marked seasonal variation in the intact plant and decline in their activity corresponding to the peak of activity of the synthetic enzymes.

Secondary product accumulation may often be increased by precursor feeding: for the lower isoprenoids the technique is necessarily limited to increasing the availability of early precursors. Using *P. fragrans* suspensions, supplementation of the growth medium with acetate or with intermediates of the citric acid cycle (viz. citrate, fumarate and malate) at concentrations of 10 mM gave rise to variable (but typically ten-fold) increases in accumulation of essential oil. Malate had little effect on growth or cell morphology, but the other additives caused pigmentation within the culture and gave rise to a marked inhibition in growth. On the other hand, the presence of up to 100 mM aspartate stimulated growth, and, when applied at levels of 10–20 mM, this additive brought about a 50-fold increase in oil accumulation. Only with fumarate as supplement did the accumulated oil exhibit the composition expected for *P. fragrans* suspension cells (Fig. 1B). Acetate, citrate and malate-containing medium gave rise to oils (Fig. 1C) with *p*-cymene and camphene as the major components, whilst an unidentified sesquiterpene hydrocarbon, limonene and 1,8-cineole were formed in the presence of aspartate (Fig. 1D). The stimulation of product observed upon treatment with organic acids is probably occasioned in part by the stress effect of these additives, although enhancement of acetyl-CoA availability cannot be ruled out.

Monoterpene accumulation may be controlled, albeit indirectly, by the manipulation of phytohormone concentrations within the medium. Suspension cultures of *P. fragrans* (maintained in a 16-h light/8-h dark photoperiod) became increasingly pigmented and aggregated upon incubation in media containing 0.5 mg l^{-1} BAP and supplemented with NAA in the range 0.5 to 0.05 mg l^{-1}. Such aggregation was accompanied by a small but significant increase in oil accumulation (Table 2), but not by the development of specialized cells or glandular hairs. Typically, nearly 80% of the essential oil that accumulated was excreted into the medium and consisted almost entirely of α-pinene and limonene (Fig. 1E). Non-pigmented and morphologically stable aggregates were formed when *P. fragrans* suspension cells were entrapped within reticulate polyurethane foam. Application of the range of phytohormone regimes indicated above did not give rise to differentiation within the immobilized aggregates, but did lead to an enhancement of essential oil accumulation (up to 18% of that produced by the intact plant) the major components of which were limonene, α-pinene and myrcene (Fig. 1F). At least 90% of the oil produced by immobilized cells was excreted into the medium.

In contrast, callus cultures responded quite differently when subjected to similar phytohormone regimes. When undifferentiated calli were grown in medium containing 0.5 mg l^{-1} BAP and 0.25 mg l^{-1} NAA differentiation, in the form of shoot organogenesis and concomitant glandular hair formation, occurred and oil accumulation increased considerably (Table 2). Shoot morphogenesis could also be induced in submerged cultures from morphogenic callus lines of *Pelargonium* variants (especially from *P. fragrans* and *P. tomentosum*). Shoot-proliferation culture, incubated in medium containing 0.5 mg l^{-1} BAP and 0.05 mg l^{-1} NAA under 16-h light/8-h dark photoperiod, eventually accumulate essential oils at levels up to 50% of those of the intact plants. Furthermore, both oil accumulation and culture morphology remain stable over prolonged periods. Menthone and isomenthone are the major components of the oil of both the intact plant and the shoot-proliferation culture of *P. tomentosum* (Figs. 2A,B). In contrast, the shoot-proliferation culture of *P. fragrans* (Fig. 1G) shows a marked reduction in the formation of the oxygenated monoterpene fenchone and a concomitant increase in accumulation of α- and β-pinene and sabinene.

Within these cultures both peltate and capitate glandular hairs are formed on both the leaf and the petiole and it is within these structures that the essential oil

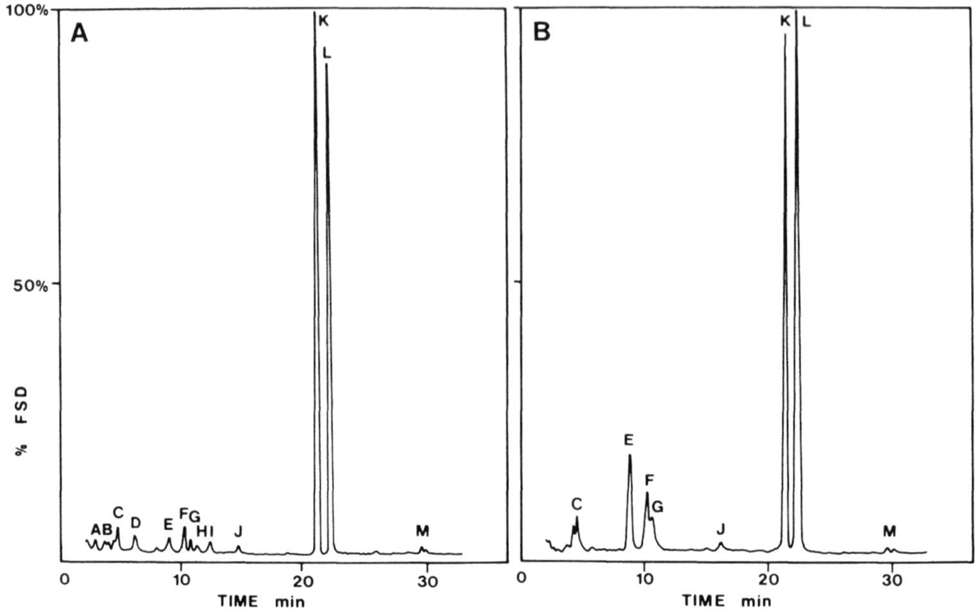

Fig. 2A,B. Oil accumulated by *Pelargonium tomentosum:* **A** intact plant; **B** shoot-proliferation culture. *E* sabinene; *F* limonene; *K* menthone; *L* isomenthone

is stored. The formation of glandular hairs is influenced by the plant growth regulator regime applied to the culture, and this in turn effects the level of oil accumulation. Figure 3 shows that there is a significant (though complex) correlation between the accumulation of essential oil in *P. fragrans* shoot-proliferation cultures and the density of the storage sites.

Treatment of both tissue and cell cultures with colchicine has been shown to increase monoterpene production and to alter the product profile. Colchicine treatment of *Mentha piperita* plant tissues gave rise to a three-fold stimulation of monoterpene production which was still detectable 2 months after treatment (Bricout et al. 1978). An approximately 80-fold increase in valepotriate accumulation in suspension cultures of *Valeriana wallichii* was reported (Becker and Chavadej 1985) after two treatments with colchicine, and the product profile varied during consecutive treatments. It was held that the increase in valepotriate accumulation could not be explained by an increase in ploidy level, since treated cultures returned to their normal chromosome status within six passages after treatment. Increase in monoterpene level may be due to gene amplification, selection by colchicine of high-producing lines or by inhibition of synthesis of steroid precursors. When suspension cells of *P. fragrans* were treated over a 5-day period with colchicine, an initial rapid reduction in growth rate (up to 50% reduction at the higher colchicine levels) was observed and this was accompanied by a progressive increase in accumulation of essential oil. After four passages, cells treated at the higher levels of colchicine exhibited extremely slow growth and even cells treated with the lowest levels of colchicine showed growth rates some ten-fold lower than those of untreated cells. Essential oil accumulation had, however,

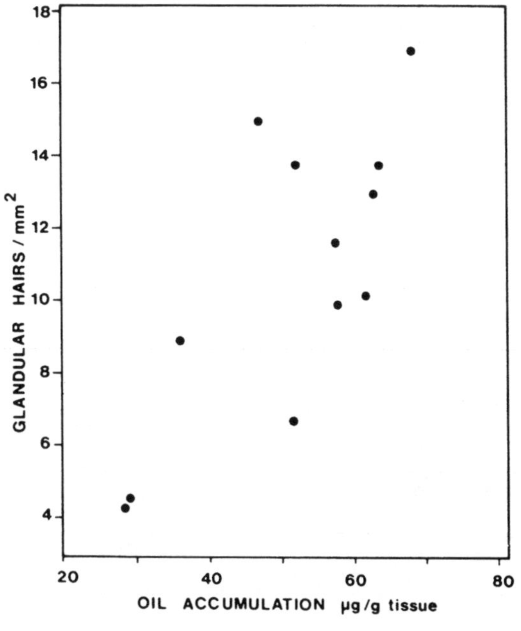

Fig. 3. Correlation between essential oil accumulation and glandular hair density in shoot-proliferation cultures of *Pelargonium fragrans* (r = 0.738:significant at ρ 0.01)

increased to 97 $\mu g\ g^{-1}$ fresh wt. in the highest accumulating line. The product profile of the oil was, however, very different from that found in either the parent plant or the untreated culture and consisted almost entirely of monoterpene hydrocarbons including α- and β-pinene, sabinene, myrcene and limonene (Fig. 1H). Colchicine treatments in the range 5 to 50 $\mu g\ l^{-1}$ had no immediate effect on the growth of shoot-proliferation cultures of either *P. fragrans* or *P. tomentosum*. After two passages in colchicine-free medium a small, but significant decrease in oil accumulation was observed at all treatment levels although the product profiles remained substantially unchanged.

It has reasonably been suggested that it may be possible to divert carbon flux through the isoprenoid pathway towards the lower isoprenoids by the use of sterol or carotenoid inhibitors. Thus, El-Keltawi and Croteau (1987) reported that the foliar application of AMO-1618 and DCPTA to *Salvia officinalis* increased essential oil accumulation with only a small reduction in plant growth. The sterol inhibitors AMO-1618, CCC, MER 29 and miconazole have all been shown to increase the production of the sesquiterpene lactone, artemisinin, in cell-free systems from shoot cultures of *Artemisia annua* (Kudakasseril et al. 1987). The effects of AMO-1618, amitrol, metflurazon, desmethy-norflurazon and 5C cycocel on *P. tomentosum* shoot-proliferation cultures are shown in Table 3. Only amitrol produced a small (but significant) increase in oil accumulation, the composition of which was identical to that of the untreated culture. The slight decrease in growth rate was accompanied by an increase in density of glandular hairs. Application of metflurazon and desmethyl-norflurazon caused progressive dedifferentiation of the tissue leading to the formation of non-accumulating aggregate cultures within nine subcultures.

Table 3. Effects of growth regulators on oil accumulation and growth of *Pelargonium tomentosum* shoot-proliferation cultures

Treatment	Subculture number	Accumulation of oil[a] (cf. control)	Growth rate[b] (cf. control)
Amitrol	1	1.45	0.97
	5	2.17	0.95
	9	0.54	0.89
5C Cycocel	1	1.29	0.99
	5	1.35	1.07
	9	0.29	0.92
Metflurazon	1	1.24	0.90
	5	0.11	0.76
	9	(*)	
Des-methyl-norflurazon	1	1.06	1.00
	5	0.03	0.79
	9	(*)	

[a] (*) After subculture 5 the culture gradually became dedifferentiated and by subculture 9 accumulated negligible oil.
[b] Growth factor (Wt-Wi)/Wi for control was 2.45: accumulation of oil by control was 34.9 $\mu g\ g^{-1}$ fresh wt.

It would appear that the ability of the shoot-proliferation culture to accumulate essential oil is closely associated with the formation of storage sites. Attempts to regulate oil production by the application of elicitors generally lead to a decrease in density of glandular hairs and a concomitant reduction in oil accumulation. A similar correlation between the formation of storage site and monoterpene accumulation has also been established in callus cultures of *Centranthus macrosiphon* (Violon et al. 1984). In this case valepotriate accumulation increased as root differentiation occurred, and eventually slow growing, root-differentiated cultures were obtained that accumulated product at a level similar to, or slightly higher than, that of the plant. We have recently obtained "hairy-root" cultures of *Fedia cornucopiae, Valeriana officinalis, Valerianella discoidea* and *Centranthus ruber* following infection of plant material with *Agrobacterium rhizogenes.* These transformed cultures exhibit very high growth rates [growth factor (Wt-Wi)/Wi up to 13 for *F. cornucopiae* incubated for 14 days under a photoperiod] in medium containing no plant growth regulators. The roots exhibit extensive lateral branching and, initially at least, synthesize agropine. Transformed roots produce similar valepotriates to those found in the parent plants (e.g. Fig. 4) but in amounts up to seven-fold greater on a dry weight basis. Unlike the shoot-differentiated cultures, product accumulation in transformed root cultures can be regulated by various external factors.

It has been demonstrated that many monoterpenes (including α- and β-pinene, limonene, carvone and isomenthone) are toxic to plant cells at concentrations as low as 0.5 mg ml^{-1} of medium (Brown et al. 1987; Cormier and Ambid 1987) and this would suggest that end-product toxicity may be the regulating factor in oil accumulation in vitro. If this is so, then it is not surprising that cultures that are able to sequester oil in the natural storage sites are also able to accumulate significantly more of the product. In such cultures regulation of accumulation is probably at the storage level rather than at the level of synthesis. In cases where oil

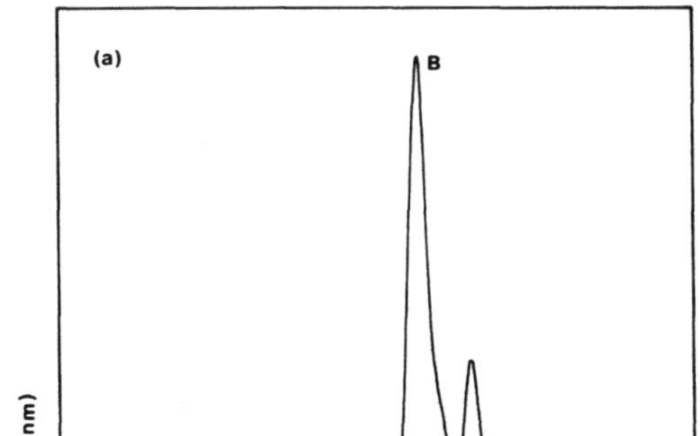

Fig. 4. Valepotriate accumulation in *Fedia cornucopiae*: **a** Roots from intact plant, **b** roots transformed with *Agrobacterium rhizogenes*. *A* Iso-valtrate; *B* valtrate

is excreted into the medium it may prove possible to reduce end-product toxicity by the continuous removal of monoterpenes from the producing cells. This has been accomplished to a limited extent by employing a second, inert and lipophilic phase (such as Miglyol) together with the culture medium, although the growth rate of the cells is often dramatically reduced (see Charlwood and Brown 1988). More success may accrue from using a more suitable lipophilic phase, such as the perfluorocarbon PCF 43, or by continuously scrubbing the medium by passing it through a column of absorbent resin.

Acknowledgement. The ownership of all registered trade names is acknowledged.

References

Arebalo RE, Mitchell ED (1984) Cellular distribution of 3-hydroxy-3-methylglutaryl-coenzyme A reductase and mevalonate kinase in leaves of *Nepeta cataria*. Phytochemistry 23:13–18

Bach TJ (1987) Synthesis and metabolism of mevalonic acid in plants. Plant Physiol Biochem 25:163–178

Banthorpe DV, Njar VCO (1984) Light-dependent monoterpene synthesis in *Pinus radiata* cultures. Phytochemistry 23:295–299

Banthorpe DV, Wirz-Justice A (1972) Terpene biosynthesis. Part VI. Monoterpenes and carotenoids from tissue cultures of *Tanacetum vulgare* L. JCS Perkin 1:1769–1772
Banthorpe DV, Bucknall GA, Gutowski JA, Rowan MG (1977) Epoxidation and hydration of prenyl pyrophosphates by plant extracts. Phytochemistry 16:355–358
Banthorpe DV, Branch SA, Njar VCO, Osborne MG, Watson DG (1986) Ability of plant callus cultures to synthesise and accumulate lower terpenoids. Phytochemistry 25:629–636
Becker H, Chavadej S (1985) Valepotriate production of normal and colchicine-treated cell suspension cultures of *Valeriana wallichii.* J Nat Prod 48::17–21
Bricout J, Garcia-Rodriguez M-J, Paupardin C (1978) Action de la colchicine sur la synthese d'huile essentielle par des tissue de *Mentha piperita* cultives in vitro. CR Acad Sci Paris Ser D 286:1585–1588
Brown JT, Charlwood BV (1986a) The control of callus formation and differentiation in scented pelargoniums. J Plant Physiol 123:409–417
Brown JT, Charlwood BV (1986b) The accumulation of essential oils by tissue cultures of *Pelargonium fragrans* (Willd.). FEBS Lett 204:117–120
Brown JT, Hegarty PK, Charlwood BV (1987) The toxicity of monoterpenes to plant cell cultures. Plant Sci 48:195–201
Charlwood BV, Brown JT (1988) The accumulation of mono- and sesquiterpenes in plant cell cultures. Biochem Soc Trans 16:61–63
Charlwood BV, Hegarty PK, Charlwood KA (1986) The synthesis of biotransformation of monoterpenes by plant cells in culture. In Morris P, Scragg AH, Stafford A, Fowler MW (eds) Secondary metabolism in plant cell cultures. Univ Press, Cambridge, pp 15–34
Cheniclet C, Carde J-P (1985) Presence of leucoplasts in secretory cells and of monoterpenes in the essential oil: a correlative study . Isr J Bot 34:219–238
Cormier F, Ambid C (1987) Extractive bioconversion of geraniol by a *Vitis vinifera* cell suspension employing a two-phase system. Plant Cell Rep 6:427–430
Curry KJ (1987) Initiation of terpenoid synthesis in osmophores of *Stanhopea anfracta* (Orchidaceae): a cytochemical study. Am J Bot 74:1332–1338
El-Keltawi NE, Croteau R (1987) Influence of herbicides and growth regulators on the growth and essential oil content of sage. Phytochemistry 26:675–679
Gleizes M, Camara B, Walter J (1987) Some characteristics of terpenoid biosynthesis by leucoplasts of *Citrofortunella mitis.* Planta 170:138–140
Heinrich G, Schultze W (1985) Composition and site of biosynthesis of the essential oil in fruits of *Phellodendron amurense* Rupr. (Rutaceae). Isr J Bot 34:205–217
Koch-Heitzmann I, Schultze W (1988) Compilation of volatile compounds found in plant cell cultures. In: Schreier P (ed) Bioflavour '87. De Gruyter, Berlin, pp 365–395
Kudakasseril GJ, Lam L, Staba EJ (1987) Effect of sterol inhibitors on the incorporation of ^{14}C-isopentenyl pyrophosphate into artemisinin by cell-free system from *Artemisia annua* tissue cultures and plants. Planta Med 53:280–284
Mettal U, Boland W, Beyer P, Kleinig H (1988) Biosynthesis of monoterpene hydrocarbons by isolated chromoplasts from daffodil flowers. Eur J Biochem 170:613–616
Stermer BA, Bostock RM (1987) Involvement of 3-hydroxy-3-methylglutaryl-coenzyme A reductase in the regulation of sesquiterpenoid phytoalexin synthesis in potato. Plant Physiol 84:404–408
Violon C, Dekegel D, Vercruysse A (1984) Relation between valepotriate content and differentiation level in various tissues from Valerianeae. J Nat Prod 47:934–940

Toward Directed Modulation of Rosmarinic Acid Production in Cultured Plant Cells

W. De-Eknamkul and B.E. Ellis[1]

1 Introduction

Hydroxycinnamoyl conjugates form a class of metabolites whose production in cultured plant cells is often remarkably high (for review, see Ellis 1985). Questions of metabolic regulation can be posed in such high production systems which are very relevant to attempts to reduce or enhance the accumulations of specific metabolites in other culture or whole plant systems. Determination of which combination of regulatory mechanism(s), genetic organization and environmental cue(s) results in accumulation of such high concentration (as much as 25% of the tissue biomass) of a single metabolite should provide important insights for plant metabolic regulation in general. More specifically, the information may suggest new approaches for rational intervention in specific plant systems where modified expression of a metabolic phenotype is desired. The particular advantage of a high-producing culture system is that the enzyme catalysts, together with their associated short- and long-term regulatory processes, clearly are present and functional within a readily manipulated tissue.

Suspension cultures of *Anchusa officinalis* L. produce large amounts of rosmarinic acid (α-O-caffeoyl 3,4-dihydroxyphenyllactic acid), to the virtual exclusion of other soluble phenylpropanoid metabolites (De-Eknamkul and Ellis 1984). Both the level of accumulation (up to ten-fold higher than whole plant tissue on a dry weight basis), and the metabolic commitment to a single pathway, are notable traits of this culture system.

Enzymes involved in specialized facets of plant metabolism typically comprise an extremely small ($<$ 0.05%) fraction of the cellular protein, even in tissues displaying elevated levels of synthesis, which makes their purification a significant technical challenge. The high expression and relatively restricted pattern of phenylpropanoid metabolism in *Anchusa* cells have therefore been important in facilitating detailed study of the rosmarinic acid pathway and its integration into the remainder of metabolism.

While this work is adding to our understanding of phenylpropanoid biosynthesis and enzyme function, definitive tests of potential regulatory mechanisms will require development of appropriate genetically engineered material in which the operation of specific metabolic loci can be modified at will. For the development of such modified cell lines, a series of discrete goals will have to be attained.

[1]Department of Chemistry and Biochemistry, University of Guelph, Guelph, Ontario NIG 2W1, Canada

Primary and Secondary Metabolism of Plant Cell Cultures, ed. by W.G.W. Kurz

These include (but are not restricted to) characterization of potential regulatory enzymes, analysis of the corresponding cDNA's and genes, and optimization of transformation procedures. This chapter outlines the progress that has been made towards these goals in the *Anchusa* system, and points out problems that still need to be dealt with.

2 Biogenesis

The biosynthetic route from the shikimate pathway to rosmarinic acid has been established by radiotracer studies in whole plants (Ellis and Towers 1970) and cell cultures (Ellis et al. 1979; Ellis 1985). As outlined in Fig. 1, L-phenylalanine is converted to cinnamic acid and its hydroxylated derivatives through the well-known general phenylpropanoid pathway. The most significant uncertainty with respect to this sequence is the mechanism by which the second ring hydroxylation

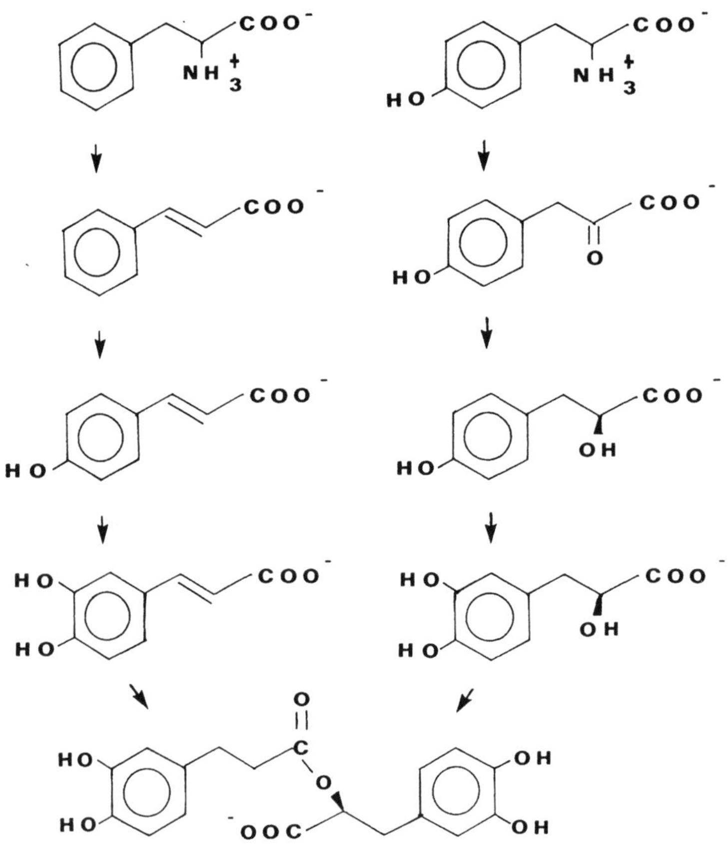

Fig. 1. Proposed biosynthetic pathway for rosmarinate formation in *Anchusa officinalis* and *Coleus blumei*. The steps of acyl activation before ester formation have been omitted

proceeds. Results from other systems have led to proposals for phenolase catalyzed hydroxylation of free 4-hydroxycinnamic acid (Vaughan and Butt 1969), phenolase catalyzed hydroxylation of 4-hydroxycinnamoyl quinate ester (Hanson and Zucker 1963) and mono-oxygenase action on 4-hydroxycinnamoyl quinate (Kühnl et al. 1987). Although the analogous ester of 4-hydroxycinnamic acid with 3,4-dihydroxyphenyllactic acid (DOPL) has not been detected in rosmarinic acid-producing cells, a small rapidly turning over pool could be functioning as the hydroxylation substrate. The recent detection of an acyltransferase activity capable of forming both 4-hydroxycinnamoyl DOPL and caffeoyl DOPL in protein extracts from *Coleus blumei* cell cultures (Petersen and Alfermann, personal communication) is consistent with such a model. The activated acyl donor in the *Coleus* system was found to be the coenzyme A ester.

The conversion of L-tyrosine to the (S)-DOPL moiety of rosmarinic acid was originally proposed to involve L-DOPA as an intermediate, based on the incorporation of label from administered ^{14}C-DOPA (Ellis and Towers 1970). It was subsequently discovered that a small pool of 4-hydroxyphenyllactate existed in *Coleus* cells, and that this was efficiently labelled from administered ^{14}C-tyrosine, whereas DOPA was not (Ellis et al. 1979). This observation led to the proposal that tyrosine was initially being converted to 4-hydroxyphenylpyruvate, followed by reduction to 4-hydroxyphenyllactate and hydroxylation to form DOPL (Fig. 1). Since the recently detected acyltransferase activity mentioned above is able to use 4-hydroxyphenyllactate efficiently as an acyl acceptor substrate (Petersen and Alfermann, personal communication), it is possible that 3-hydroxylation of the phenyllactate moiety also requires the ester as substrate. In this regard, it may be relevant that labelled free 4-hydroxyphenyllactate was not an efficient precursor of rosmarinic acid in *Coleus* cell cultures (Ellis et al. 1979).

3 Enzymology

The deamination of tyrosine through transamination was found to be much more active in *Anchusa* cultures than was the corresponding oxidative deamination (De-Eknamkul and Ellis 1987a). The involvement of a tyrosine aminotransferase in rosmarinic acid synthesis, and the presumed occurrence of the same reaction as the terminal step in tyrosine biosynthesis, suggested that the aminotransferase enzyme might play a different role at different points in the *Anchusa* culture cycle. The reversibility of aminotransferase reactions would be well-suited for such a metabolic arrangement. Early in the purification of the tyrosine aminotransferase (TAT) from *Anchusa*, however, the activity was resolved into four discrete peaks during anion exchange chromatography (De-Eknamkul and Ellis 1987b). The third of these proved to be far more active in aspartate transamination than with tyrosine, and was not pursued further.

The slowest eluting peak also displayed good activity as a glutamate: oxaloacetate aminotransferase, but, in addition, it was quite active in the transamination of prephenate to arogenate (De-Eknamkul and Ellis 1988). The latter reaction is one of a triad of reactions within the shikimate pathway which provide an alternative route for formation of phenylalanine and tyrosine in many organisms, particularly photoautotrophic species (Bonner and Jensen 1985; Jensen 1986). The *Anchusa* prephenate aminotransferase shared a number of features with the partially characterized enzymes from *Nicotiana* (Bonner and Jensen 1985) and *Sorghum* (Siehl et al. 1986), including resistance to heat denaturation and a tendency to substrate inhibition. In contrast to the total TAT activity in the *Anchusa* cultures, which reached a peak on day 9 coincident with the time of most rapid rosmarinic acid synthesis, the prephenate aminotransferase activity peaked at day 5 and declined sharply thereafter. This pattern is in accord with a primary role in generation of phenylalanine and tyrosine for protein synthesis during the period of rapid cell growth and division early in the culture cycle.

After purification to apparent homogeneity, the *Anchusa* prephenate aminotransferase yielded approximately equal amounts of two polypeptides (44 and 57 kD) on SDS-PAGE. Since the native size was estimated at 220 kD from calibrated gel exclusion chromatography, the prephenate aminotransferase appears to have an $\alpha_2\beta_2$ structure. This is the first reported complete purification of this activity from any organism, so no comparisons with other data are available.

The existence of a prephenate aminotransferase activity in *Anchusa* suggests that tyrosine and phenylalanine may be synthesized via arogenate in this species. If this is the case, the presence of a specific tyrosine aminotransferase could be viewed in a different light, since the only known role for such an activity would now be as an entry point enzyme for the rosmarinic acid biosynthetic pathway. The temporal pattern of expression across the culture cycle and the high specificity for tyrosine as an amino acid substrate are both consistent with such a role. In addition, both tyrosine aminotransferase activities were inhibited by the product of the tyrosine-derived pathway, 3,4-dihydroxyphenyllactic acid (DOPL) (De-Eknambul and Ellis 1987b), a trait which suggests the possibility of metabolic feedback loops as a fine control mechanism.

Purification of the phenylalanine ammonia-lyase activity from 7-day-old *Anchusa* cells resulted in >300-fold enrichment but the final preparation still yielded three polypeptide bands on SDS-PAGE. Whether the lower M_r bands represent contaminants, or the degradation products observed during PAL purification in other systems (Bolwell et al. 1986) is not known. It is worth noting that only a single form of PAL could be resolved during chromatofocussing. In *Phaseolus* cells, multiple forms are resolved by chromatofocussing (Bolwell et al. 1986) and it may be that these represent the products of different PAL genes, each of which could be transcribed at a rate determined by the specific signals received by the cell. If this were a general phenomenon, the apparent presence of a single PAL species in *Anchusa* cells may reflect the commitment of this culture to rosmarinic acid production as its sole phenylpropanoid product.

Considerably more information is needed concerning the subcellular localization, protein-protein organization and metabolic channelling characteristics of the rosmarinic acid pathway before a definitive model of its regulation will be available. Although TAT and PAL can be extracted as "soluble" enzymes, preliminary results from sucrose density gradient fractionation of carefully disrupted *Anchusa* protoplasts indicate that TAT and PAL enter the gradient as discrete particle-bound fractions with different buoyant densities. The TAT-containing fractions coincided with the position of marker enzymes for rough endoplasmic reticulum. This approach therefore shows promise as a means of establishing the macromolecular organization behind rosmarinic acid formation.

Production of murine monoclonal antibodies to purified TAT-1 protein has recently yielded two antibodies which have a high affinity for native TAT in ELISA and on Western blots, but which also react strongly with commercial ovalbumin. These preliminary results, together with other observations, suggest that TAT-1 could be a glycoprotein. Such a post-translational modification may have important implications for the synthesis and organization of TAT within *Anchusa* cells.

4 Future Directions

The absence of significant pools of biosynthetic intermediates, and the rapidity of labelling (< 5 min) of rosmarinic acid from exogenously supplied amino acids, are indications of a highly organized, and possibly channelled, reaction sequence. Efforts to characterize the individual enzymes involved will have to be complemented by isolation of functional complexes and by identification of other proteins and metabolites whose participation is important to pathway organization and regulation. Unravelling the role of each component, whether catalytic, structural or modulatory, may eventually require directed suppression or enhancement of specific gene products in transformed tissues. *Anchusa* cells are killed by 100 $\mu g\ ml^{-1}$ kanamycin or 10 $\mu g\ ml^{-1}$ hygromycin (Fig. 2), which offers the possibility of developing a transformation system based on Ti vectors carrying antibiotic resistance as a selectable marker. To date, several fast-growing colonies

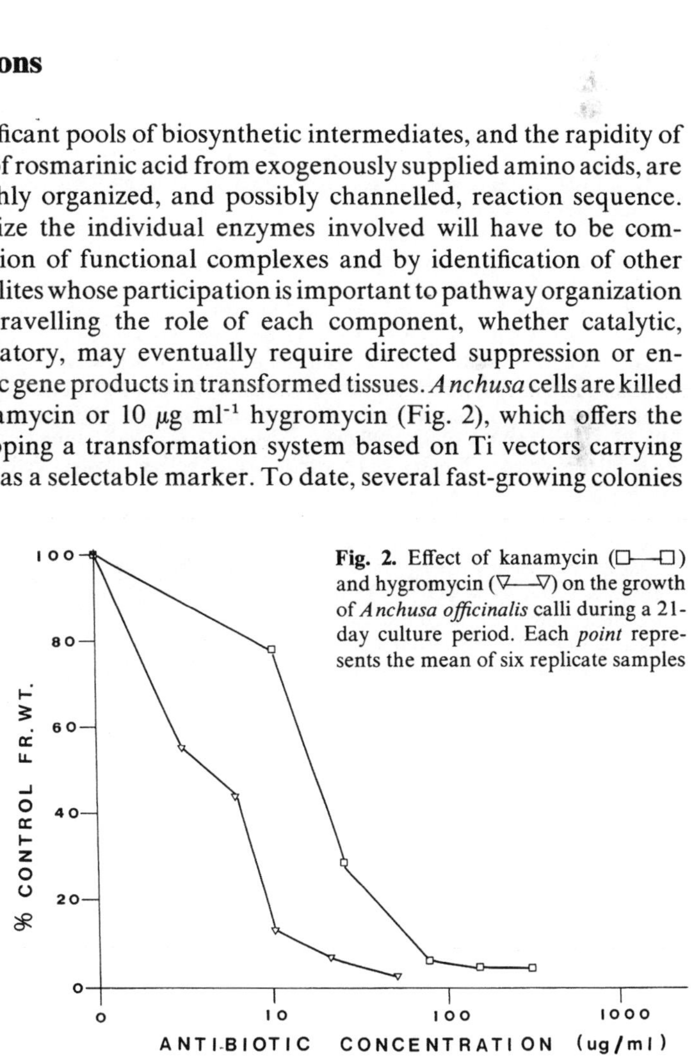

Fig. 2. Effect of kanamycin (□—□) and hygromycin (▽—▽) on the growth of *Anchusa officinalis* calli during a 21-day culture period. Each *point* represents the mean of six replicate samples

of *Anchusa* cells have been recovered from 100 $\mu g\ ml^{-1}$ kanamycin plates spread with 4-day-old *Anchusa* suspension cultures pre-incubated with *Agrobacterium tumefaciens* carrying a disarmed Ti plasmid (Kan^r). These recovered calli are presently being evaluated for the level and stability of the apparent antibiotic resistance.

Acknowledgements. Our studies of the *Anchusa* system have been financially supported by the Natural Sciences and Engineering Research Council of Canada. We would also like to acknowledge the many helpful discussions with other members of the Plant Biochemistry Group, the technical assistance of Lauretta Williams, Cathy Fuchs and Denise McClellan, the advice on antibody production from Dr. J. Lam and the assistance with manuscript preparation from Kris Hladun. Dr. Maike Petersen (Universität Düsseldorf) kindly provided us with unpublished results from her laboratory.

References

Bonner CA, Jensen RA (1985) Novel features of prephenate aminotransferase from cell cultures of *Nicotiana silvestris*. Arch Biochem Biophys 238:237–246

Bolwell GP, Sap J, Cramer CL, Lamb CJ, Schuch W, Dixon RA (1986) L-phenylalamine ammonia-lyase from *Phaseolus vulgaris*: partial degradation of enzyme subunits in vitro and in vivo. Biochem Biophys Acta 881:210–221

De-Eknamkul W, Ellis BE (1984) Rosmarinic acid production and growth characteristics of *Anchusa officinalis* cell suspension cultures. Planta Med 51:346–350

De-Eknamkul W, Ellis BE (1987a) Tyrosine aminotransferase: the entry point enzyme of the tyrosine-derived pathway in rosmarinic acid biosynthesis. Phytochemistry 26:1941–1946

De-Eknamkul W, Ellis BE (1987b) Purification and characterization of tyrosine aminotransferase activities from *Anchusa officinalis* cell cultures. Arch Biochem Biophys 257:430–438

De-Eknamkul W, Ellis BE (1988) Purification and characterization of prephenate aminotransferase from *Anchusa officinalis* cell cultures. Arch Biochem Biophys (in press)

Ellis BE (1985) Metabolism of caffeoyl derivatives in plant cell cultures. In: Neuman K-H, Barz W, Reinhard E (eds) Primary and secondary metabolism of plant cell cultures. Springer, Berlin Heidelberg New York Tokyo, p 164

Ellis BE, Towers GHN (1970) Biogenesis of rosmarinic acid in *Mentha*. Biochem J 118:291–297

Ellis BE, Remmen S, Goeree G (1979) Interactions between parallel pathways during biosynthesis of rosmarinic acid in cell suspension cultures of *Coleus blumei*. Planta 147:163–167

Hanson KR, Zucker M (1963) The biosynthesis of chlorogenic acid and related conjugates of the hydroxycinnamic acids. J Biol Chem 238:1105–1115

Jensen RA (1986) Tyrosine and phenylalanine biosynthesis: relationship between alternative pathways, regulation and subcellular location. In: Conn EE (ed) Recent advances in phytochemistry, vol 20. The shikimic acid pathway. Plenum, New York London, p 57

Kühnl T, Koch U, Heller W, Wellmann E (1987) Chlorogenic acid biosynthesis: characterization of a light-induced microsomal 5-0-(4-coumaroyl)-D-quinate/shikimate 3′-hydroxylase from carrot (*Daucus carota* L.) cell suspension cultures. Arch Biochem Biophys 258:226–232

Siehl DL, Connelly JA, Conn EE (1986) Tyrosine biosynthesis in *Sorghum bicolor*: characteristics of prephenate aminotransferase. Z Naturforsch 41C: 79–86

Vaughan PFT, Butt VS (1969) The hydroxylation of p-coumaric acid by an enzyme from leaves of spinach beet (*Beta vulgaris* L.) Biochem J 113:109–115

Regulation of Chalcone Synthase with Regard to Anthocyanin Biosynthesis

H.U. Seitz and J. Gleitz[1]

1 Introduction

Cell cultures of an Afghan cultivar of *Daucus carota* L. ("black carrot") propagated under dark conditions accumulate large amounts of cyanidin 3-(sinapoyl-xylosylglucosylgalactoside) as the major anthocyanin (Harborne et al. 1983; Hopp and Seitz 1987). The biosynthesis of anthocyanin is inhibited by physiological concentrations of gibberellic acid (GA_3; 30 μM). In the presence of GA_3 added 7 days after the onset of cultivation, the key enzyme of the anthocyanin pathway, the chalcone synthase (CHS), is inhibited by 60–80%. This inhibition of the catalytic activity of CHS takes place within 24 h. Subsequent to the inhibition of the CHS activity by GA_3, anthocyanin biosynthesis could be restored by the addition of precursors of the anthocyanin to the culture medium (naringenin, eriodictyol, or dihydroquercetin). The best supplementation was achieved by feeding dihydroquercetin, a precursor which is very close to cyanidin (Hinderer et al. 1984). These feeding experiments clearly demonstrate that the physiological block is located at the chalcone synthase. Inhibition of anthocyanin biosynthesis by GA_3 was also described for a cell culture derived from *Daucus carota* L. cv. Kurodagosun in a culture medium lacking 2,4-D (Ozeki and Komamine 1986).

Cell cultures are not homogeneous with regard to anthocyanin accumulation. A cell culture of *Catharanthus roseus* consists of about 10% anthocyanin-accumulating cells (Hall and Yeoman 1986). Dougall et al. (1983) isolated cell clones of the wild carrot differing in anthocyanin accumulation. The carrot cell cultures used throughout the present study are also heterogeneous with regard to anthocyanin production.

2 Results and Discussion

2.1 Anthocyanin-Containing and Anthocyanin-Free Cell Lines

The experiments described here were carried out with two cell lines. The cell line designated as DC_b is of a deep violet colour and accumulates anthocyanin under dark conditions. As mentioned above, it consists of anthocyanin-accumulating and non-accumulating cells. Anthocyanin-free parts were selected from the heterogeneous

[1]Botanisches Institut, Allgemeine Botanik und Pflanzenphysiologie, Universität Tübingen, Auf der Morgenstelle 1, D-7400 Tübingen, FRG

Primary and Secondary Metabolism of Plant Cell Cultures, ed. by W.G.W. Kurz

callus propagated on solid medium and subcultivated for about 1 year. The resulting cell cultures is anthocyanin-free and is designated as DC_s (s = selected).

2.2 Light-Induced Changes in DC_b and DC_s Cells

The best-known system is the flavonoid-accumulating cell culture of *Petroselinum hortense,* where phytochrome and a blue-light photoreceptor are involved in UV-induced flavonoid biosynthesis (Wellmann 1971; Duell-Pfaff and Wellmann 1982). During this induction process, enzymes are regulated within two coordinated sequences designated as group I and group II. The transient changes in enzyme synthesis depend on the regulation of mRNA transcription (for review, see Hinderer and Seitz 1988).

Dark-grown DC_b and DC_s cells were irradiated continuously with UV light (λ_{max} = 354 nm, flux rate: 6.3 Wm^{-2}). Then the catalytic activities of various enzymes of the general phenylpropanoid pathway (group I) and of the flavonoid pathway (group II) were assayed. The results of this set of experiments are summarized in Figs. 1 and 2.

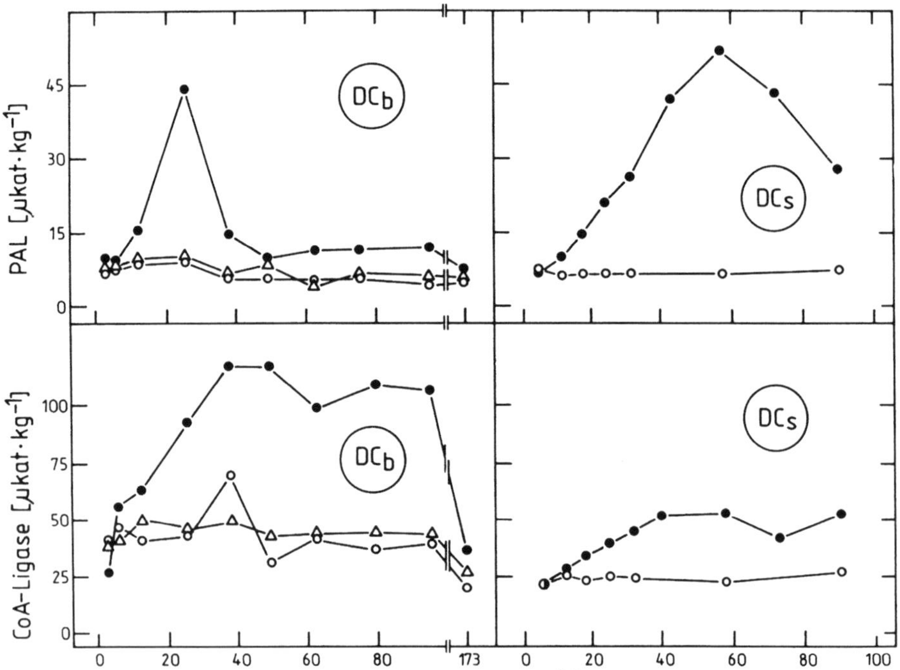

Fig. 1. UV-induced changes in the catalytic activities of phenylalanine ammonia-lyase (PAL) and 4-coumarate:CoA ligase (4CL) in anthocyanin-containing (*DC_b*) and anthocyanin-free (*DC_s*) cell suspensions of an Afghan cultivar of *Daucus carota* L. ("black carrot"). The specific activities were determined spectrophotometrically. The cells were irradiated continuously with UV-A (λ_{max} = 354 nm; 6.3 W m^{-2}). The onset of irradiation was at time zero at day 10 after onset of cultivation. Additionally, pulse irradiation of 30 min at time zero was carried out. ●—● Continuous irradiation; △—△ pulse irradiation; ○—○ constant darkness

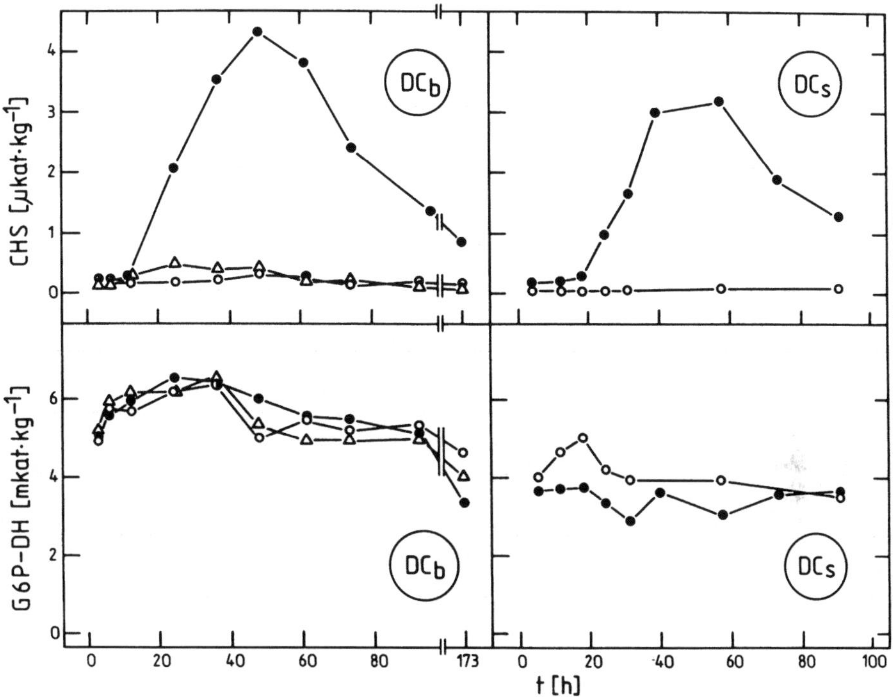

Fig. 2. UV-induced changes in the catalytic activities of chalcone synthase (CHS) and glucose 6-phosphate dehydrogenase (Glc 6-P-dh) in anthocyanin-containing (*DC_b*) and anthocyanin-free (*DC_s*) cell lines of *Daucus carota* L. CHS activity was assayed using ^{14}C labelled malonyl CoA; glucose 6-phosphate dehydrogenase activity was determined spectrophotometrically. For further conditions see Fig. 1. ●—● Continuous irradiation; △—△ pulse irradiation; ○—○ constant darkness

In Fig. 1 the activity changes of enzymes of the general phenylpropanoid pathway are shown. In DC_b cells phenylalanine ammonia-lyase (PAL; EC 4.3.1.5) exhibits a rapid and transient increase. In DC_s cells the maximum activity level of PAL is about the same but the induction is longer-lasting, resulting in a broader peak. The induction of the 4-coumarate:CoA ligase (4CL; EC 6.2.1.12) in DC_b as well as in DC_s cells is not as prominent as it is in the case of PAL, but in both cell lines the activity of 4CL is enhanced by the same factor (about two-fold).

In Fig. 2 the activity curves of chalcone synthase (CHS) upon irradiation are shown. The catalytic activity of CHS is induced in the violet-coloured DC_b cells as well as in DC_s cells. The latter result is highly surprising since no anthocyanin is accumulated by these cells. As shown in Fig. 2 the primary metabolism represented by glucose 6-phosphate dehydrogenase (EC 1.1.1.49) is not affected.

Figure 3 summarizes the results after the continuous irradiation of DC_b cells with UV light. The relative PAL activity (representative for group I enzymes) reaches its maximum approximately 25 h after onset of irradiation, whereas the catalytic activity of CHS reaches its maximum after 48 h. Anthocyanin accumulation is closely correlated to CHS activity, demonstrating once more the key role of this enzyme in anthocyanin biosynthesis, as already demonstrated by feeding experiments (Hinderer et al. 1984). The half-lives of PAL and CHS were

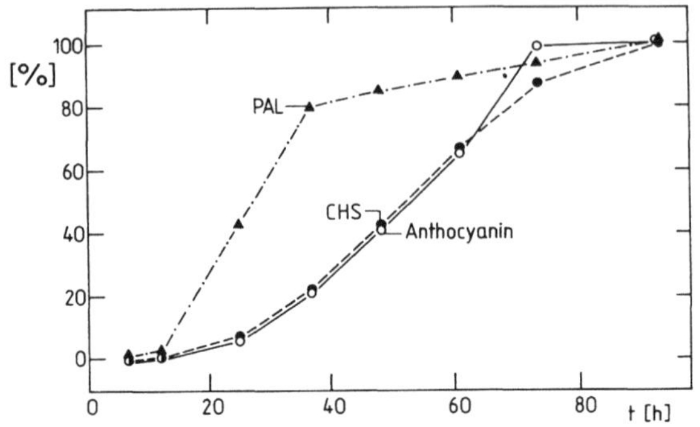

Fig. 3. Time course of anthocyanin accumulation and the integrated enzyme activities of phenylalanine ammonia-lyase and chalcone synthase in anthocyanin-containing DC_b cells upon continuous irradiation with UV-A. The onset of irradiation was at time zero (day 10 of subculture). The relative anthocyanin content (A_{537} g^{-1} fresh weight) was determined by extracting approximately 1 g of cells (fresh weight) in 10 ml of 1% HCl in methanol. Anthocyanin was determined spectrophotometrically at 537 nm. The maximum absorbance in extracts from dark-grown cells was A_{537} = 0.74 g^{-1} and A_{537} = 7.74 g^{-1} from irradiated cells. The maximum absolute values for the specific activities for PAL were 44.6 μkat kg^{-1} and 4.3 μkat kg^{-1} for CHS. ○——○ Anthocyanin content; ●——● specific CHS activity; ▲ —·—▲ specific PAL activity

calculated. In DC_b cells the half-life is 5.5 h for PAL and 21 h for CHS. The half-lives in DC_s cells were slightly higher: 6.3 for PAL and 26 h for CHS.

2.3 Protein Subunits of CHS (Western Blotting)

In order to gain information concerning the CHS protein after the irradiation of DC_b cells with UV light, SDS-PAGE, Western blotting and immunostaining were performed. The cells were harvested after various illumination periods. The crude protein extracts were separated by SDS-PAGE and subsequently blotted on nitrocellulose sheets. The Western blots were incubated with an antiserum against chalcone synthase from parsley (a gift from K. Hahlbrock, Köln). In a second step the CHS subunits immobilized on the nitrocellulose were treated with goat anti-rabbit IgG carrying alkaline phosphatase; then the subunits were stained using BCIP (5-bromo-4-chloro-3-indolyl phosphate).

In Fig. 4 the subunit patterns of dark-grown and UV-irradiated DC_b cells are compared. In dark-grown cells only one type of CHS subunit is visible on the immunoblot. Its molecular weight is 43 kD and it co-migrates with the CHS subunits from parsley. Its intensity did not change with time. Immunoblots from SDS-PAGE revealed that in irradiated DC_b cells a second protein appeared exhibiting a transient increase in the intensity of staining. This time course parallels the time course of the catalytic activity of CHS (see Fig. 2). In both cases the maxima are at about 50 h. The molecular weight of this additional protein was determined to be 40 kD; it is clearly separated from the 43 kD subunit. Ozeki et al. (1985) purified a carrot CHS which also had a molecular weight of 40 kD.

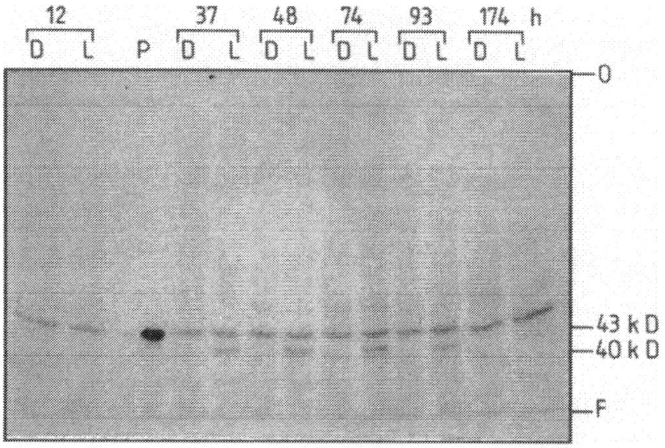

Fig. 4. Immunoblotting of crude extracts from anthocyanin-containing DC_b cells. The Western blot was treated with an antiserum against CHS from *Petroselinum hortense*. The samples were taken from the experiment depicted in Fig. 2. Alternately, protein samples from dark-grown (*D*) and continuously irradiated (*L*) cells were loaded onto a 10% SDS-PAGE gel, separated by electrophoresis and blotted to a nitrocellulose membrane. Each lane contains the same amount of protein. The Western blot was incubated with CHS antiserum, goat anti-rabbit IgG conjugated with alkaline phosphatase and developed with BCIP (5-bromo-4-chloro-3-indolyl phosphate). CHS subunits emerge as blue-stained bands. The molecular weights were determined using the low molecular weight calibration kit from Pharmacia (Uppsala, Sweden). The standard proteins were directly stained with Ponceau S. The samples were taken at times indicated at the top of the Western blot. *F* Front; *O* origin; *P* CHS subunits from *Petroselinum hortense*

In anthocyanin-free cells (DC_s) only one species of subunits with a molecular weight of 43 kD is present on the immunoblot (Fig. 5), although the catalytic activity is induced by UV light to the same extent (see Fig. 2). Preliminary experiments with non-denatured CHS revealed a shift of the isoelectric point of the 43 kD CHS upon irradiation. The IP was found to be 5.25, compared to 3.85 in dark-grown cells. The IP of the 40 kD CHS in DC_b cells was determined to be 6.2. These results strongly indicate the existence of two different types of CHS which might be regulated by different mechanisms.

2.4 Specificity of the Immunoreaction

In order to determine whether the immunoreaction of CHS is specific or not, an immunotitration of CHS activity was carried out (Fig. 6). The catalytic activity of CHS in extracts from DC_b and DC_s cells was inhibited by an antiserum prepared against soybean CHS (a gift from R. Welle, Freiburg i.Br.; Welle and Grisebach 1987). The enzyme activity was inhibited in extracts from both DC_b and DC_s cells. These results show that the 43 kD-type protein in DC_s cells carries catalytic activity. This result was confirmed by determining the CHS activity in the supernatant of a protein A-Sepharose precipitation of the antibody-enzyme complex (results not shown).

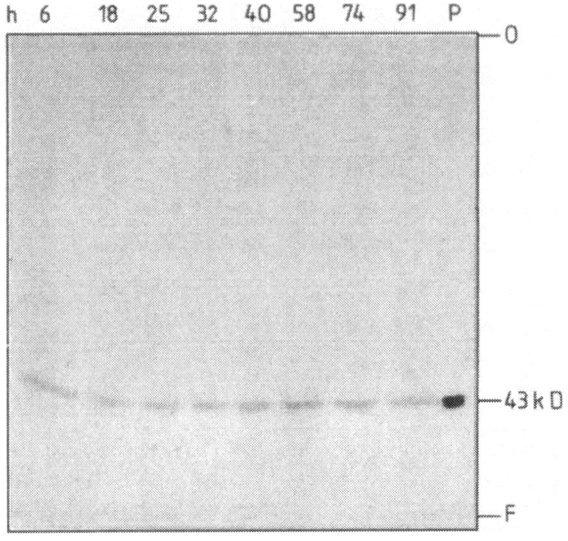

Fig. 5. Immunoblotting of chalcone synthase of anthocyanin-free DC_s cells. The samples were taken from the experiment depicted in Fig. 2. For conditions, see Fig. 4

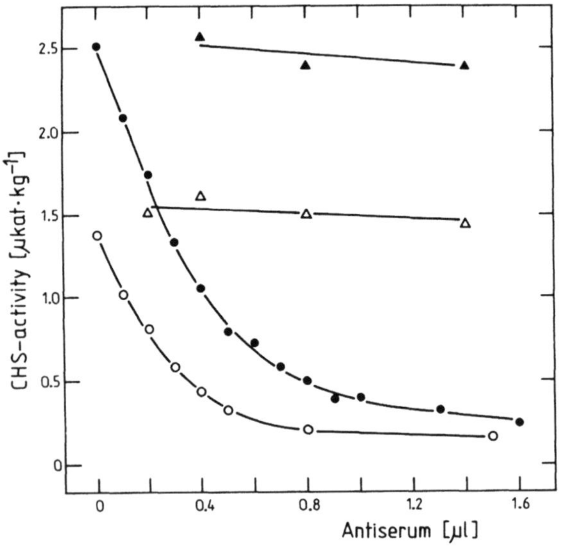

Fig. 6. Immunotitration of chalcone synthase activity from anthocyanin-containing (●—●, DC_b) and anthocyanin-free (○—○, DC_s) cells with an antiserum against CHS from *Glycine max*. 30 µl (about 50 µg protein) of crude protein extracts (DC_b; 2.5 µkat kg^{-1}; DC_s: 1.3 µkat kg^{-1}) were mixed with 170 µl of 0.1 M potassium phosphate buffer (containing 1 mg ml^{-1} bovine serum albumin) which contained the indicated amounts of the antiserum. The mixture was incubated at 25°C for 15 min and assayed for enzyme activity. Controls were performed with pre-immunserum (▲—▲, DC_b; △—△, DC_s)

The polyvalent antibody against CHS from soybean does not recognize the 40 kD subunit under the immunoblotting conditions applied here, but reacts with the 43 kD protein from both cell lines (Fig. 7).

These results establish that the CHS subunit (43 kD) is stained specifically, thus confirming the view that two types of subunits are present in our carrot cells.

2.5 The Putative Function of the CHS Proteins

The catalytic activity of CHS in DC_b cells is closely correlated with anthocyanin biosynthesis (Fig. 3), and may account for the anthocyanin production in this cell line. In order to get a better picture of what is happening, additional experiments were performed with extracts from anthocyanin-free and anthocyanin-containing petals of the umbel of the European wild carrot (*Daucus carota* L. ssp. *carota*). In Fig. 8 the extracts from white and dark purple petals from the center of the inflorescence are compared on a Western blot. White petals do not have any catalytic CHS activity (Hinderer et al. 1983), nor is there any CHS protein. By contrast, the dark purple petals contained a high CHS activity. On the immunoblot only the 40 kD subunit is present and it is clearly separated from the 43 kD subunit from parsley used as a control. These results lend additional support to the view that the 40 kD protein is involved in anthocyanin biosynthesis.

The function of the 43 kD protein is not yet understood. In *Sinapis alba* seedlings the upper epidermis contains flavonols, whereas the lower epidermis is coloured by anthocyanins (Beggs et al. 1987). Presumably there are tissue-specific CHS activities; we hypothesize in our case that the 43 kD-type CHS could be involved in the biosynthesis of other flavonoids which remain to be identified. The different responses to UV treatment are of particular interest with regard to the different modes of regulation of these two types of CHS molecules. In DC_b cells the CHS concentration (40 kD subunit) changes subsequent to irradiation with UV light and is closely correlated with changes in the catalytic activity of CHS, whereas in DC_s cells the amount of the 43 kD subunit remains constant, though the catalytic activity is transiently induced to the same extent as in anthocyanin-containing DC_b cells. This prompted us to assume a post-translational mechanism for the regulation of the 43 kD-type CHS.

Evidence for different types of CHS was obtained with various systems and techniques. One line of evidence comes from cloning experiments. In *Petunia hybrida* two non-allelic CHS genes were observed (Reif et al. 1985). In *Phaseolus vulgaris* a gene family consisting of six to eight CHS genes was detected (Ryder et al. 1987). The molecular weight of the polypeptides ranged from 42–43 kD with different isoelectric points. The other line of evidence comes from purified CHS of *Spinacea oleracea*. The two types of CHS differ in their isoelectric points and were separated by means of ion exchange chromatography (Beerhues and Wiermann 1985, 1988). Both subunits have the same molecular weight (41 kD) and both are located in the cytosol (Beerhues et al. 1988).

2.6 Effect of GA_3 on Both Types of CHS from *Daucus carota*

As indicated above, the treatment of anthocyanin-producing DC_b cells with 30 μM gibberellic acid resulted in a substantial reduction of anthocyanin and a loss of the 40 kD band on the corresponding Western blot (Figs. 9 and 10). We assume that the remaining catalytic CHS activity in dark-grown cells after treatment with GA_3 (about 65%) is due to the 43 kD-type CHS species.

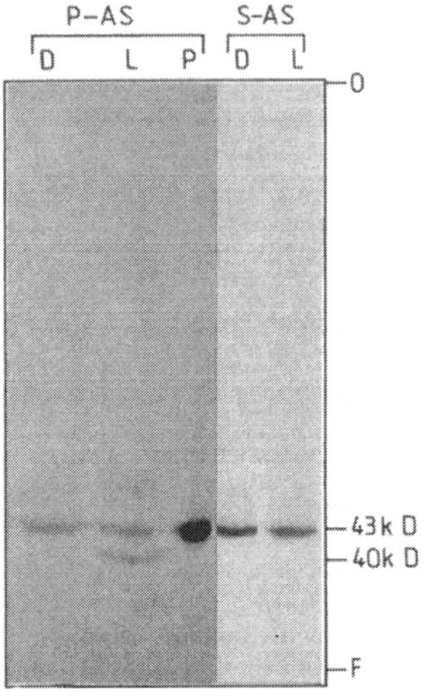

Fig. 7. Immunoblotting of crude extracts from anthocyanin-containing (DC_b) cells. The immunoblot was treated with an antiserum against CHS from *Glycine max* (*S-AS*) and *Petroselinum hortense (P-AS)*. Lanes *D* (constant darkness) and *L* (continuously irradiated) contain the same amounts of protein. Lane *P* represents the subunits from *Petroselinum hortense*. For conditions, see Fig. 4

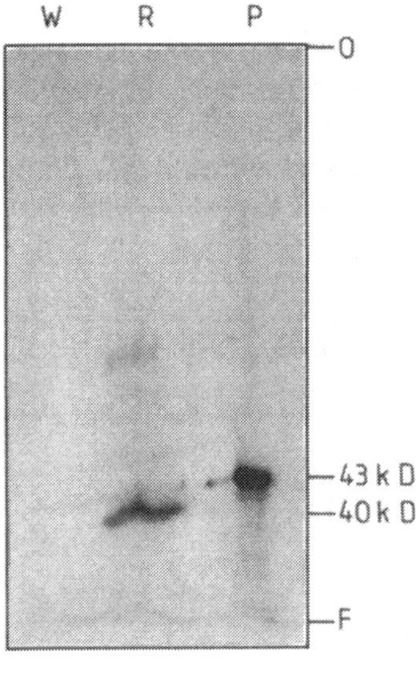

Fig. 8. Immunoblotting of crude extracts derived from petals of the European wild carrot (*Daucus carota* L. ssp. *carota*). The Western blot was incubated with an antiserum against parsley. On lane *R* (= red) crude protein extracts from dark purple petals from the center of the inflorescence were separated. Lane *W* (= white) represents the white petals from the margin of the inflorescence. *P* = CHS subunits from parsley. For conditions, see Fig. 4

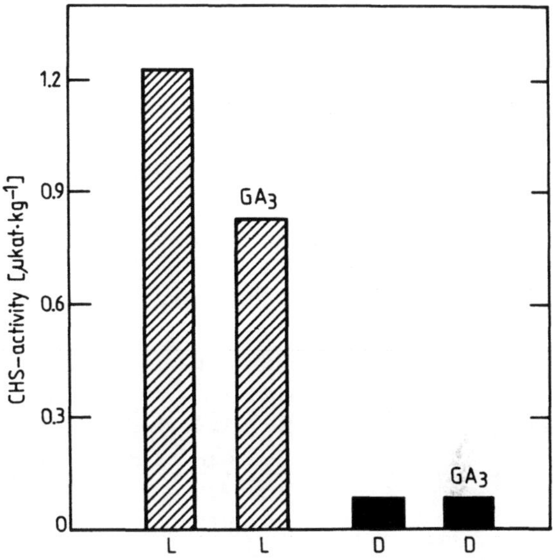

Fig. 9. Catalytic CHS activity after treatment of anthocyanin-containing cells (DC_b) with 30 μM GA_3. The hormone was applied at time zero. After 10 days the cell suspension was irradiated with UV-A for an additional 48 h then the CHS activity was determined (L/GA_3). Controls: GA_3 under constant darkness (D/GA_3), light only (*L*) and darkness only (*D*)

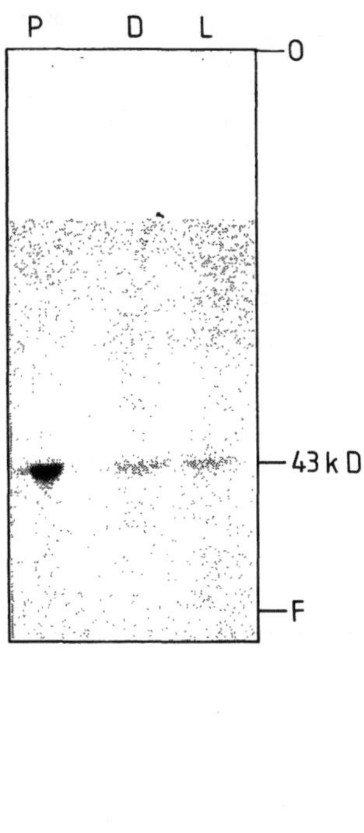

Fig. 10. Immunoblotting of samples from the experiment depicted in Fig. 9 with an antiserum against CHS from parsely. For conditions, see Fig. 4. Both samples (*D* = constant darkness; *L* = continuous irradiation with UV-A) were treated with GA_3

2.7 CHS Activity and Anthocyanin Accumulation in DC_b Cells Treated with a Fungal Elicitor

Besides light and GA_3, elicitors are of special interest in flavonoid biosynthesis (for review, see Smith and Banks 1986). Partially hydrolyzed cell walls from the oomycete *Pythium aphanidermatum* added to the culture medium of anthocyanin-producing DC_b cells inhibit the anthocyanin biosynthesis as well as the catalytic activity of the key enzyme CHS and there is consequently no 40 kD subunit present on the corresponding Western blot. However, the activities of two enzymes of the general phenylpropanoid pathway (PAL and 4CL) are induced dramatically. The activities of these enzymes lead to an enhancement of phenols of unknown structure which are deposited in the wall of the carrot cells (J.-P. Schnitzler, unpublished results). Since GA_3 and the fungal elicitor give rise to very similar physiological effects, one may hypothesize that at least parts of a common signal sequence are involved in the induction and inhibition of the enzymes.

3 Conclusions

The data presented here show that two different types of CHS proteins are present in carrot cells. The subunits exhibit different molecular weights. The 40 kD-type CHS seems to account for anthocyanin biosynthesis. The function of the other CHS species is still unknown. There are various indications in favour of different modes of regulation of these two types of CHS. By means of single cell cloning it would be possible to show whether both forms of CHS are present within one cell or are distributed to producing and non-producing cells.

Acknowledgements. We are grateful to K. Hahlbrock (Cologne) and R. Welle (Freiburg i.Br.) for providing us antisera against chalcone synthase. The investigation was supported by a grant of the Deutsche Forschungsgemeinschaft and a scholarship (J.G.) of the Friedrich-Ebert-Stiftung.

References

Beerhues L, Wiermann R (1985) Two different chalcone synthase activities from spinach. Z Naturforsch 40c:160–165

Beerhues L, Wiermann R (1988) Chalcone synthases from spinach (*Spinacea oleracea* L.) I. Purification, peptide patterns, and immunological properties of different forms. Planta 173:532–543

Beerhues L, Robenek H, Wiermann R (1988) Chalcone synthases from spinach (*Spinacea oleracea* L.) II. Immunofluorescence and immunogold localization. Planta 173:544–553

Beggs JB, Kuhn K, Böcker R, Wellmann E (1987) Phytochrome-induced flavonoid biosynthesis in mustard (*Sinapis alba* L.) cotyledons. Enzymic control and differential regulation of anthocyanin and quercetin formation. Planta 172:121–126

Dougall DK, LaBrake S, Whitten GH (1983) Growth and anthocyanin accumulation rates of carrot suspension cultures grown with excess nutrients after semicontinuous culture with different limiting nutrients at several dilution rates, pHs, and temperatures. Biotechnol Bioeng 25:581–594

Duell-Pfaff N, Wellmann E (1982) Involvement of phytochrome and a blue light photoreceptor in UV-B induced flavonoid synthesis in parsley (*Petroselinum hortense* Hoffm.) cell suspension cultures. Planta 156:213–217

Hall RD, Yeoman MM (1986) Temporal and spatial heterogeneity in the accumulation of anthocyanins in cell cultures of *Catharanthus roseus* (L.) G. Don. J Exp Bot 37:48–60

Harborne JB, Mayer AM, Bar-Nun N (1983) Identification of the major anthocyanin of carrot cells in tissue culture as cyanidin 3-(sinapoylxylosylglucosylgalactoside) Z Naturforsch 38c: 1055–1056

Hinderer W, Seitz HU (1988) Flavonoids. In: Vasil I, Constabel F (eds) Cell culture and somatic cell genetics of plants, vol 5. Phytochemicals in cell cultures. Academic Press, New York London, pp 29–48

Hinderer W, Noé W, Seitz HU (1983) Differentiation of metabolic pathways in the umbel of *Daucus carota*. Phytochemistry 22:2417–2420

Hinderer W, Petersen M, Seitz HU (1984) Inhibition of flavonoid biosynthesis by gibberellic acid in cell suspension cultures of *Daucus carota* L. Planta 160:544–549

Hopp W, Seitz HU (1987) The uptake of acylated anthocyanin into isolated vacuoles from a cell suspension culture of *Daucus carota*. Planta 170:74–85

Ozeki Y, Komamine A (1986) Effects of growth regulators on the induction of anthocyanin synthesis in carrot suspension cultures. Plant Cell Physiol 27:1361–1368

Ozeki Y, Sakano K, Komamine A, Tanaka Y, Noguchi H, Sankawa U, Suzuki T (1985) Purification and some properties of chalcone synthase from a carrot suspension culture induced for anthocyanin synthesis and preparation of its specific antiserum. J Biochem (Tokyo) 98:9–17

Reif HJ, Niesbach U, Deumling B, Saedler H (1985) Cloning and analysis of two genes for chalcone synthase from *Petunia hybrida*. Mol Gen Genet 199:208–215

Ryder TB, Hedrick SA, Bell JN, Liang X, Clouse SD, Lamb CJ (1987) Organization and differential activation of a gene family encoding the plant defense enzyme chalcone synthase in *Phaseolus vulgaris*. Mol Gen Genet 210:219–233

Smith DA, Banks SW (1986) Biosynthesis, elicitation and biological activity of isoflavonoid phytoalexins. Phytochemistry 25:979–995

Welle R, Grisebach H (1987) Purification and properties of chalcone synthase from cell suspension cultures of soybean. Z Naturforsch 42c: 1200–1206

Wellmann E (1971) Phytochrome-mediated flavone glycoside synthesis in cell suspension cultures of *Petroselinum hortense* after preirradiation with ultraviolet light. Planta 101:283–286

Induction and Suppression of Anthocyanin Synthesis in Carrot Suspension Cell Cultures Regulated by 2,4-D

Y. Ozeki[1], M. Matsuoka[2], Y. Ohashi[2], Y. Kano-Murakami[3], N. Yamamoto[4], and Y. Tanaka[5]

1 Introduction

In most plant cell cultures under dedifferentiated conditions, secondary metabolism activities expressed in parent plants are strongly repressed, or, in some cases, completely disappear. Secondary metabolism in intact plants is expressed in differentiated tissue and organs in which cell division ceases, suggesting that the expression of secondary metabolism has a close correlation with the differentiation of cells. It is thought that the repression of secondary metabolism in most cultured cells may be caused by the dedifferentiated state of cultured cells.

In order to elucidate the correlation between the expression of secondary metabolism and differentiation, we established a system in which anthocyanin synthesis was induced under the same conditions as the induction of embryogenesis in carrot suspension cultures (Ozeki and Komamine 1981). In this system, cells were subcultured in a medium containing 2,4-dichlorophenoxyacetic acid (2,4-D) in which undifferentiated growth occurred. When cells were fractionated by Ficoll density gradient centrifugation and transferred from a medium containing 2,4-D (referred to as +2,4-D medium) to a medium lacking 2,4-D (-2,4-D medium), cells of higher density induced embryogenesis as reported by Fujimura and Komamine (1979), and those of lower density induced anthocyanin synthesis. From the physiological point of view, the time course experiment of inductions and the effects of plant growth regulators have suggested a close correlation between the induction of embryogenesis, a morphological differentiation, and that of anthocyanin synthesis, a metabolic differentiation (Ozeki and Komamine 1981, 1986). In order to elucidate the mechanism of differentiation regulated by 2,4-D in this system, we studied the regulation of enzyme expression involved in anthocyanin synthesis.

[1]Department of Biology, College of Arts and Sciences, The University of Tokyo, Komaba, Meguro ku, Tokyo 153, Japan
[2]National Institute of Agrobiological Resources, Tsukuba City, Ibaraki 305, Japan
[3]Fruit Tree Research Station, Tsukuba City, Ibaraki 305, Japan
[4]Forestry and Forest Products Research Institute, Tsukuba City, Ibaraki 305, Japan
[5]National Institute of Agro-Environmental Sciences, Tsukuba City, Ibaraki 305, Japan

Primary and Secondary Metabolism of Plant Cell Cultures, ed. by W.G.W. Kurz

2 Regulation of Expression of Phenylalanine Ammonia-Lyase and Chalcone Synthase

Changes in the enzymes involved in phenylpropanoid and flavonoid metabolism were investigated in the induction of anthocyanin synthesis in this system (Ozeki and Komamine 1985a, Ozeki et al. 1987). When cells were transferred to a fresh medium, activity of phenylalanine ammonia-lyase (PAL) was rapidly transient irrespective of the presence or absence of 2,4-D in the medium. Such a rapid induction of PAL is well known as the "dilution effect" or "transfer effect". In these carrot cells, the main products of phenylpropanoid metabolism were caffeic and chlorogenic acid, both of which were accumulated in cells cultured in either -2,4-D or +2,4-D medium. The function of PAL induced by the dilution effect in vivo was unknown. In the +2,4-D medium, activity of PAL decreased thereafter and was repressed at a low level. The activities of other enzymes, cinnamate-4-hydroxylase, 4-coumarate:CoA ligase and chalcone-flavanone isomerase, maintained low levels as well. Especially chalcone synthase (CHS) activity could not be detected in cells cultured in the +2,4-D medium. On the other hand, when cells were transferred to the -2,4-D medium, anthocyanin synthesis was induced 4 or 5 days after the transfer and activity of CHS was detectable at this time. All of the activities of the enzymes involved in anthocyanin synthesis increased and reached a maximum at day 7 or 8 when anthocyanin was most rapidly synthesized and decreased thereafter (Ozeki and Komamine 1985a, Ozeki et al. 1987; see Fig. 1). The above result suggested that the induction of CHS activity has an important role in the induction of anthocyanin synthesis in this system.

When 2,4-D was added to the cells synthesizing anthocyanin in a medium lacking 2,4-D at day 6, anthocyanin synthesis immediately ceased and activities of CHS and PAL were rapidly repressed. At this time anthocyanin-synthesizing cells were diluted and transferred to a fresh medium containing 2,4-D, PAL activity was induced by the dilution effect despite the presence of 2,4-D, but accumulation of anthocyanin did not increase and CHS activity decreased with 2,4-D. Therefore,

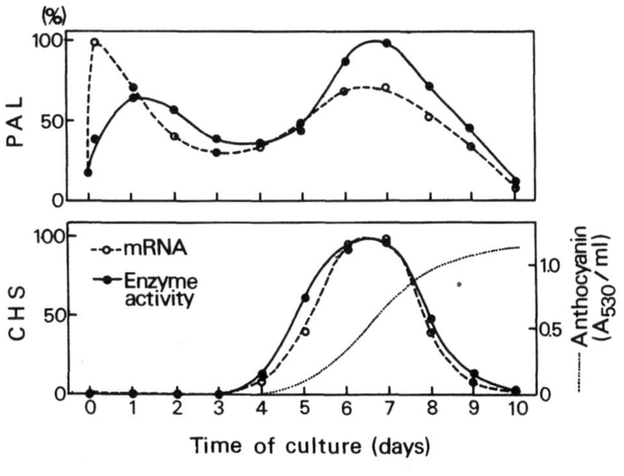

Fig. 1. Typical changes in mRNAs and enzyme activities of PAL (*upper*) and CHS (*lower*) and induction of anthocyanin synthesis of carrot suspension cultures in the –2,4-D medium. - - O - -, mRNA; —•—, activities of enzymes; ......, anthocyanin accumulation

the above results suggested that CHS plays a most important role in the regulation of anthocyanin synthesis by 2,4-D in this system and an increase of PAL activity induced by the dilution effect could not result in an increase of anthocyanin accumulation in vivo (Table 1). It appears that PAL induced by the dilution effect did not contribute to anthocyanin synthesis.

In order to examine the regulatory mechanisms of CHS and PAL, CHS and PAL proteins were purified from carrot cultured cells and antisera for CHS and PAL were prepared (Ozeki et al. 1985 and Tanaka Y, unpublished data). Using Western blotting and immunodetection, we investigated changes in protein amounts of PAL and CHS. The amounts of PAL increased and attained a peak 1 day after the transfer to a fresh medium but declined thereafter in a medium containing 2,4-D. In the same medium, CHS could not be detected above background level. CHS protein was detected 4 to 5 days after the transfer to the −2,4-D medium and the amounts of PAL and CHS gradually increased. These changes correlated closely with the changes in activities, suggesting that 2,4-D regulated the synthesis of CHS and PAL proteins (Ozeki and Komamine 1985b).

Changes in mRNAs for PAL and CHS were examined by Northern blotting using cDNAs for CHS and PAL described below as probes and using an in vitro translation system of wheat germ and immunoprecipitation. The amounts of mRNA for PAL were very high in cultures at day 0. These cultures were subjected to the fractionation procedure of cells. The increase in the mRNA level of PAL at day 0 after the transfer is due to the transfer effect, resulting in the induction of

Table 1. Effect of 2,4-D and dilution on anthocyanin accumulation and the activities of mRNAs and enzymes of PAL and CHS anthicyanin-synthesizing cells[a]

	Relative activity (%)		
	6 Days	6 Days[b] 6 h −2,4-D	6 Days[c] 6 h +2,4-D
PAL			
mRNA	100	123 (203)[d]	75 (165)[d]
Enzyme	100	142 (233)	72 (175)
CHS			
mRNA	100	181(116)	75(36)
Enzyme	100	181(66)	79(44)
	Relative amounts (%)		
Anthocyanin accumulation	100	122 (118)	93 (98)

[a] Activities of mRNAs and enzymes of PAL and CHS of anthocyanin-synthesizing cells cultured in the −2,4-D medium for 6 days were represented by 100%.
[b] Cells were continuously cultured for more than 6 h.
[c] 2,4-D was added at day 6 and cells were cultured for 6 h.
[d] In parentheses, cells were transferred to the fresh −2,4-D or +2,4-D medium at half the cell density of the original and cultured for 6 h.

protein synthesis. The level of mRNA of PAL decreased thereafter in the cells in the +2,4-D medium. mRNA for CHS was below the detectable level in the cells cultured in the +2,4-D medium. mRNA for CHS began to be detectable 4 to 5 days after the transfer to the -2,4-D medium and increased thereafter, coincident with the increase of mRNA for PAL (Fig. 1). mRNAs for CHS and PAL were at maximum 7 or 8 days after the transfer and decreased thereafter. These changes in mRNA levels for CHS and PAL were coordinated with changes in enzyme activities of CHS and PAL. All the results mentioned above suggested that the amounts of mRNA for PAL and CHS reflected those of the enzyme proteins and consequently those in their activities. It was suggested that 2,4-D regulated anthocyanin synthesis at transcriptional levels of PAL and CHS genes.

When 2,4-D was added to the cells which were cultured in -2,4-D medium for 6 days and anthocyanin synthesis and mRNA for CHS and PAL were induced, the levels of mRNA for CHS and PAL decreased (Table 1). However, when anthocyanin-synthesizing cells at day 6 were transferred to a fresh +2,4-D medium, mRNA for PAL increased by the dilution effect irrespective of the presence of 2,4-D. The mRNA level and enzyme activity of PAL were also increased by the transfer to the fresh -2,4-D medium and activity of CHS was observed, but the level of accumulated anthocyanin did not increase (Table 1). Although the function of PAL by the dilution effect is unknown to affect the phenylpropanoid pathway, it is certain that there are two types of PALs in this system: one PAL type was slowly induced in the -2,4-D medium during anthocyanin synthesis, which took about 3 days from the beginning of induction to reach a maximum, and the other PAL type was rapidly induced by the dilution effect, which took within a day to reach a maximum. No difference could be observed immunologically between them in Western blotting and immunoprecipitation of the products synthesized according to mRNA by an in vitro translation system.

3 cDNA Preparation for PAL and CHS: Difference in the cDNA Structure of PAL Induced by the Dilution Effect and Anthocyanin Synthesis

In order to reveal the regulatory mechanism of expressions of PAL and CHS genes, we attempted to obtain cDNA for PAL and CHS from the cDNA library of cells of carrot suspension cultures. In particular, PAL was induced by two triggers: the dilution effect and -2,4-D, resulting in anthocyanin synthesis. Although we cannot detect a difference in the properties of the two PAL proteins immunologically, it was thought that the regulation of expression of PAL gene(s) might be different in rapidly induced PAL through the dilution effect and in slowly induced PAL through -2,4-D contributed to anthocyanin synthesis. Therefore, mRNAs were prepared not only from cells cultured in the -2,4-D medium for 7 days, in which anthocyanin synthesis was induced by the inductions of PAL, CHS and other enzymes involved in anthocyanin synthesis, but also from cells cultured in the fresh +2,4-D medium 6 h after transfer, in which PAL was rapidly induced by the dilution effect. Using the method developed by Gubler and Hoffman (1983), cDNAs for PAL were independently prepared with both mRNAs and that for CHS was prepared with mRNA derived from cells cultured in -2,4-D medium for 7 days. cDNAs were size-separated by gel filtration and introduced into λgt11. The cDNA clone for CHS was detected using a cDNA probe for CHS of cultured parsley cells derived by Prof. Hahlbrock (Reimold et al. 1983) and that for PAL was detected by a cDNA probe for PAL of sweet potato. When cDNA libraries were prepared using mRNAs size-separated only once by sucrose density gradient centrifugation, no positive clone was obtained from more than 10^6 clones, because the amounts of mRNAs for CHS and PAL were small in total mRNAs, the major part of which were housekeeping mRNAs containing the cultured carrot cells used in our experiments. Therefore, mRNAs for CHS and PAL were further purified by sucrose density gradient centrifugation under different separating conditions. Using more purified mRNA, cDNA libraries were prepared and cDNAs for CHS and PAL were surveyed using the cDNA probes described above.

We could detect about 200 positive plaques for CHS cDNA and 12 plaques strongly hybridized to the cDNA probe were purified. The lengths of the inserts of ten clones were about 1.4 kbp and those of the other two clones were 0.8 and 1.2 kbp. All these inserts were hybridized to cDNA for CHS of cultured parsley cells. The cDNA sequence of the 1.4 kbp insert had more than 70% homology to cDNA for CHS of parsley. The expressed fusion protein of β-galactosidase and the cDNA insert by the λgt11 system was cross-reacted with CHS antiserum prepared against suspension cultured cells of carrot. These results confirmed that the cDNA obtained here is cDNA for CHS of cultured carrot cells which was induced during anthocyanin synthesis in -2,4-D medium.

More than 500 positive plaques were found in each cDNA library from either the mRNA of cells in which PAL was induced by the dilution effect in +2,4-D medium or that of cells in which anthocyanin synthesis was induced in -2,4-D medium. Among the positive plaques, nine and eight of the strongly hybridized plaques for PALs induced by the dilution effect and -2,4-D respectively, were picked up. All nine and eight cDNA inserts were hybridized with cDNA for PAL of sweet potato, and fused protein produced by λgt11 were cross-reacted with an antiserum against PAL derived from carrot suspension cultured cells, confirming cDNAs for PALs. Remarkable differences were found between cDNAs for PAL induced by the dilution effect and that induced by -2,4-D contributed to anthocyanin synthesis. All nine cDNAs for PAL induced by the dilution effect had lengths of more than 1.9 kbp, whereby the 2.4 kbp lengths of two clones were the longest. cDNAs less than 1.9 kbp could not be obtained due to the size separation of the cDNAs by gel filtration and the selection of strength in hybridization signals. On the other hand, none of the eight clones of cDNA for PAL induced by -2,4-D showed an insert longer than 1.45 kbp by Eco RI from λgt11 arms. Three clones showed two bands by Eco RI cutting, in which one is 1.45 kbp and the others have lengths from 0.6 kbp to 0.95 kbp, that is 2.05 to 2.4 kbp in total lengths. Figure 2 shows one of the cDNAs for PAL induced by the dilution effect and that for PAL induced by -2,4-D which has an Eco RI site in cDNA. These results showed that all of the cDNA for PAL induced by -2,4-D had an Eco RI site at 1.45 kbp upstream from the 3′ end. It was confirmed that eight cDNAs for PAL were picked up from another cDNA library independently prepared with mRNA from other cultured

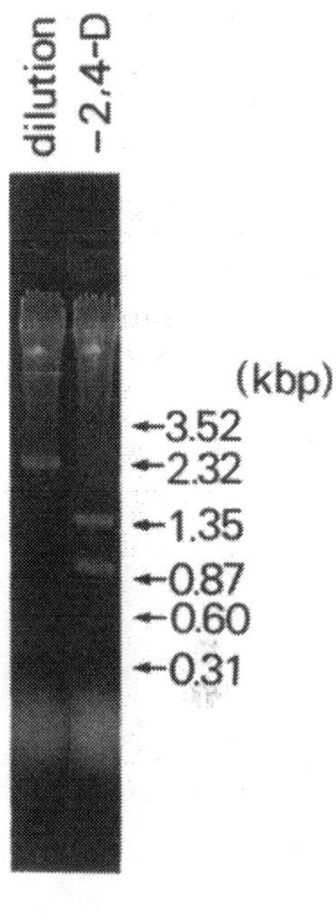

Fig. 2. cDNA for PAL induced by the −2,4-D medium associated with anthocyanin synthesis (*right*) and that induced by dilution effect (*left*)

cells in the −2,4-D medium and all eight cDNAs for PAL induced had an Eco RI site 1.45 kbp from the 3′ end. On the other hand, all of the cDNAs for PAL induced by the dilution effect have no Eco RI site in the inserts. The sequences of cDNAs for the two PALs, induced by the dilution effect and −2,4-D, were compared. The sequences of cDNAs have a high homology of over 80% in the coding region. But the sequences were less homologous, below 50%, in their non-coding region. These results showed that sequences of cDNA for PALs were different, suggesting that mRNA for PAL rapidly induced by the dilution effect might be transcribed from different PAL genes which, expressed in the −2,4-D medium, is associated with anthocyanin synthesis (Fig. 3). It was supposed that there are two or more genes for PAL in the carrot genome; one expresses rapidly by the dilution effect but the translated PAL does not contribute to an increase of anthocyanin, while the other expresses slowly by −2,4-D and the translated PAL contributes to synthesis of anthocyanin.

4 Conclusions

In this carrot cell suspension culture system, 2,4-D regulated induction and suppression of anthocyanin synthesis which has a close correlation with the induction of embryogenesis. Induction and suppression of anthocyanin synthesis were caused by the induction and suppression of CHS and PAL regulated by 2,4-D. We could obtain cDNAs for CHS associated with anthocyanin synthesis and PALs which were induced by the dilution effect as well as by the −2,4-D medium during

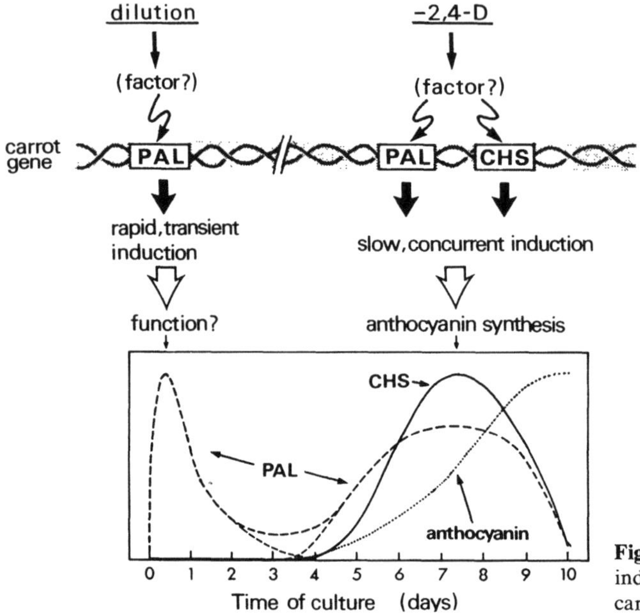

Fig. 3. Proposed scheme of the induction of PAL and CHS in carrot suspension cultures

anthocyanin synthesis. cDNA for CHS was 1.4 kbp in length and both of the cDNAs for PAL were 2.4 kbp each. The cDNA sequences for PALs were different, suggesting the possibility that the mRNA for PAL induced by the dilution effect may be transcribed from different genes of PAL induced by the -2,4-D medium associated with anthocyanin synthesis. The function of the PAL transient induced by the dilution effect was unknown in contrast to the other PAL associated with anthocyanin synthesis.

Further studies are required to focus on the regulatory mechanisms of CHS and PAL genes associated with anthocyanin synthesis in order to elucidate the induction and suppression of anthocyanin synthesis regulated by 2,4-D, and to reveal the function and regulation of PAL induced by the dilution effect.

References

Fujimura T, Komamine A (1979) Synchronization of somatic embryogenesis in a carrot suspension culture. Plant Physiol 64:162–164

Gubler U, Hoffman BJ (1983) A simple and very efficient method for generating for cDNA libraries. Gene 25:263–269

Ozeki Y, Komamine A (1981) Induction of anthocyanin synthesis in relation to embryogenesis in a carrot suspension culture: correlation of metabolic differentiation with morphological differentiation. Physiol Plant 53:570–577

Ozeki Y, Komamine A (1985a) Changes in activities of enzymes involved in general phenylpropanoid metabolism during the induction and reduction of anthocyanin synthesis in a carrot suspension culture as regulated by 2,4-D. Plant Cell Physiol 26:903–911

Ozeki Y, Komamine A (1985b) Induction of anthocyanin synthesis in relation to embryogenesis in a carrot suspension culture – a model system for the study of expression and repression of secondary metabolism. In: Neumann KH, Barz W, Reinhard E (eds) Primary and secondary metabolism of plant cell cultures. Springer, Berlin Heidelberg New York Tokyo, pp 99–106

Ozeki Y, Komamine A (1986) Effects of growth regulators on the induction of anthocyanin synthesis in carrot suspension cultures. Plant Cell Physiol 27:1361–1368

Ozeki Y, Sakano K, Komamine A, Tanaka Y, Noguchi H, Sankawa U, Suzuki T (1985) Purification and some properties of chalcone synthase from a carrot suspension culture induced for anthocyanin synthesis and preparation of its specific antiserum. J Biochem 98:9–17

Ozeki Y, Komamine A, Noguchi H, Sankawa U (1987) Changes in activities of enzymes involved in flavonoid metabolism during the initiation and suppression of anthocyanin synthesis in carrot suspension cultures regulated by 2,4-dichlorophenoxyacetic acid. Physiol Plant 69:123–128

Reimold U, Kroger M, Kreuzaler F, Hahlbrock K (1983) Coding and 3' non-coding nucleotide sequence of chalcone synthase mRNA and assignment of amino acid sequence of the enzyme. EMBO J 2:1801–1805

Digitoxin 12β-Hydroxylase, a Cytochrome P-450-Dependent Enzyme from Cell Cultures of *Digitalis lanata* Ehrh

M. Petersen[1], H. U. Seitz[2], and A. W. Alfermann[1]

1 Introduction

Since the publication of the famous book *An Account of the Foxglove, and Some of its Medical Uses: with Practical Remarks on Dropsy, and other Diseases* by William Withering in 1785, cardiac glycosides of *Digitalis* species became a widely used drug against heart diseases. The most frequently used compounds today are derivatives of digoxin, which are isolated from *Digitalis lanata*. This plant is native to south-eastern Europe, but is cultivated today in many countries all over the world.

The molecular structure of the most important cardenolides of *Digitalis* species is shown in Fig. 1. It is derived from a sterol with *cis*-linked rings C and D and an unsaturated, five-membered lactone ring at C-17. These two characteristics are very important for the medicinal effectiveness of the compounds. At C-3 there is a sugar side chain composed of three digitoxose molecules. In contrast to digitoxin, digoxin has an additional hydroxyl function at C-12. The glucose and the acetyl group, which can be bound to the last digitoxose of the molecule in the plant, are often lost during the isolation of the cardenolides.

The 12β-hydroxyl group, which makes the difference between digitoxin and digoxin, influences the pharmacokinetic behaviour of the cardiac glycosides in humans. This results in an easier handling of digoxin derivatives for the treatment of the patients, therefore, they are used more frequently in medicinal therapy today. Digitoxin derivatives can be isolated from several *Digitalis* species, whereas a technical isolation of digoxin derivatives is only possible from *Digitalis lanata*. However, only about two-thirds of the total cardenolides isolated from *Digitalis lanata* are digoxin derivatives, the rest being digitoxin and others. For this reason, the biotransformation process of digitoxin to digoxin derivatives, which has been worked out by Reinhard and co-workers (Reinhard 1974; Heins 1978; Alfermann et al. 1983, 1985), has become of importance. Cell suspension cultures of *Digitalis lanata* are able to transform β-methyldigitoxin to β-methyldigoxin by hydroxylation at C-12, although they are not able to synthesize cardiac glycosides *de novo*. Since the reaction product appeared in the medium and could easily be isolated, this process seemed to be suited for a biotechnological production of digoxin derivatives and was scaled up to a level of a 1000-l pilot plant.

[1]Institut für Entwicklungs- und Molekularbiologie der Pflanzen, Heinrich-Heine-Universität Düsseldorf, Universitätsstr. 1, D-4000 Düsseldorf 1, FRG
[2]Institut für Biologie I, Universität Tübingen, Auf der Morgenstelle 1, D-7400 Tübingen, FRG

Primary and Secondary Metabolism of Plant Cell Cultures, ed. by W.G.W. Kurz

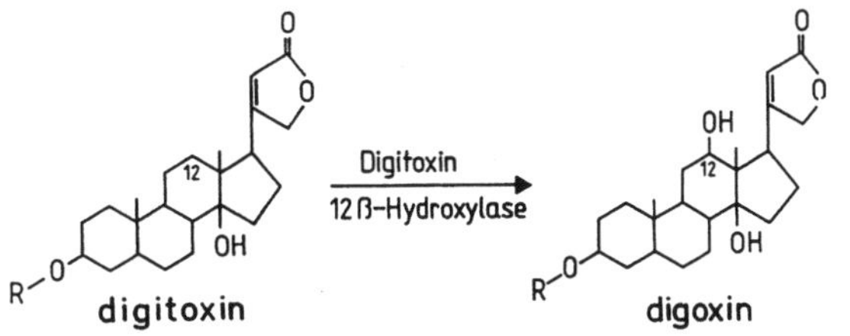

Fig. 1. Molecular structure of the cardenolides, digitoxin and digoxin, and the reaction of the enzyme digitoxin 12β-hydroxylase. *R* Digitoxose-digitoxose-digitoxose

2 Characteristics of Digitoxin 12β-Hydroxylase

The enzyme catalyzing the 12β-hydroxylation of digitoxin to digoxin is a membrane-bound cytochrome P-450-dependent monooxygenase. The digitoxin 12β-hydroxylase is strictly dependent on molecular oxygen. It can be isolated from cell cultures of *Digitalis lanata* by preparation of microsomes by $MgCl_2$ precipitation or centrifugation at 100 000 g (Petersen and Seitz 1985; Petersen et al. 1988). The enzyme assay for the digitoxin 12β-hydroxylase consists of 2.5 mg of microsomal protein, 20 μM of β-methyldigitoxin and an NADPH-regenerating system consisting of 1 mM $NADP^+$, 5 mM glucose-6-phosphate and 8.4 nkatal ml^{-1} glucose-6-phosphate dehydrogenase in a total volume of 5 ml 0.1 M Tris/HCl buffer containing 0.6 M mannitol, 1 mM EDTA and 1 mM DTT at a pH of 7.5. After an incubation time of 4 h at 26°C the cardiac glycosides were extracted with chloroform, purified by a petroleum benzine (bp. 40–60°C) treatment and analyzed by HPLC (Petersen and Seitz 1985). An HPLC chromatogram of the extracted cardiac glycosides of an enzyme assay is shown in Fig. 2.

The digitoxin 12β-hydroxylase is located in the endoplasmic reticulum. This was shown by sucrose density gradient fractionation and marker enzyme analysis (Petersen et al. 1988). The enzyme could be reconstituted from the partially or totally purified cytochrome P-450, the NADPH-cytochrome c (P-450) reductase and microsomal lipids (Petersen and Seitz 1988). The cytochrome P-450 is responsible for the substrate binding and specificity and for the activation of molecular oxygen. The electrons are transferred from NADPH via the flavine-dependent NADPH-cytochrome c (P-450) reductase to the cytochrome P-450. As a matrix, phospholipids are essential. The main microsomal phospholipids of *Digitalis lanata* were identified as phosphatidylethanolamine and phosphatidylcholin (Petersen and Seitz 1988). Therefore, the composition of the digitoxin 12β-hydroxylase is similar to the composition of mammalian liver monooxygenases, but it is not dependent on cytochrome b_5.

Cytochrome P-450 is present in *Digitalis lanata* microsomes at a level of 110 to 140 pmol mg^{-1} microsomal protein. This can be calculated from CO-difference spectra of sodium dithionite-reduced microsomes using a millimolar extinction coefficient of 91 mM^{-1} cm^{-1} between 450 and 490 nm (Omuta and Sato 1964). The dependence of a hydroxylation reaction on cytochrome P-450 can be determined by the inhibition of the reaction by CO and the reversal of this inhibition by illumination with 450 nm light (West 1980). The digitoxin 12β-hydroxylase activity is inhibited by 40% in an atmosphere containing 15% CO besides 15% O_2 and N_2,

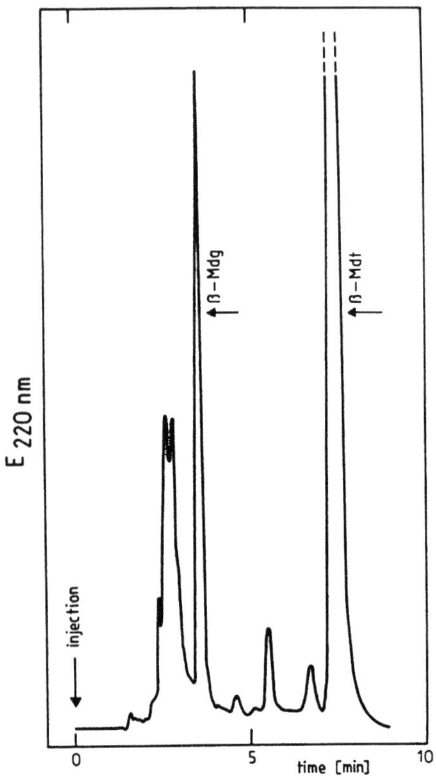

Fig. 2. HPLC chromatogram of the cardiac glycosides extracted from a digitoxin 12β-hydroxylase assay. Column: Shandon Hypersil ODS 5 μm, 250×4.6 mm; eluent: 80% methanol, 20% water; flow rate: 1 ml min^{-1}; detection at 220 nm. *β-Mdt*, β-methyldigitoxin; *β-Mdg*, β-methyldigoxin

but this inhibition can be completely reversed by the illumination with light of the wavelength 450 nm (Table 1). This is strong evidence for the participation of cytochrome P-450 in the 12β-hydroxylation of cardiac glycosides.

Digitoxin 12β-hydroxylase is strictly dependent on NADPH, which can be either supplied directly or by a regenerating system. The regenerating system has the advantage of permanent high levels of NADPH and low levels of $NADP^+$, which was shown to inhibit the digitoxin 12β-hydroxylase competitively (Petersen et al. 1988). The inhibition constant of $NADP^+$ was determined to be at 1.1 mM. The saturation concentration for NADPH was at 1 mM and a Km value of 26 μM was determined. NADH could not replace NADPH and no synergistic effect was observed when NADH and NADP(H) were supplied together in the presence or absence of ATP. This demonstrates that no transhydrogenases are present in the microsomal preparation, which would transfer the hydrogen from NADH to $NADP^+$. The missing synergistic effect of a stimulation of the hydroxylation reaction by simultaneous addition of NADH and NADPH shows that cytochrome b_5, which is also present in *Digitalis lanata* microsomes (Petersen et al. 1988), does not participate in the electron transport chain of this enzyme. In mammalian microsomal liver monooxygenases NADH can supply electrons to the cytochrome P-450 via cytochrome b_5, too. This kind of electron transport could not be proven for plant cytochrome P-450 systems up to now (West 1980).

Table 1. Influence of different gas compositions on the activity of digitoxin 12β-hydroxylase (100% = 124.3 nkat kg^{-1})

Gas composition	Hydroxylation activity (%)
Air	100
100% N_2	10
15% O_2 + 85% N_2	117
15% O_2 + 15% CO + 70% N_2	60
15% O_2 + 15% CO + 70% N_2 + 450 nm light	98

The digitoxin 12β-hydroxylase accepts all cardenolides with the aglycone structure of digitoxigenin as substrates. The number and kind of sugar molecules attached at C-3 of the digitoxigenin is not important for the reaction. In standard assays for the digitoxin 12β-hydroxylase the semi-synthetic β-methyldigitoxin was used as a substrate. In enzyme assays the saturation concentration for β-methyldigitoxin was at 20 μM, and a Km value of 7.1 μM was determined. Very similar values were observed for digitoxin, which may be the natural substrate of the enzyme.

The digitoxin 12β-hydroxylase is strongly inhibited by cytochrome c, which also accepts electrons from the NADPH-cytochrome c (P-450) reductase. Therefore, the inhibition is of the competitive mode. Other inhibitors are Hg^{2+} and Cu^{2+} and p-hydroxymercuribenzoate. On the other hand, KCN, which is a typical inhibitor of heme enzymes but not for cytochrome P-450, has no inhibiting but a stimulating effect (Petersen and Seitz 1985). This is further strong evidence for the dependence of the digitoxin 12β-hydroxylase reaction on cytochrome P-450.

3 Characterization of a Cell Culture of *Digitalis lanata* During a Culture Period

Suspension cultures of *Digitalis lanata* were characterized during a culture period of 10 days by observing growth, medium parameters and activities of enzymes of primary and secondary metabolism. The *Digitalis lanata* cells were grown in a Murashige and Skoog medium (1962) with 1 mg l^{-1} IAA and 2 mg l^{-1} kinetin. They were subcultured every 7 days by transferring 12 ml of the old culture to 50 ml of fresh medium in a 300-ml Erlenmeyer flask. The cultures were cultivated in the dark at 26°C on a gyratory shaker (100 rpm).

After the transfer of the cells to fresh culture medium the fresh weight of the cells and the water content strongly decline, probably due to the high difference in the osmolarity of the old and the new medium (Fig. 3a, c). Cell growth starts immediately without any lag phase, as can be seen in the graphs of fresh and dry weight (Fig. 3a) and packed cell volume and cell number (Fig. 3b). All four parameters show a parallel increase up to day 6 or 7 of the culture period and then

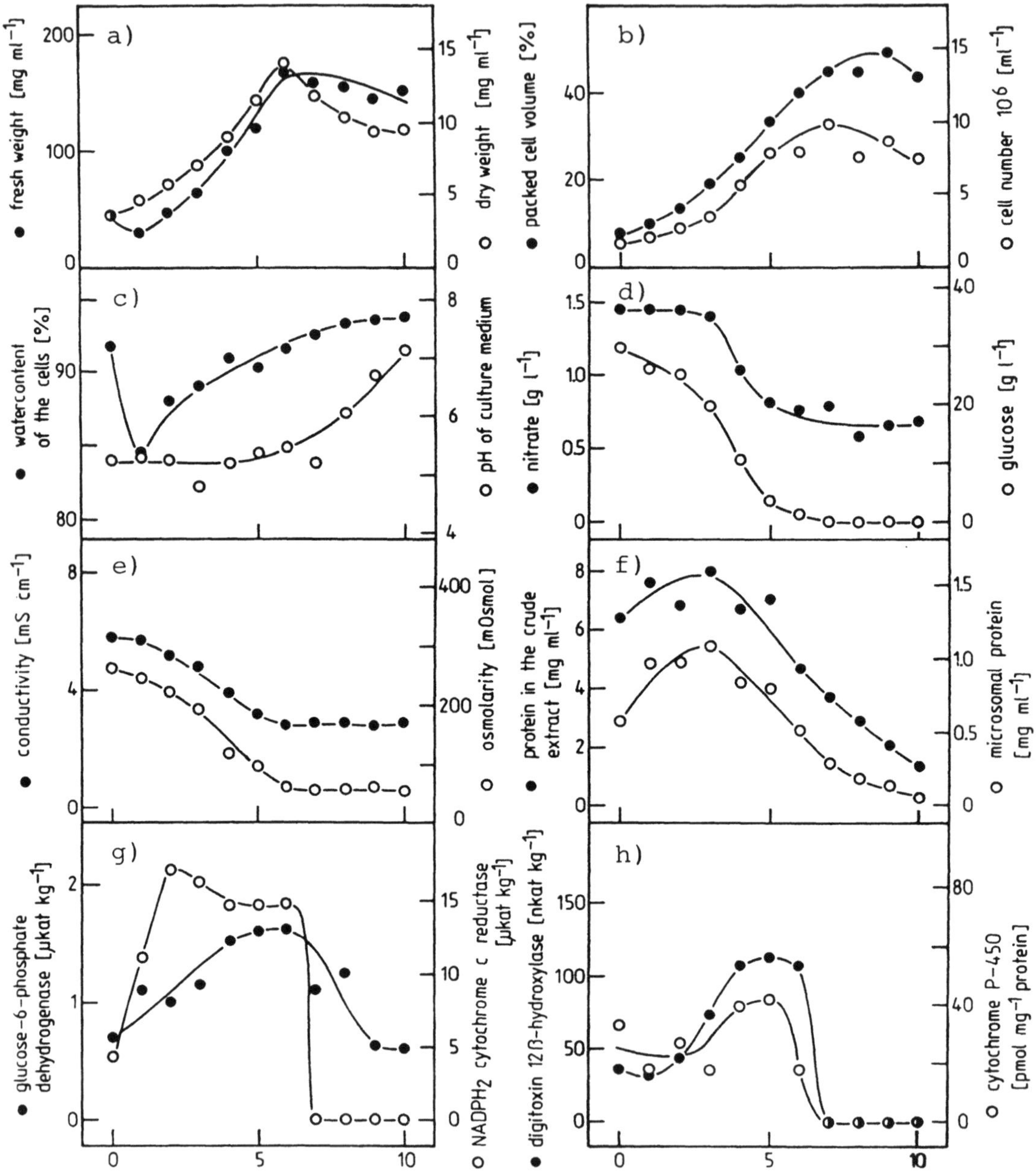

Fig. 3a-h. Characterization of a cell culture of *Digitalis lanata* during a culture period of 10 days. **a** (●) fresh weight, (○) dry weight; **b** (●) packed cell volume, (○) cell number; **c** (●) water content of the cells, (○) pH of the medium; **d** (●) nitrate content in the medium, (○) glucose concentration; **e** (●) conductivity and (○) osmolarity of the medium; **f** protein content of the crude cell extract (●) and the microsomal fraction (○); **g** specific activity of glucose-6-phosphate dehydrogenase (●) and NADPH-cytochrome c (P-450) reductase (○) **h** (●) specific digitoxin 12β-hydroxylase activity, (○) cytochrome P-450 in the microsomal fraction decline without a stationary phase. Dry weight and cell number drop more rapidly than fresh weight and packed cell volume. This coincides with the increasing water content of the cells during the culture period (Fig. 3c). Therefore, the cell cultures of *Digitalis lanata* show only two prominent phases, the growth and death phase. The pH value of the culture medium is rather constant at pH 5.2 during the growth phase of the cells, but increases up to pH 7.2 at the end of the culture period (Fig. 3c). Fresh medium contains 30 g glucose l^{-1}. The glucose is consumed completely by the cells during 6 days (Fig. 3d). The decrease in the glucose content is paralleled by a decrease in the nitrate content, which drops to about one-half of the original nitrate concentration (Fig. 3d). In the same way conductivity and osmolarity of the medium decrease until day 6 or 7 of the culture period (Fig. 3e). These two parameters mainly show the variation of nitrate as the dominant ion and of glucose as the main osmotically active component of the medium. Probably the glucose content is limiting for the cell growth. After the cells have stopped growing, also other medium components can no longer be consumed.

The protein content in the crude cell extract and in the microsomal fraction shows similar changes (Fig. 3f). It rises until day 3 of the culture period and then decreases constantly. The microsomal protein content is nearly zero at day 10. Therefore, the intracellular membrane system seems to have characteristic changes in its expression. This can be seen in context to the graphs of the digitoxin 12β-hydroxylase activity, the cytochrome P-450 content and the NADPH-cytochrome c (P-450) reductase activity (Fig. 3g, h). These parameters seem to be tightly coupled with a high protein content and possibly a distinct intracellular membrane system. Cytochrome P-450 and digitoxin 12β-hydroxylase activity show parallel curves. Both rise from day 3 to day 5 and then drop steeply until they are at zero on day 7. The main part of the measurable cytochrome P-450 seems to belong to the digitoxin 12β-hydroxylase. On the other hand, the activity of the NADPH-cytochrome c (P-450) reductase has a high level from day 2 to 6 of the culture period, but then also drops to zero at day 7. Probably the reductase works with several species of cytochrome P-450 with different substrates and therefore is a component of several different cytochrome P-450-dependent enzyme systems. These other systems probably do not show such prominent changes in the cytochrome P-450 content as the digitoxin 12β-hydroxylase. This enzyme, which catalyzes one of the last steps of cardiac glycoside biosynthesis, is only active in the last part of the growth phase. This step of secondary metabolism seems to be dependent on sufficient nutrition of the cells, since all parameters concerning secondary metabolism drop to zero shortly after the glucose is exhausted. An enzyme of primary metabolism, glucose-6-phosphate dehydrogenase, shows a different pattern. The activity rises up to day 7 of the culture period and then declines, but never reaches zero. Primary metabolism is active during the entire time of the culture period, even in the death phase.

Digitoxin 12β-hydroxylase is not a constitutive enzyme. It is probably regulated through the provision of the cytochrome P-450, which is specific for cardiac glycosides. The highest levels of digitoxin 12β-hydroxylase activity and cytochrome P-450 content can be measured during the last days of the growth phase shortly before the cells die. A very similar curve can be observed for the in vivo biotransformation of β-methyldigitoxin to β-methyldigoxin in suspension cultures of *Digitalis lanata*. The digitoxin 12β-hydroxylase is not a main regulation point of cardiac glycoside biosynthesis, because this enzyme is active in cell cultures which no longer synthesize cardiac glycosides. The biosynthetic pathway of the cardiac glycosides and its regulation in cell cultures of *Digitalis lanata* still have to be investigated in the future.

Acknowledgements. This work was supported by a grant of the Deutsche Forschungsgemeinschaft and by a scholarship to MP from the Studienstiftung des deutschen Volkes.

References

Alfermann AW, Bergmann W, Figur C, Helmbold U, Schwantag D, Schuller I, Reinhard E (1983) Biotransformation of β-methyldigitoxin to β-methyldigoxin by cell cultures of *Digitalis lanata*. In: Mantell SH, Smith H (eds) Plant biotechnology. Univ Press, Cambridge, pp 67–74

Alfermann AW, Spieler H, Reinhard E (1985) Biotransformation of cardiac glycosides by *Digitalis* cell cultures in airlift reactors. In: Neumann KH, Barz W, Reinhard E (eds) Primary and secondary metabolism of plant cell cultures. Springer, Berlin Heidelberg New York Tokyo, pp 316–322

Heins M (1978) Screening of *Digitalis lanata* plants and cell cultures for hydroxylating activity. In: Alfermann AW, Reinhard E (eds) Production of natural compounds by cell culture methods. Ges Strahlen Umweltforsch, München, pp 39–42

Murashige T, Skoog F (1962) A revised medium for the rapid growth and bioassays with tobacco tissue cultures. Physiol Plant 15:473–497

Omura T, Sato R (1964) The carbon monoxide-binding pigment of liver microsomes I. Evidence for its hemoprotein nature. J Biol Chem 239:2370–2378

Petersen M, Seitz HU (1985) Cytochrome P-450-dependent digitoxin 12β-hydroxylase from cell cultures of *Digitalis lanata*. FEBS Lett 188:11–14

Petersen M, Seitz HU (1988) Reconstitution of cytochrome P-450-dependent digitoxin 12β-hydroxylase from cell cultures of *Digitalis lanata* Ehrh. Biochem J 252:537–543

Petersen M, Seitz HU Reinhard E (1988) Characterization and Localization of digitoxin 12β-hydroxylase from cell cultures of *Digitalis lanata* Ehrh. Z Naturforsch 43c:199–206

Reinhard E (1974) Biotransformations of plant tissue cultures. In: Street HE (ed) Tissue culture and plant science. Academic Press, New York London, pp 433–459

West CA (1980) Hydroxylases, monooxygenases, and cytochrome P-450. In: Davies DD (ed) The biochemistry of plants, vol 2. Academic Press, New York London, pp 317–367

The Integration of Cardenolide Biosynthesis in the Differentiation Program "Somatic Embryogenesis" of *Digitalis lanata*

M. Luckner and B. Diettrich[1]

1 Introduction: The Phase-Dependent Expression of Secondary Metabolism

The restriction to certain specialized cells, tissues, or organs, as well as to certain developmental stages is the most outstanding characteristic of secondary metabolism. In several experimental systems it was shown that this "phase dependence" is caused by the synthesis of the enzymes forming the secondary compounds shortly before or during the expression of the biosynthesis of secondary products. This demonstrated that secondary metabolism is the result of differential gene expression [1–3].

Under normal physiological conditions formation of the enzymes of secondary pathways is integrated in the programmed expression of genes during cell specialization as well as tissue and organ development. In extreme physiological situations, however, the genes encoding secondary metabolism may also be expressed independently of these programs [1–3], a feature probably caused by multiple regulatory sites.

Most plant cell cultures are derived from explants which contain only specialized cells which are unable to divide. These cells have to be reprogrammed during the initiation of the primary callus in a way which reduces the extent of their specialization and allows them to enter the cell division cycle. Reprogramming usually is brought about by using a growth regulator with strong auxin activity. The decrease of specialization is accompanied by the loss of parameters characterizing the specialized cells, including secondary metabolism. Many types of cultured cells are therefore unable to form secondary products even if kept under very different conditions. The biosynthesis may be triggered, however, in these "recalcitrant" cells using the natural way, i.e., launching a program of cell differentiation or organ development in which the biosynthesis of the desired secondary products in nature is integrated [4]. This possibility will be discussed in the following with regard to the formation of cardenolides in cell and organ cultures of *Digitalis lanata.*

D. lanata is cultivated widely for its cardiac glycosides, secondary products used in the treatment of heart disease. In several laboratories callus and suspension cultures were proven unable to build cardiac glycosides. They formed, however, cardiac glycosides after triggering certain developmental programs, the most important of which is somatic embryogenesis (the respective literature is summarized in [5] and [6]).

[1]Martin-Luther-University Halle-Wittenberg, Section of Pharmacy, DDR-4050 Halle, Weinbergweg 15, GDR

Primary and Secondary Metabolism of Plant Cell Cultures, ed. by W.G.W. Kurz

2 The Sites of Synthesis and Storage of Cardenolides in the Plant *Digitalis lanata*

Information is necessary about the types of cells that build and accumulate the secondary products in question before attempts can be made to express the formation of these cells in vitro. With respect to the cardiac glycosides of *D. lanata* the leaves were known to contain the highest amounts and therefore were used as raw material in the preparation of these compounds in the pharmaceutical industry. Detailed investigations with sensitive radioimmunoassays showed that all green organs of the plant (and the seed coat) are rich in cardenolides [7, 8].

The investigation of individual tissues revealed that the mesophyll cells and similar types of chloroplast-containing "green" cells contained most of the cardiac glycosides. In contrast, epidermis cells, pith cells, different types of root cells, and the cells of the vascular bundles were nearly free of these compounds [7, 8]. In addition, it was shown that mesophyll protoplasts of *D. lanata* (like those of *D. purpurea* [9]) were able to synthesize cardenolides from administered radioactive labelled precursors [4]. These results proved the mesophyll cells as the cell type most important for cardenolide formation and accumulation in *D. lanata*. It could therefore be expected that the formation of mesophyll-like cells in *D. lanata* cultures would trigger cardenolide biosynthesis in vitro.

3 Reprogramming of Specialized Cells During the Establishment of *Digitalis lanata* Cell Cultures

As yet cell and tissue cultures were obtained from only a few of the many kinds of specialized cells making up the plant *D. lanata*. Most important were cell types which, due to a high degree of specialization, were unable to divide under normal physiological conditions. Despecialization of these cell types was brought about by treatment of the explants with 2,4-D. Cell types forming primary callus under these conditions were:

1. The xylem parenchyma cells of filaments, leaves, stem, and roots [10]. (Probably the same type of cells formed primary callus also in leaf explants of *D. purpurea* [11]).
2. Unripe pollen grains [unpublished results] (This type of cell yielded haploid callus. The haploid state, however, was not stable and after some time diploid cultures were obtained.)
3. The ovules [unpublished results].

(Results described in the literature regarding the formation of callus from mesophyll protoplasts could not be substantiated in the laboratory of the authors.)

In addition, shoot tip meristems yielded a mass of durable, nodular, green tissue after wounding or administration of high concentrations of cytokinins, e.g., benzyl adenine [unpublished results]. No addition of any auxin was necessary because these cells are able to produce indole-3-acetic acid. Since meristem cells already divide in the mother plant, no 2,4-D must be added to reduce the degree of specialization or to initiate cell division.

The administration of high concentrations of benzyl adenine depressed the tendency of the meristems to form leaves and shoots. The rapid formation of shoots in media with low cytokinin activity provides the base for recently developed methods for the micropropagation of valuable *D. lanata* genotypes by shoot tip cultures [12–15].

4 Triggering Morphogenetic Programs in Cell Cultures of *Digitalis lanata*

The callus obtained from shoot tip meristems was shown to be fixed with respect to its morphogenetic potential. Like in the intact plant it formed only leaves and shoots. Using many different cultural conditions, no other morphogenetic programs could be triggered [unpublished results]. In contrast, reprogramming of the more specialized cells by 2,4-D yielded cell lines with a potential for the expression of a much broader variety of morphogenetic programs. The 2,4-D-treated cells thus obviously were more despecialized than the shoot tip meristem cells and had a wider option to react.

The morphogenetic potential of 2,4-D-treated cells was extensively investigated with cell lines derived from xylem parenchyma cells [10]. In nutrient media with high auxin activity these cell lines grew as colonies of thin-walled, vacuolated cells, which resembled parenchyma cells. But in contrast to the true parenchyma cells, these cells where able to divide. In the following this cell type will therefore be referred to as "parenchymoid".

The parenchymoid cells were free of cardiac glycosides. In agreement similar cell types in the mother plant, e.g., pith cells, they also contained only traces of cardenolides [8]. Parenchymoid cells were, however, able to transform added cardenolides, e.g., by glucosylation [16, 17]. The cardiac glucosides built were stored in the vacuoles [18, 19]. These results indicate that the final reactions of cardenolide formation and the processes involved in accumulation do not limit the ability of the parenchymoid cells for cardiac glycoside biosynthesis in vitro.

Glucosylation obviously is a prerequisite for cardenolide accumulation, since only the glucosylated compounds undergo transport through the tonoplast membrane [20, 21]. Since the vacuoles are the site of cardiac glycoside storage also in whole plants, as shown for instance with *Convallaria majalis* [22], all bulk cardenolides occurring in *D. lanata* carry a terminal glucose residue.

After lowering the auxin-cytokinin ratio in the nutrient medium, different programs of morphogenesis were triggered [10]. Thereby the selected cell lines reacted quite differently, some building only a hard nodular tissues, whereas others formed shoots, roots, or somatic embryos. If the process of morphogenesis was carried out upon irradiation with white light, several of the morphological entities formed became green and contained chloroplasts.

Substantial amounts of cardiac glycosides were found in the green shoots and the green somatic embryos [10]. In contrast, the green, nodular, nonembryoid tissue which developed in the cultures of some lines was nearly free of cardiac glycosides. This agrees with results of Reinhard et al. [23] on *D. lanata* and Hagimori et al.

[9, 24, 25] on *D. purpures* which in similar green, nonmorphogenetic cultures also found no cardenolide formation.

The morphological potential of cell lines obtained from root, shoot, stem, and filament explants was statistically different though in all cases the primary callus was derived from despecialized xylem parenchyma cells [10]. In agreement with earlier results [26, 27], cell lines forming large amounts of somatic embryos (embryogenic cell lines) were obtained from filament xylem parenchyma cells only.

5 The Integration of Cardenolide Formation in the Developmental Program "Somatic Embryogenesis"

The somatic embryos formed were much more suitable for a detailed investigation of the relations between cardenolide biosynthesis and morphological development than shoots and leaves because of their simple morphological structure. Cardenolide biosynthesis became measurable during the development of the early globular embryo structures. Figure 1 shows that these structures consist of only two cell types, cortex and core cells. During further development xylem cells were built within the core and the embryos became bipolar showing the shoot and the root pole (late stages of embryo development).

Two types of globular embryos may be distinguished which represent succesive stages in embryo development:

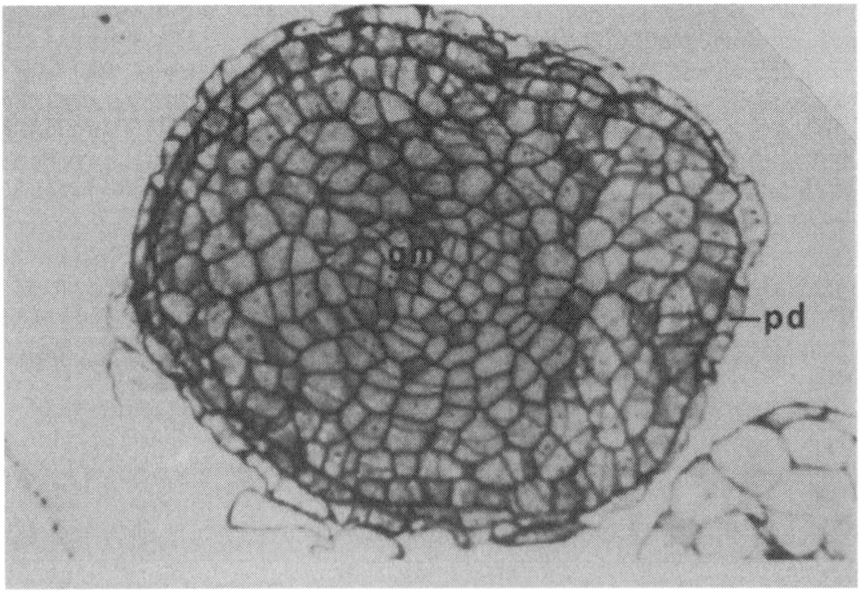

Fig. 1. Cross-section through a somatic embryo (stage I globule) of *D. lanata*. *gm* Matrix of isodiametric cells; *pd* periderm

1. Stage I globules, which are nearly unpigmented and are UV-A sensitive, and
2. Stage II globules which are yellow (the nature of the pigment is still unknown), UV-A insensitive and able to turn green upon illumination.

Whereas stage I globules contained only a small base level of cardenolides, illuminated stage II globules accumulated $>$ 1 μmol digitoxin equivalents g^{-1} dry weight.

Stage I globules slowly multiplied by adventive embrogenesis, but did not further develop to stage II globules and late embryo structures if grown at high density in a medium containing a low auxin-cytokinin ratio [28]. The high density cultures (Fig. 2) provided morphologically synchronized material for experiments on further embryo development. The transition of the stage I globules to the stage II globules was initiated by dilution of the embryo suspension. It may be divided into the following phases:

1. A first period (5 days) during which the embryos formed the yellow pigment and became insensitive to UV-A irradiation;
2. A second period (10 days) during which the embryos underwent a competence period with respect to the influence of visible light on cardiac glycoside and chlorophyll accumulation; and
3. A third period (20 days) during which the embryos grew rapidly, started biosynthesis and accumulation of cardiac glycosides, formed chloroplasts, and began to build roots and shoots.

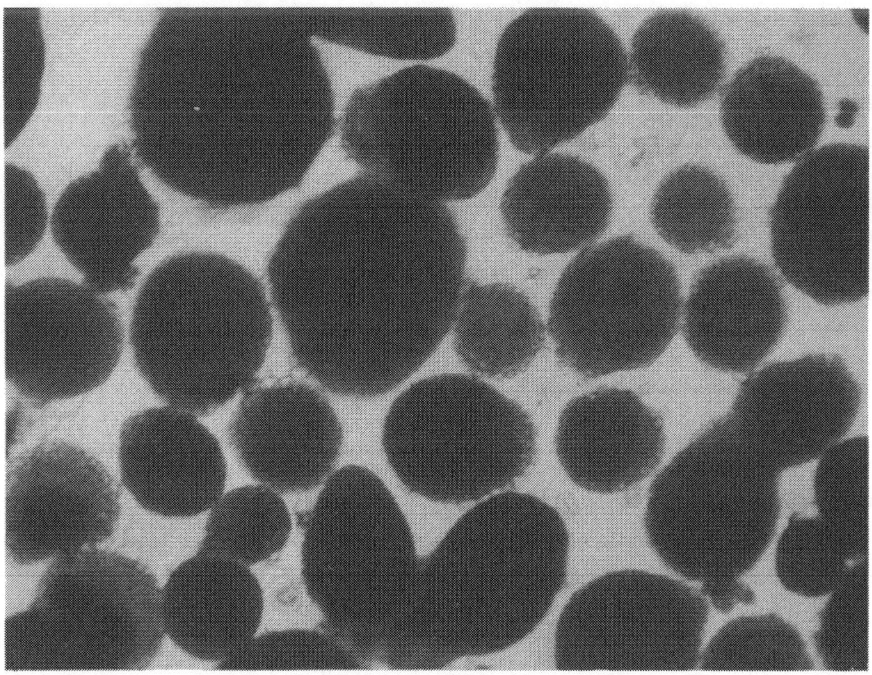

Fig. 2. Suspension of stage I globules obtained by secondary embryogenesis

The embryo structures grown in the dark contained a base level of cardiac glycosides which was about ten times that of the parenchymoid cells. This value was increased by two orders of magnitude upon illumination. Detailed experiments showed that the accumulation of the cardenolides was influenced by the energy flux, the photoperiod, and the wavelength of the light used for irradiation [29]. The main cardenolide accumulated was identified as glucodigifucoside by physicochemical methods and comparison with an identical specimen [30, 31]. It was recently shown to occur also in another embryogenic cell strain of *D. lanata* [32].

The highest cardenolide content was always found in structures with fully developed chloroplasts. However, there was no direct relation between chloroplast formation and cardenolide accumulation. On the one hand, in chloroplast-forming, nonembryoid cell lines no cardenolides were synthesized [10] and, on the other hand, an embryogenic cell line was found which was unable to form chloroplasts, but accumulated considerable amounts of cardiac glycosides [33]. In addition, the treatment of the somatic embryos described in this Chapter with different light regimes [29] or inhibitors of chloroplast development [34] had a different influence on both processes.

In summary, cardiac glycoside biosynthesis seems to be regulated in *D. lanata* by two principles:

1. At a given stage of a suitable developmental program, like somatic embryogenesis, the genes involved in cardiac glycoside biosynthesis become competent for expression and a certain base level of cardenolides is formed; and
2. By illumination of the competent structures, the genes become fully expressed.

Further experiments are necessary to show whether the chloroplasts provide precursors or cosubstrates necessary for cardiac glycoside biosynthesis, or whether they contain several enzymes catalyzing this pathway.

6 Is There a Possibility to Trigger the Biosynthesis of Cardiac Glycosides Independently of the Program "Somatic Embryogenesis"?

Due to growth stress, in vitro secondary products may be formed in the parenchymoid cells of *D. lanata* cultures, although in the whole plant, their biosynthesis is part of other differentiation programs (cf. Sect. 1). The biosynthesis of the oxygenated carotenoids, which give the parenchymoid cells a yellow tint [unpublished results], and of cinnamic acid derivatives, like caffeic acid, which after transformation to polyphenols cause the deep brown color of the old parenchymoid cells [unpublished results], indeed shows that the parenchymoid cells are able to form certain types of secondary products.

In addition, it was demonstrated recently that parenchymoid *D. lanata* cells transferred from suspension cultures to the surface of a solidified nutrient medium showed transient synthesis and accumulation of cardiac glycosides [4]. The accumulation was different in individual cell colonies ranging from < 0.001 to 0.75 μmol digitoxin equivalents g^{-1} dry weight. This diversity probably reflected genetic heterogeneity in the cell strain used [35]. After cessation cardenolide biosynthesis could be retriggered if the cycle of submerged/surface cultivation was repeated. The accumulation was somewhat higher in cell colonies obtained from protoplasts compared with normally grown cell colonies. In all experimental variants cardenolide biosynthesis was independent of illumination.

These results indicate that the beginning of cultivation at the surface of a solidified nutrient medium is a stress which can trigger cardenolide biosynthesis directly, i.e., independently from morphological programs. Protoplast formation and the growth of cell colonies from protoplasts obviously adds to this stress and therefore causes higher cardenolide values. It is the aim of further experiments to optimize this system, e.g., by the selection of better stress factors and the isolation of cell lines which express cardenolide biosynthesis as a stable characteristic apart from morphological development.

References

1. Luckner M, Nover L, Boehm H (1977) Secondary metabolism and cell differentiation (Molecular biology, biochemistry and biophysics, vol 23). Springer, Berlin Heidelberg New York
2. Luckner M (1980) Expression and control of secondary metabolism. In: Bell EA, Charlwood B (eds) Encyclopedia of plant physiology, New Ser, vol 8: Secondary plant products. Springer, Berlin Heidelberg New York, pp 23–63
3. Luckner M (1984) Secondary metabolism in microorganisms, plants, and animals. Fischer, Jena; Springer, Berlin Heidelberg New York
4. Luckner H, Diettrich B (1987) Biosynthesis of cardenolides in cell cultures of *Digitalis lanata* – the result of a new strategy. In: Green CE, Somers DA, Hackert WP, Biesboer DD (eds) Plant tissue and cell culture. Liss, New York, pp 187–197
5. Luckner M, Diettrich B (1985) Formation of cardenolides in cell and organ cultures of *Digitalis lanata*. In: Neumann K-H, Barz W, Reinhard E (eds) Primary and secondary metabolism of plant cell cultures. Springer, Berlin Heidelberg New York Tokyo, pp. 154–163
6. Luckner M, Diettrich B (1989) Cardenolides. In: Constabel F (ed) Cell culture and somatic cell genetics of plants, vol 5: Phytochemicals in plant cell cultures. Academic Press, New York London, pp 193–212
7. Weiler EW, Zenk MH (1976) Radioimmunoassay for the determination of digoxin and related compounds in *Digitalis lanata*. Phytochemistry 15:1537–1545
8. Vogel E, Luckner M (1981) Distribution of cardenolides in *Digitalis lanata*. Planta Med 41:161–165
9. Hagimori M, Matsumoto T, Mikami Y (1984a) Digitoxin biosynthesis in isolated mesophyll cells and cultured cells of *Digitalis*. Plant Cell Physiol 25:947–953
10. Diettrich B, Steup C, Neumann D, Scheibner H, Reinbothe C, Luckner M (1986) Morphogenetic capacity of cell strains derived from filament, leaf and root explants of *Digitalis lanata*. J Plant Physiol 124:441–453
11. Rücker W, Jentzsch K, Wichtl M (1976) Wurzeldifferenzierung und Glykosidbildung bei in vitro kultivierten Blattexplantaten von *Digitalis purpurea* L. Z Pflanzenphysiol 80:323–335
12. Breuel K, Poetter H, Luckner M, Diettrich B, Springer M, Oertel C (1984) Verfahren zur virusfreien vegetativen Vermehrung und Erhaltung von *Digitalis*-Hochleistungspflanzen. GDR-Patent DD 207 731
13. Luckner M, Diettrich B, Springer M, Breuel K, Oertel C (1984) Verklonung von *Digitalis-lanata*-Hochleistungspflanzen durch Sproßspitzenkultur. Int Vortragstag Methoden und Verfahren der Züchtung, des Anbaues, der Sammlung und der industriellen Verarbeitung von Arznei- und Gewürzpflanzen, Artern, 18. – 22. 6. 1984. Vortragstexte, Pt 1. Pharmazeutisches Werk, Halle, pp 113–127
14. Schöner S, Reinhard E (1986) Long-term cultivation of *Digitalis lanata* clones propagated in vitro: cardenolide content and the regeneration of plants. Planta Med 52:478–481

15. Diettrich B, Mertinat H, Luckner M, Breuel K, Dauth C (1987) Die Gewinnung von Klonlinien aus *Digitalis-lanata*-Hochleistungspflanzen durch Sproßspitzenkultur. Wiss Z Univ Halle 36M/5:90–102
16. Alfermann AW, Reinhard E (1980) Biotransformation by plant tissue cultures. In: Sala F, Parisi B, Cella R, Ciferri O (eds) Plant cell cultures: results and perspectives. Elsevier/North Holland Biomedical Press, Amsterdam, pp 399–404
17. Reinhard E, Alfermann AW (1980) Biotransformation by plant cell cultures. Adv Biochem Eng 16:49–83
18. Pfeiffer B, Roos W, Luckner M (1982) Accumulation of *purpurea* glycoside A in vacuoles of *Digitalis lanata* cells cultivated in vitro. Planta Med 45:154
19. Kreis W, Reinhard E (1987) Selective uptake and vacuolar storage of primary cardiac glycosides by suspension-cultured *Digitalis lanata* cells. J Plant Physiol 128:311–326
20. Diettrich B, Aster U, Greidziak N, Roos W, Luckner M (1987) Glucosylation of digitoxin and other cardenolides in cell cultures of *Digitalis lanata*. Biochem Physiol Pflanzen 182:245–255
21. Kreis W, Reinhard E (1985) Uptake, metabolism, and storage of cardenolides by *Digitalis lanata* cells. Pharm Z 130:2315–2316
22. Loeffelhardt W, Kopp B, Kubelka W (1979) Intracellular distribution of cardiac glycosides in leaves of *Convallaria majalis*. Phytochemistry 18:1289–1291
23. Reinhard E, Boy M, Kaiser F (1975) Umwandlung von Digitalisglykosiden durch Zellsuspensionskulturen. Planta Med Suppl 163–168
24. Hagimori M, Matsumoto T, Obi Y (1982) Studies of the production of *Digitalis* cardenolides by plant tissue culture. II. Effect of light and plant growth substances on digitoxin formation by undifferentiated cells and shoot-forming cultures of *Digitalis purpurea* L. grown in liquid media. Plant Physiol 69:653–656
25. Hagimori M, Matsumoto T, Mikami Y (1984b) Photoautotrophic culture of undifferentiated cells and shoot-forming cultures of *Digitalis purpurea* L. Plant Cell Physiol 25:1099–1102
26. Nover L, Luckner M, Tewes A, Garve R, Vogel E (1980) Cell specialization and cardiac glycoside formation in cell cultures of *Digitalis* species. Acta Hortic 96:65–74
27. Tewes A, Wappler A, Peschke E-M, Garve R, Nover L (1982) Morphogenesis and embryogenesis in long-term cultures of *Digitalis*. Z Pflanzenphysiol 106:311–324
28. Scheibner H, Diettrich B, Schulz U, Luckner M (1989b) Somatic embryos of *Digitalis lanata*. Synchronization of development and cardenolide biosynthesis. Biochem Physiol Pflanzen (in press)
29. Scheibner H, Bjoerk L, Schulz U, Diettrich B, Luckner M (1987) Influence of light on cardenolide accumulation in somatic embryos of *Digitalis lanata*. J Plant Physiol 130:211–219
30. Hering F, Lehmann T, Luckner M (1987) Glucodigifucosid und Odorobiosid G – Hauptglycoside von somatischen Embryonen und jungen in-vitro-Pflanzen von *Digitalis lanata*. Pharmazie 42:215–216
31. Beale JM, Floss HG, Lehmann T, Luckner M (1988) Glucodigifucoside (digitoxigenin-3b-O-[b-D-fucopyranosyl-4′-b-D-glucopyranoside]), the main cardenolide of somatic embryos of *Digitalis lanata*. Phytochemistry 27:3143–3146
32. Seidel S, Reinhard E (1987) Major cardenolide glycosides in embryogenic suspension cultures of *Digitalis lanata*. Planta Med 53:308–309
33. Ohlson AB, Bjoerk L, Gatenbeck S (1983) Effect of light on cardenolide production by *Digitalis lanata* tissue cultures. Phytochemistry 22:2447–2450
34. Scheibner H, Bjoerk L, Schulz U, Neumann D, Diettrich B, Luckner M (1989a) The influence of inhibitors of chloroplast differentiation on chlorophyll and cardenolide accumulation in somatic embryos of *Digitalis lanata*. Biochem Physiol Pflanzen 184:63–67
35. Rhodes MJC, Robins RJ (1987) The use of plant cell cultures in studies of metabolism. In: Davies DD (ed) The biochemistry of plants, vol 13: Methodology. Academic Press, New York London, pp 65–125

Compartmentation of Flavour Precursor Synthesis in the Onion

H.A. Collin[1], D. Musker[2], and G. Britton[2]

The spatial separation of metabolites and enzymes within plants is a major factor in regulating metabolism; this subject has been extensively reviewed recently (Luckner 1984). Secondary products in plants are often stored or biosynthesized within specialized cells e.g. isoquinoline alkaloids in laticifer cells of *Papaver somniferum* or in specific organs e.g. nicotine biosynthesis in roots of *Nicotiana tabacum*. Biosynthetic pathways may be localized in specialized compartments within cells, either in specific organelles, e.g. the shikimic acid pathway in chloroplasts of *Pisum sativum* (Mousedale and Coggins 1985), or in particular micro-environments within the cytoplasmic endomembrane system, as is the case with polymethylated flavonal glucoside biosynthesis in *Chrysosplenium americanum* (Ibrahim et al. 1987).

The enzyme-mediated release of flavour in *Allium* is an interesting example of compartmentation. Undamaged tissue of *Allium* has only a slight odour. Immediately upon tissue damage, however, there is a rapid development of an odour which is characteristic of each *Allium* species: onion, in addition, produces a powerful lachrymator. Early work established that on disruption of onion tissue S-*trans*-prop-1-enyl-L-cysteine sulphoxide (propenylcysteine sulphoxide) is hydrolyzed by an enzyme, alliinase. The initial products of the reaction are ammonia, pyruvate and propenylsulphenic acid (Schwimmer 1968). Propenyl sulphenic acid is the volatile lachrymator characteristic of the onion. Propenylsulphenic acid is chemically unstable, and spontaneously rearranges to form propanal sulphoxide which is the stable end product of the reaction. The milder components of the odours in onion and other species of *Allium* are formed by the hydrolysis of the other flavour precursors, namely propyl-, allyl- and methyl-cysteine sulphoxides. In these reactions the initial products are thiosulphinates which rearrange to form volatile disulphides. These products do not have a lachrymatory effect.

The spatial separation of alliinase, the degradative enzyme, from the flavour precursors in intact onion cells is necessary to prevent the premature release of volatiles into healthy tissue. It has been suggested that flavour precursors are localized within the vacuole, whilst alliinase is lysozymal (Becker and Schupan 1975). However, when onion bulb protoplasts were prepared and the intact vacuoles subsequently isolated, alliinase was present solely within the cell vacuole fraction, whereas propenylcysteine sulphoxide was found in the cytoplasm

[1]Department of Botany, University of Liverpool, Nicholson Building, P.O. Box 147, Liverpool, L69 3BX, UK
[2]Department of Biochemistry, University of Liverpool, Liverpool, L69 3BX, UK

Primary and Secondary Metabolism of Plant Cell Cultures, ed. by W.G.W. Kurz

(Lancaster and Collin 1981); small vesicles, bounded by a single membrane were implicated in the storage of propenylcysteine sulphoxide (Turnbull et al. 1981).

Normally, undifferentiated tissue cultures of onion do not accumulate the flavour precursor, propenylcysteine sulphoxide (Selby and Collin 1976), although the alliinase is present at an activity comparable to that in the onion bulb tissue (Selby et al. 1979). Differentiation of the tissue culture into plantlets does reactivate the production of flavour precursor (Turnbull et al. 1981), indicating that the capacity for flavour production is still present in the tissue culture, but not expressed. Attempts at stimulating precursor synthesis by feeding intermediates of the primary and secondary pathways leading to precursor synthesis were partially successful. Granroth (1970) suggested that propenylcysteine sulphoxide was derived from the primary pathway of valine oxidation and cysteine. However, the addition of valine, methacrylic acid or cysteine to the tissue culture medium had no effect on accumulation of propenylcysteine sulphoxide in callus. When the intermediates on the secondary pathway, S-2-carboxypropyl-L-cysteine and S-*trans*-prop-l-enyl-L-cysteine, were added to the medium propenylcysteine sulphoxide accumulated in the calus (Selby et al. 1980; Turnbull et al. 1980). Clearly the secondary pathway was able to function but lacked a suitable supply of intermediates from the primary pathway, or a site at which the synthesis or accumulation could occur.

In order to determine the effect of limited differentiation on the production of flavour precursor, i.e. cell differentiation as opposed to organ formation, tissue cultures were redifferentiated in a much more controlled way. The herbicide Picloram (4-amino-3,5,6-trichloropicolinic acid) was found to be very effective as an auxin source at initiating differentiation in onion tissue cultures (Phillips and Luteyn 1983). Onion tissue cultures were transferred for one subculture onto a control medium containing 2,4-dichlorophenoxyacetic acid (0.55 mgl^{-1}), which maintained the callus in an undifferentiated state, low Picloram (0.5 mgl^{-1}), which caused the production of shoot initials, and high Picloram (5.0 mgl^{-1}), which maintained the callus in an undifferentiated state. These treatments allowed the distinction to be made between the effect of differentiation and Picloram on flavour precursor formation (Musker et al. 1988). The high Picloram medium suppressed all tissue differentiation and reduced growth of the callus, whereas callus on low Picloram showed signs of early differentiation but no obvious leaf, root or shoot formation, or greening. HPLC analysis of the callus on 2,4-D, high and low Picloram showed very low levels of flavour precursor, propenylcysteine sulphoxide, initially in all treatments, but during a period of subculture the level of flavour precursor increased in the Picloram treatments (Table 1). The effect of high Picloram was to suppress differentiation but still stimulate production of propenylcysteine sulphoxide, whereas low Picloram stimulated early differentiation and also production of propenylcysteine sulphoxide.

The cellular structure showed cells from culture maintained on a 2,4-D medium with large vacuoles and only small amounts of cytoplasm, whereas callus on both high and low Picloram medium contained cells which had the characteristics of more metabolically active cells (Gunning and Steer 1986). In addition, the Picloram-treated cells had many cytoplasmic vesicles. The presence of the cytoplasmic vesicles was correlated therefore with the accumulation of flavour

Table 1. The effect of incubation on a low and a high Picloram medium on propenylcysteine sulphoxide accumulation in callus

Propenylcysteine sulphoxide (mg g^{-1} f.wt.)[a]			
Time after inoculation (weeks)	Low Picloram	High Picloram	2,4-D
0	n.d.	n.d.	n.d.
4	0.12	0.51	n.d.
8	0.32	0.55	n.d.

[a] n.d. Not detected.

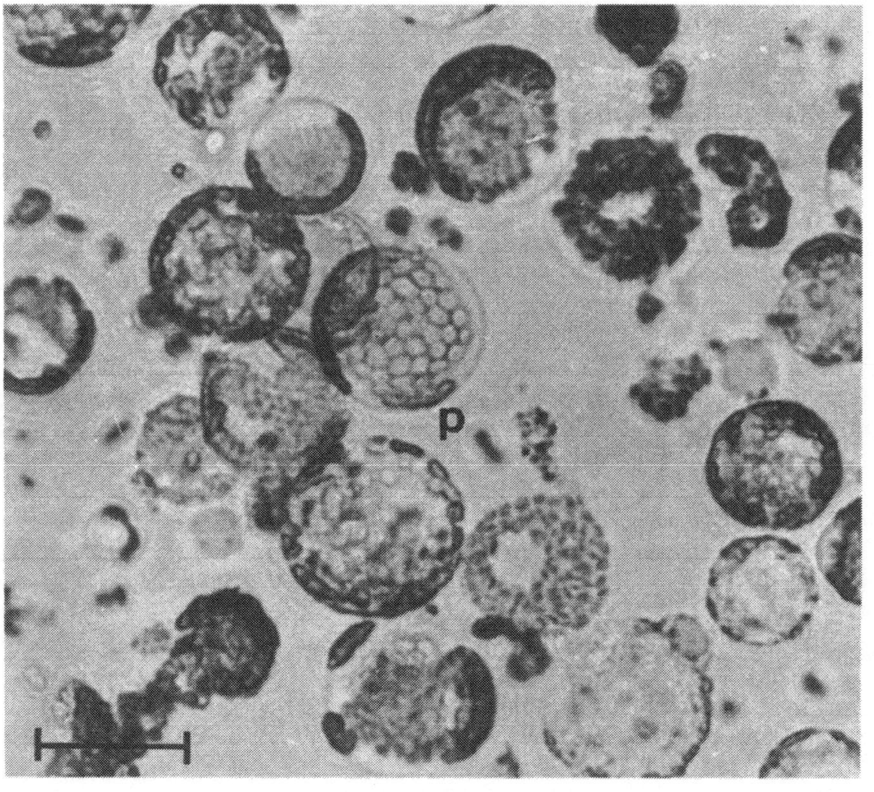

Fig. 1. Protoplast preparation from leaf blades of onion (*Allium cepa* L.) *Bar* = 10 μm; *p* protoplast

precursors in tissue cultures. The role of the vesicles as a site of accumulation of flavour precursor was investigated in more detail by fractionating the cytoplasm of the intact onion plant rather than the onion tissue culture, since the plant was biosynthetically more active and more easily available.

Leaf mesophyll protoplasts (Fig. 1) were lysed and the cytoplasm organelles separated on a continuous sucrose gradient (Musker 1988). The fractions were analyzed for markers, i.e. sucrose, chlorophyll, protein, rotenone insensitive NADH-cytochrome c reductase (a marker specific for endoplasmic reticulum) and propenylcysteine sulphoxide. The fractions were also examined by electron microscopy.

Propenylcysteine sulphoxide was detected in all fractions (Fig. 2a) but fractions 2 and 13 were particularly enriched. The maximum concentration of chlorophyll and protein occurred in fractions 8, 9 and 10 (Fig. 2a,b). A further peak of chlorophyll in fraction 12 was considered to be due to thylakoid membrane released from the chloroplasts during homogenization of protoplasts. Activity of NADH-cytochrome c reductase was detected in fractions 6 to 17, with fractions 8 and 13 having the highest specific activities (Fig. 2b). The distribution of the enzyme was distorted due to the presence of chloroplast protein in fraction 8. Examination of fractions 2 and 13 by electron microscopy showed that fraction 2 contained single-membrane-bounded vesicles 0.5–1.0 μm in diameter (Fig. 3a,b). These vesicles were aggregated and contained electron dense fibrillar material. In fraction 13 single-membrane-bounded vesicles were also evident 0.5–0.75 μm in diameter and membranous fragments believed to be derived from endoplasmic reticulum (Fig. 3b).

The fact that on the sucrose density gradient the distribution of propenylcysteine sulphoxide was not homogeneous suggested that propenylcysteine sulphoxide was stored within a specific compartment, and not freely soluble in the cytoplasm. Some of the gradient fractions enriched in propenylcysteine sulphoxide were associated with NADH-cytochrome c reductase. For example, this enzyme was present largely in fraction 8 (rough ER) and fraction 13 (smooth ER). Propenylcysteine sulphoxide was detected in fraction 13 and is therefore associated with sER membranes, whereas propenylcysteine sulphoxide in fraction 2 was not associated with NADH-cytochrome c reductase activity. The presence of propenylcysteine sulphoxide and reductase activity in fraction 13 suggests that the flavour precursors could be compartmented within membranous vesicles derived from or closely associated with the sER. This hypothesis is supported by the electron micrographs of membrane obtained from fraction 13. Electron micrographs of sedimented material from fraction 2 showed aggregated, single-membrane-bound vesicles, containing electron dense fibrillar material. These structures could be the product of vesicle maturation or possibly of membrane aggregation.

Heterogeneity of storage vesicles (vacuoles) has often been reported as a feature of maturation for further packaging and storage of metabolites (Ali et al. 1985, 1986; Griffing and Fowke 1985). An extra-vacuolar, single-membrane-bounded vesicle storage site for propenylcysteine sulphoxide in bulb cells of *A. cepa* has been proposed (Turnbull et al. 1981). The work here concerns leaf blades of *A. cepa* which are the most active site of synthesis of propenylcysteine sulphoxide (Granroth 1970). Vesicles similar in structure to those observed in bulb cells were evident in shoot cells and callus cells. The storage of propenylcysteine sulphoxide may therefore be similar in both leaf blade, bulb and bulb tissue and cultured tissue.

In summary, the results presented support the concept of a GERL-mediated mechanism for the storage of propenylcysteine sulphoxide in leaf blades of *A. cepa*

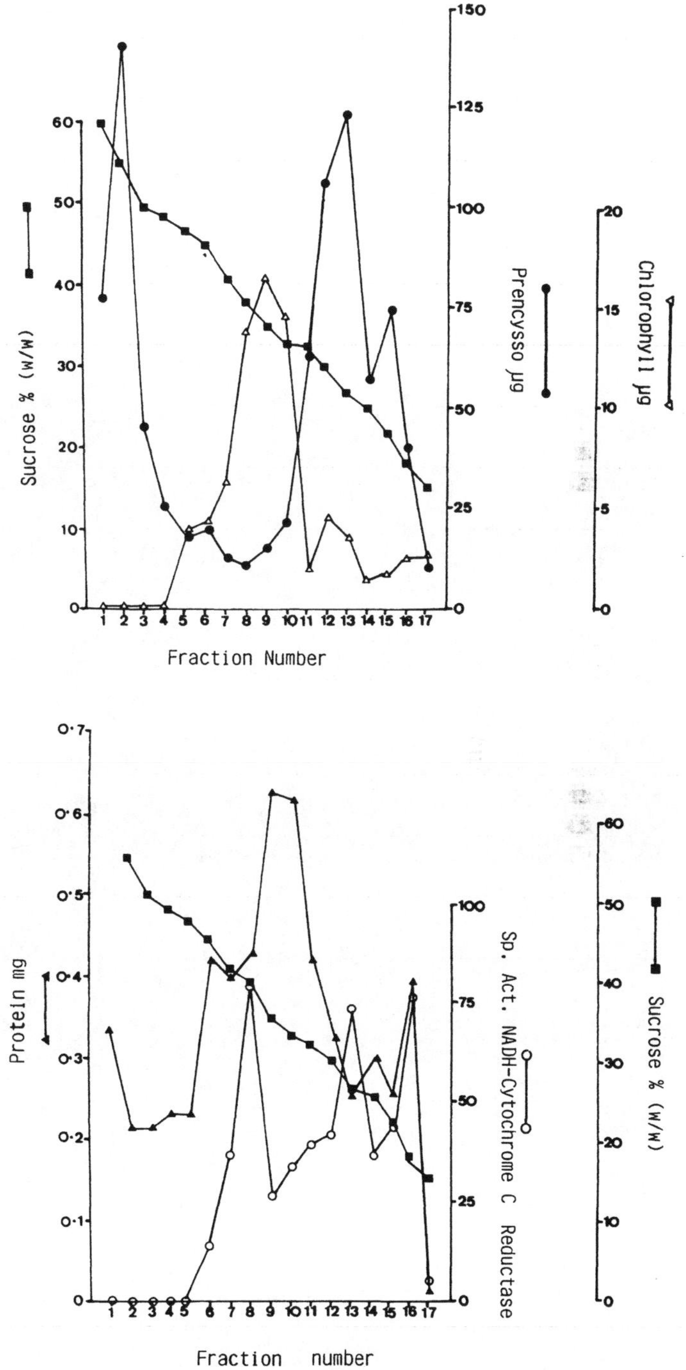

Fig. 2a,b. Distributions of markers in fractions of protoplast lysate separated by isopycnic centrifugation on continuous sucrose gradients. **a** Sucrose, propenylcysteine sulphoxide and chlorophyll; **b** sucrose, NADH cytochrome c reductase activity and protein

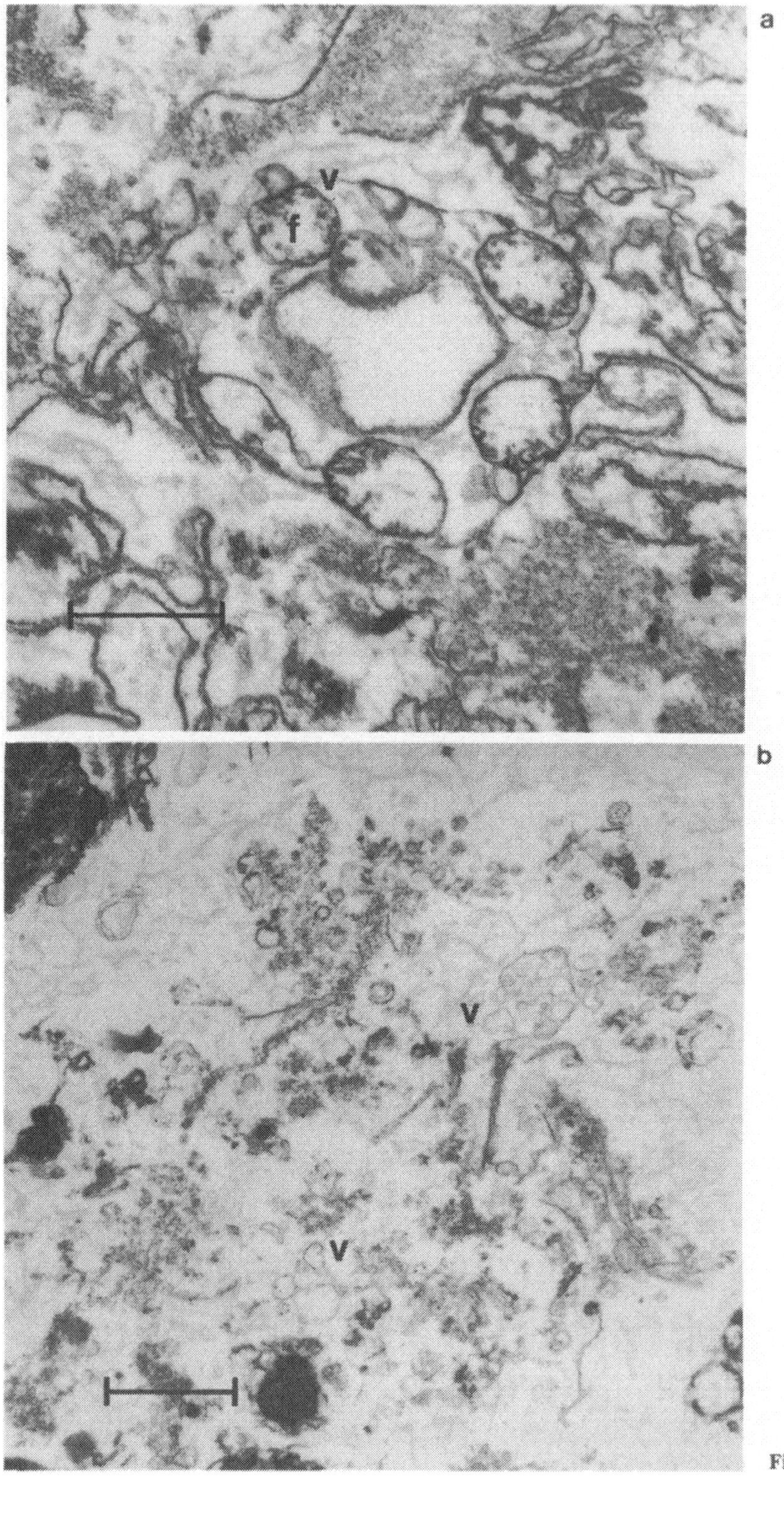

Fig. 3a,b

and possibly in tissue cultures. It is suggested that propenylcysteine sulphoxide is stored within single-membrane-bounded vesicles (0.5–1 μm in diameter which are closely associated with or derived from sER membrane. The site of synthesis of propenylcysteine sulphoxide may be the rER. The cell differentiation required for the accumulation of propenylcysteine sulphoxide is initiated by Picloram. The presence of vesicles in the Picloram-treated tissue cultures confirmed the role of these vesicles in the accumulation of flavour precursors in the onion cell.

References

Ali MS, Nishimura N, Mitsui T, Akazawa T, Kojima K (1985) Isolation and characterisation of Golgi membranes from suspension cultured cells of Sycamore (*Acer pseudoplatanus* L.). Plant Cell Physiol 26:1119–1133

Ali MS, Mitsui T, Akazawa T (1986) Golgi-specific localisation of transglycosylases engaged in glycoprotein biosynthesis in suspension cultured cells of Sycamore (*Acer pseudoplatanus* L.). Arch Biochem Biophys 251:421–431

Becker A, Schupan W (1975) Ein Beitrag zur Biogenese und Biochemie antimicrobiell wirkender ätherischer Öle der Küchenzwiebel (*Allium cepa* L.). Qual Plant 25:107–169

Drew SW, Demain AL (1977) Effect of primary metabolites on secondary metabolism. Annu Rev Microbiol 31:343–356

Granroth B (1970) Biosynthesis and decomposition of cysteine derivatives in the onion and other *Allium* species. Ann Acad Sci Fennica Ser A2 154:1–71

Griffing LR, Fowke LC (1985) Cytochemical localisation of peroxidase in soya bean suspension culture cells and protoplasts. Intracellular vacuole differentiation and presence of peroxidase in coated vesicles and multivesicular bodies. Protoplasma 128:22–30

Gunning BES, Steer MW (1986) Plant cell biology: an ultrastructural approach. M W Steer Dublin, Eire

Ibrahim RK, De Luca V, Khouri H, Latchinion L, Brisson L, Charest PM (1987) Enzymology and compartmentation of polymethylated flavonal glucosides in *Chrysoplenium americanum*. Phytochemistry 26:1237–1245

Luckner M (1984) Secondary metabolism in micro-organisms, plants and animals. Springer, Berlin Heidelberg New York

Mousedale DM, Coggins JR (1985) Subcellular localisation of the common shikimate pathway enzymes in *Pisum sativum* L. Planta 163:241–249

Musker D (1988) Secondary product biosynthesis in plant cell cultures. Ph D Thesis, Univ Liverpool, UK

Musker D, Collin HA, Britton G, Ollerhead G (1988) Propenylcysteine sulphoxide biosynthesis in callus cultures in *Allium cepa* L. In: Rhodes M, Robbins R (eds) Manipulating secondary metabolism in culture. (in press)

Phillips GC, Luteyn KJ (1983) Effects of picloram and other auxins on onion tissue cultures. J Am Soc Hortic Sci 108:948–953

Saunders JA, Conn EE (1978) Presence of the cyanogenic glucoside dhurrin in isolated vacuoles from *Sorghum*. Plant Physiol 61:154–157

Saunders JA, Conn EE, Lin CH, Shimada M (1977) Localisation of cinnamic acid 4-monooxygenase and the membrane bound enzyme system for dhurrin biosynthesis in *Sorghum* seedlings. Plant Physiol 60:629–634

Schwimmer S (1968) Enzymic conversion of trans-(+)-S-1-propenyl-L-cysteine sulphoxide to the bitter and odour bearing components of onion. Phytochemistry 7:401–404

Fig. 3a,b. Electron micrograph of material isolated from a linear sucrose gradient (*Bar* = 1 μm). **a** Fraction 2; **b** fraction 13. *v* Single membrane-bound vesicle; *f* fibrillar material

Selby C, Collin HA (1976) Clonal variation in growth and flavour production in tissue cultures of *Allium cepa* (L.) Ann Bot (London) 40:911–918

Selby C, Galpin IJ, Collin HA (1979) Comparison of the onion plant (*Allium cepa*) and onion tissue culture. I. Alliinase activity and flavour precursor compounds. New Phytol 83:351–359

Selby C, Turnbull A, Collin HA (1980) Comparison of the onion plant (*Allium cepa*) and onion tissue culture. II. Stimulation of flavour precursor synthesis in onion tissue culture. New Phytol 84:307, 312

Turnbull A, Galpin IJ, Collin HA (1980) Comparison of the onion plant (*Allium cepa*) and onion tissue culture. III. Feeding of 14C labelled precursors of the Stauour precursor compounds. New Phytol 85:483–487

Turnbull A, Galpin IJ, Smith JL, Collin HA (1981) Comparison of the onion plant (*Allium cepa*) and onion tissue culture. IV. Effect of shoot and root morphogenesis on flavour precursor synthesis in onion tissue culture. New Phytol 87:257–268

Biosynthesis of Tropane Alkaloids

Y. Yamada, T. Hashimoto, and T. Endo[1]

1 Cultured Roots

As with other secondary products, the biosynthesis of tropane alkaloids in plants has been extensively investigated by the use of labelled precursors (for a review, see [11]). Recent advances in the techniques of alkaloid analysis and enzyme purification, however, have now made it possible to isolate and characterize biosynthetic enzymes in the pathway, and earlier results using precursor feeding must be validated and may necessitate a revision of the biosynthetic pathways proposed. A prerequisite to such studies is a sufficient supply of cell material with high biosynthetic capability, thus, with high enzyme activities.

Tropane alkaloids are not produced in large amounts by cultured, undifferentiated cells. Screening of prospective plant species, repeated selection of high-producing cell lines and optimization of production medium did not result in stable and efficient formation of tropane alkaloids in cell cultures (e.g. [2, 3]). Differentiation of root tissues seems to be essential for the active biosynthesis of these alkaloids [4, 5]. Fortunately, excised roots of several solanaceous species grow vigorously when the culture medium is supplemented with appropriate concentrations of auxin [6, 7]. Plant species that give cultured roots with poor growth even in the presence of auxin may be transformed by *Agrobacterium rhizogenes* to promote rapid growth (for a review, see [8]). These cultured roots, natural or transformed, actively synthesize tropane alkaloids characteristic of the species, and thus are rich sources of biosynthetic enzymes.

2 Biosynthesis of Scopolamine

The first reaction in the biosynthetic pathway from hyoscyamine to scopolamine is catalyzed by an enzyme, hyoscyamine 6β-hydroxylase [9] (Fig. 1). The hydroxylase requires hyoscyamine, 2-oxoglutarate, ferrous ions and molecular oxygen. Ascorbate stimulated this reaction. A 1:1 stoichiometry was shown between the hyoscyamine-dependent formation of $^{14}CO_2$ from 2-oxo[1-^{14}C]glutarate and the hydroxylation of hyoscyamine as determined by GLC analysis. Therefore, the enzyme, a 2-oxoglutarate-dependent dioxygenase (EC 1.14.11.11), incorporates

[1]Research Center for Cell and Tissue Culture, Faculty of Agriculture, Kyoto University Sakyo-ku, Kyoto 606, Japan

Primary and Secondary Metabolism of Plant Cell Cultures, ed. by W.G.W. Kurz

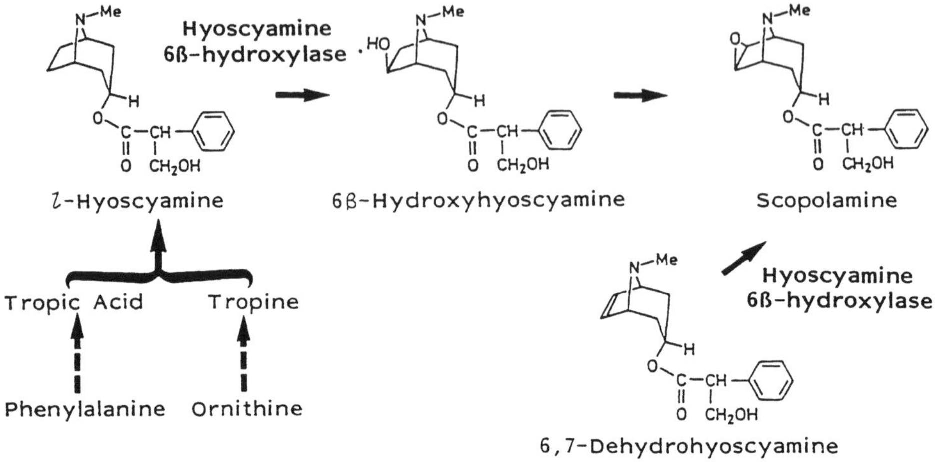

Fig. 1. Metabolism of hyosajamine to scopolamine

one atom of molecular oxygen into the tropane moiety at the C-6 of hyoscyamine, forming 6β-hydroxyhyoscyamine. This reaction is coupled with the simultaneous incorporation of the other atom of molecular oxygen into 2-oxoglutarate, thereby decarboxylating the α-keto acid to succinate and CO_2.

Various cultures were tested for the presence of tropane alkaloids and hydroxylase activity. Root cultures of tobacco and non-solanaceous species, and shoot cultures of *Duboisia* and *Hyoscyamus* produced no alkaloids. There was no enzyme activity present in these cultures, except for weak activities detected in shoot cultures of *D. myoporoides* and *H. niger*. Several *Datura* root cultures and calluses of *H. niger* produced only hyoscyamine, no enzyme activity being found.

Root cultures of *Atropa belladonna, Duboisia leichhardtii*, two *Datura* species and five *Hyoscyamus* species produced both hyoscyamine and scopolamine; 6β-hydroxyhyoscyamine was present in several of them. Hydroxylase activity was generally found in the crude enzyme preparations from these root cultures. Transfer of the root cultures to auxin-free medium generally produced marked increases in the alkaloid contents and in the activity of the extractable hydroxylase. Among them, root cultures of *Hyoscyamus niger, H. gyorffi, H. muticus* and *Duboisia leichhardtii* had the highest enzyme activities.

Hyoscyamine 6β-hydroxylase was purified from root cultures of *H. niger* [10]. It has a pI of 4.5 and a molecular weight of 41000 (gel filtration) or 38000 (SDS-PAGE), indicating that the enzyme is a monomer. Maximum activity was observed at pH 7.8. The respective Km values of (S)-hyoscyamine and 2-oxoglutarate are 35 and 43 μM. 2-Oxoglutarate was not replaced by any of the ten other oxo acids tested, nor was Fe^{2+} by any of the nine other divalent cations tested. The enzyme activity was inhibited strongly by nitroblue tetrazolium and the divalent cations Mn^{2+}, Co^{2+}, Zn^{2+} and Hg^{2+}. Pyridine 2,4-dicarboxylate and 3,4-dihydroxybenzoate are competitive inhibitors of 2-oxoglutarate, having respective Ki values of 9 and 90 μM.

Several alkaloids that have structures similar to (S)-hyoscyamine were hydroxylated at the C-6 position of their tropane moieties. The N-methyl group of the tropane moiety and the hydroxymethylene group of the tropic acid moiety of hyoscyamine are not important for the 6β-hydroxylation reaction. The hydroxylase accepts as a substrate only esters with tropane-3α-ol, but not esters with tropane-3β-ol. The spatial configuration around the 2′-C of the tropic acid moiety may also be critical for hydroxylase activity because (R)-hyoscyamine is inactive. That tropine is not hydroxylated by the enzyme is consistent with the currently accepted biosynthetic pathway of scopolamine, in which epoxidation of the tropane moiety occurs after tropine and tropic acid are condensed to form hyoscyamine.

Hyoscyamine 6β-hydroxylase also epoxidized 6,7-dehydrohyoscyamine to scopolamine (Km, 10 μM) (Fig. 1). 6,7-Dehydrohyoscyamine was suggested to be a precursor of scopolamine because the unsaturated alkaloid added by puncture to scions of *Datura ferox* was converted to scopolamine; but, there has been no report of the isolation of this hypothetical alkaloid from plants.

To determine whether 6,7-dehydrohyoscyamine is a true precursor of scopolamine in vivo, we made use of the fact that the proposed dehydration scheme from 6β-hydroxyhyoscyamine to scopolamine via 6,7-dehydrohyoscyamine would result in the release of hydroxyl oxygen, whereas intramolecular epoxidation of 6β-hydroxyhyoscyamine would not. [6-^{18}O]6β-Hydroxyhyoscyamine was prepared from (S)-hyoscyamine and $^{18}O_2$ using the hyoscyamine 6β-hydroxylase obtained from root cultures of *H. niger.* Shoot cultures of *Duboisia myoporoides* were used for the transformation because they not only are capable of biosynthesis, but do not accumulate alkaloids which would complicate quantitation. When [6-^{18}O]6β-hydroxyhyoscyamine was fed to the shoot cultures, the labelled alkaloid was converted to scopolamine. Analysis of the transformed scopolamine by GC-MS revealed that all the ^{18}O in the hydroxyl group of 6β-hydroxyhyoscyamine was retained in the epoxide oxygen of scopolamine after biotransformation in vivo [11]. Therefore, 6β-hydroxyhyoscyamine is converted in vivo to scopolamine without a dehydration step.

[6β, 7β-$^2H_2$] (S)-Hyoscyamine was synthesized by hydrogenation of 6,7-dehydrohyoscyamine with $^2H_2$ in the presence of Pd/C. The deuterated hyoscyamine then was converted to [7β-^{2}H]6β-hydroxyhyoscyamine by the reaction of hyoscyamine 6β-hydroxylase. When the deuterated 6β-hydroxyhyoscyamine was fed to the shoot cultures of *D. myoporoides,* it was converted to scopolamine which retained no deuterium (T. Hashimoto, Y. Yamada, unpublished work). It is concluded that 6β-hydroxyhyoscyamine is converted in vivo to scopolamine by a *cis*-dehydrogenation reaction.

The in vitro conversion of 6β-hydroxyhyoscyamine to scopolamine was demonstrated using the partially purified enzyme preparations from *H. niger* root cultures (T. Hashimoto, J. Kohno, Y. Yamada, unpublished work). The reaction requires the cofactors for 2-oxoglutarate-dependent dioxygenases and proceeds with retention of the hydroxyl oxygen and with a loss of 7β-hydrogen. At present, the epoxidase activity is weak and our attempts to further purify the enzyme have been unsuccessful. Still unknown factors may be involved in this reaction.

3 Somatic Hybrids Between Tobacco and *Duboisia*

Attempts were made to establish somatic hybrids of *Nicotiana tabacum* which produces nicotine alkaloids and *Duboisia* spp. which produce both tropane and nicotine alkaloids. These novel somatic hybrids will become potentially interesting materials for the study of the regulation of nicotine and tropane alkaloid biosynthesis. Fusion of protoplasts from *N. tabacum* mesophyll cells and protoplasts from *D. hopwoodii* suspension cultures was induced by polyethylene glycol treatment [12]. Heterokaryons were identified microscopically on the basis of the presence of both green chloroplasts (from mesophyll protoplasts) and thick cytoplasmic strands (from protoplasts of cell suspension cultures). They were isolated with a micropipette and were cultured individually in small nylon mesh chambers, around which mesophyll protoplasts of tobacco were cultured as nurse cells. Culture medium can permeate into the chamber freely, but the tobacco protoplasts remained outside of the chamber. One putative intergeneric hybrid cell line was obtained, and the hybrid nature of this line was confirmed by biochemical and cytological investigations. Indirect evidence was provided by the studies of the electrophoretic patterns of peroxidase and acid phosphatase, and the thin-layer chromatography of fluorescent compounds. The callus culture derived from the fused cell showed intermediate patterns of both parental species. Observations of metaphase chromosomes directly indicated the hybrid nature of this cell line. Chromosomes of *Duboisia* are small and therefore easily distinguished from the larger chromosomes of tobacco. The putative hybrid cells contained both types of chromosomes. The chromosome numbers in the hybrid cells varied from ca. 70 to 240, indicating genetic instability in this line. The relative nuclear DNA content of the hybrid line and both parental species were also investigated. Callus cultures of *N. tabacum* and *D. hopwoodii* had a 4c nuclear DNA content, which suggests that the culture conditions we used favoured the doubling of the nuclear DNA content. The mean value of the DNA content in the hybrid line was much larger than the sum of the 2c value for *N. tabacum* (mesophyll) and the 4c value for *D. hopwoodii* (callus), indicating that this hybrid was a product of multiple fusion. This high DNA content, however, might have been the result of somatic polyploidization after fusion.

The genetic instability of the intertribal hybrid cell line was further studied [13]. Ten subclones of calluses derived from this hybrid cell line have been cultured for 3 years, and their genetic makeup has been clarified as to nuclear DNA content, chromosome constitution and peroxidase isozymes.

Nuclear DNA content differed in each subclone. For most subclones the mean DNA contents were smaller than the mean DNA content for the original hybrid cell line determined 1 year after fusion. This decrease in DNA content is partly attributable to the elimination of tobacco chromosomes which took place in all the subclones. The extent of the elimination of tobacco chromosomes varied among the subclones. No tobacco type (large) chromosomes were found in two subclones. Other subclones had fewer than 20 tobacco-type chromosomes per metaphase. The chromosome elimination occurred slowly since no such tendency for specific elimination of chromosomes was evident 1 year after fusion. Multinuclear cells were common in all the subclones. A cell containing 22 or 23 nuclei was found in one subclone. The peroxidase isozyme analysis indicated the loss of a tobacco-specific isozyme, confirming the result of chromosome analysis.

Attempts to regenerate plants during the early stage of culture were unsuccessful until 2 years after fusion, when shoots regenerated from two hybrid subclones. These shoots were heterogeneous in morphology and nuclear DNA content. When these shoots were treated for 3 days with 3-indolebutyric acid (10^{-4}M) then cultured on hormone-free medium, adventitious roots formed at a high frequency. The growth of those plantlets did not, however, go beyond root initiation.

These results show that the *Duboisia* + *Nicotiana* hybrid has morphogenetic potential in spite of considerable genetic instability.

Recently, we have obtained new hybrid cell lines by electric fusion of mesophyll protoplasts of *N. tabacum* and protoplasts of a *D. leichhardtii* suspension culture which was derived from a root culture producing a large amount of scopolamine (T. Endo, Y. Yamada, unpublished work). Alkaloid biosynthesis of these hybrids are now under investigation.

References

Endo T, Yamada Y (1985) Alkaloid production in cultured roots of three species of *Duboisia*. Phytochemistry 24:1233–1236

Endo T, Komiya T, Masumitsu Y, Morikawa M, Yamada Y (1987) An intergeneric hybrid cell line of *Duboisia hopwoodii* and *Nicotiana tabacum* by protoplast fusion. J Plant Physiol 129:453–459

Endo T, Komiya T, Mino M, Nakanishi K, Fujita S, Yamada Y (1988) Genetic diversity among sublines originating from a single somatic hybrid cell of *Duboisia hopwoodii* + *Nicotiana tabacum*. Theor Appl Genet (in press)

Hamill JD, Parr AJ, Rhodes MJ, Robins RJ, Walton NJ (1987) New routes to plant secondary products. Bio/Technology 5:800–804

Hashimoto T, Yamada Y (1983) Scopolamine production in suspension cultures and redifferentiated roots of *Hyoscyamus niger*. Planta Med 47:195–199

Hashimoto T, Yamada Y (1986) Hyoscyamine 6β-hydroxylase, a 2-oxoglutarate-dependent dioxygenase, in alkaloid-producing root cultures. Plant Physiol 81:619–625

Hashimoto T, Yamada Y (1987) Effects of culture conditions on tropane alkaloid formation in *Hyoscyamus niger* suspension cultures. Agric Biol Chem 51:2769–2774

Hashimoto T, Yamada Y (1987) Purification and characterization of hyoscyamine 6β-hydroxylase from root cultures of *Hyoscyamus niger* L Eur J Biochem 164:277–285

Hashimoto T, Yukimune Y, Yamada Y (1986) Tropane alkaloid production in *Hyoscyamus* root cultures. J Plant Physiol 124:61–75

Hasimoto T, Kohno J, Yamada Y (1987) Epoxidation in vivo of hyoscyamine to scopolamine does not involve a dehydration step. Plant Physiol 84:144–147

Leete E (1979) Biosynthesis and metabolism of the tropane alkaloids. Planta Med 36:97–112

Yamada Y, Endo T (1984) Tropane alkaloid production in cultured cells of *Duboisia leichhardtii*. Plant Cell Rep 3:186–188

Yamada Y, Hashimoto T (1982) Production of tropane alkaloids in cultured cells of *Hyoscyamus niger*. Plant Cell Rep 1:101–103

Secondary Metabolites in Cell Cultures of *Tabernaemontana* Species

R. Verpoorte[1], R. van der Heijden[1], J. Schripsema[1], M. Sierra[1], E.J.M. Pennings[1], F. van Iren[2], and H.J.G ten Hoopen[3]

1 Introduction

The pantropical genus *Tabernaemontana* comprises about 120 species. Many species are used in traditional medicine. The most often reported uses are as an antimicrobial agent, curing diseases caused by parasites and protozoa, against warts, as analgetic, as stimulant and as febrifuge. Despite these activities, until recently little work has been done on the isolation of active compounds. Most of the work on this genus was pure phytochemical: the isolation of indole alkaloids. For extensive reviews the reader is referred to van Beek et al. (1984) and Danieli and Palmisano (1986).

Characteristic for the genus is the occurrence of indole alkaloids, particularly the ibogàn (e.g. coronaridine), corynanthean (e.g. vobasine) and aspidospermatan (e.g. tubotaiwine, apparicine) class of alkaloids. Dimeric alkaloids, formed from the vobasine-type of alkaloids with the ibogan class of alkaloids or with apparicine, are commonly found in *Tabernaemontana* species. Altogether more than 250 different alkaloids have been isolated from plants of this genus. Dimers were found to have antimicrobial activity (van Beek et al. 1985, Verpoorte 1987).

The occurrence of interesting biological active compounds in the plants, the dimeric alkaloids and the close relationship with *Catharanthus* were reasons to start cell cultures of *Tabernaemontana*. Here, we will briefly review some of the results from our studies on cell cultures of this genus.

2 Initiation of Cell Cultures

With eight different species a variety of media was tested for callus formation on leaf explants (van der Heijden et al. 1986 c). Murashige Skoog medium containing 2,4-D (1 ppm) and zeatin (1 ppm) proved to be most suitable. To improve friability of the callus B5 medium (1 ppm 2,4-D and 0.2 ppm kinetin) should be preferred. All callus cultures obtained, produced alkaloids.

Projectgroup Plant Cell Biotechnology, Biotechnology Delft Leiden (BDL), The Netherlands
[1]Department of Pharmacognosy, Center for Bio-Pharmaceutical Sciences, University of Leiden, P.O. Box 9502, 2300RA Leiden, The Netherlands
[2]Department of Plant Molecular Biology, University of Leiden, The Netherlands
[3]Department of Biochemical Engineering, University of Delft, The Netherlands

Primary and Secondary Metabolism of Plant Cell Cultures, ed. by W.G.W. Kurz

A more detailed comparison of a callus culture of *T. elegans* and the whole plant showed clear differences (van der Heijden et al. 1986a,b). Apparicine was the major alkaloid from the callus material. In the plant dimeric alkaloids of the tabernaelegantine type were the major compounds. The callus also contained small amounts of dimers, i.e. 3R/S-hydroxy-conodurine and monogagaine. The latter was not isolated from the plant.

From several of the callus cultures cell suspension cultures were initiated. These cultures have been studied further for their growth and alkaloid production under various conditions.

3 Phytochemical Characterization

After extraction from the cells the alkaloids were isolated by using a combination of ion-pair droplet, countercurrent chromatography (van der Heijden et al. 1987b) and preparative TLC and subsequently identified by their spectral data (UV, MS, NMR). In Table 1 the results are summarized for various cell lines (van der Heijden

Table 1. Alkaloids identified in suspension cultures of *Tabernaemontana*

	T. africana	*T. elegans*	*T. divaricata*	*T. orientalis*
Tryptamine		+	+	+
5-Hydroxy-tryptamine				+
Tetrahydroharman				+
5-Hydroxy-tetra-hydroharman				+
Norharman				+
Apparicine	+	+	+	+
Pericine			+	
Tubotaiwine		+	+	+
Condylocarpine			+	
Vallesamine			+	
O-acetyl-vallesamine		+	+	
16-Hydroxy-16,22-dihydro-apparicine			+	
Voaphylline			+	+
Voaphylline hydroxy-indolenine			+	+
Vobasine		+	+	
Vobasinol		+		
Perivine		+	+	
Tabernaemontanine			+	
Pericyclivine			+	
Isositsirikine		+		
Coronaridine			+	
19S-Heyneanine			+	
Catharanthine[a]			+	
12-Hydroxy-akuammicine[a]			+	

[a] Pawelka and Stoekigt (1983).

et al. 1987a,b,c, 1988d,e) The *T. divaricata* cell cultures produced similar alkaloids as reported by Pawelka and Stoeckigt (1983), except for catharanthine, which could not be detected in any of our cultures. Comparing the alkaloids found in the various cultures, it can be noted that they contain a similar spectrum of alkaloids; apparicine, vallesamine and tubotaiwine being present in almost all cultures. Also the vobasine type of alkaloids are rather common. On the other hand, the ibogan class of alkaloids are present only at low concentration, if at all (e.g. they are not present in *T. elegans*). This could explain why we have not been able to detect dimeric alkaloids (containing an ibogan moiety) with certainty, if they are present it is as trace components only.

In general, one may conclude that the alkaloid mixtures isolated from the cell cultures are far less complex than those obtained from the plant. Whether this is due to a difference in the biosynthetic capacity of the cell culture and the plant or to (photo)chemical conversions in the plant during prolonged storage of the alkaloids needs further study on the level of enzyme activities.

The alkaloid production of a *T. divaricata* cell line has been followed over a prolonged period in order to learn more about the stability of the alkaloid production. As this cell line was grown on a medium containing 2,4-D, a growth regulator often said to be detrimental for alkaloid production (e.g. Zenk et al. 1977, Knobloch et al. 1982), it was also transferred to a medium containing NAA. The growth and alkaloid production on this medium were also followed during the same period. In Fig. 1 the results are summarized. The alkaloid production of the cell line grown on 2,4-D containing medium remained constant during the whole period. The cell line transferred to an NAA-containing medium showed a rapid increase of the alkaloid production during the first four subcultures, then the production declines to a stable value still higher than on the 2,4-D medium (Sierra et al. unpublished results). Whether this is reproducible is presently being studied.

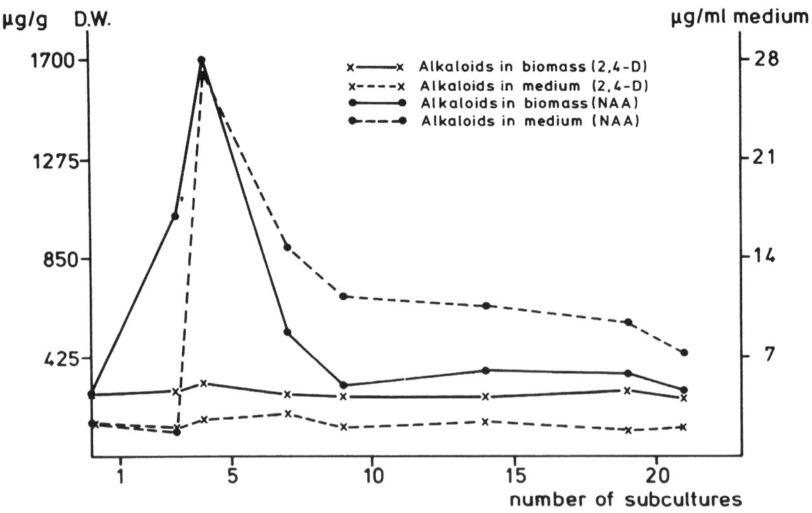

Fig. 1. Alkaloid production (µg ml^{-1} medium) in a *Tabernaemontana divaricata* cell suspension culture (line 6) during 21 subcultures after changing from a 2,4-D- to a NAA-containing medium, compared with the production of the original cell culture during the same period

4 Growth and Production Characteristics

A typical growth curve of a *Tabernaemontana* cell culture is given in Fig. 2. The sugar utilization is similar to that found in *Catharanthus*. Sucrose is extracellularly rapidly converted into glucose and fructose. Glucose is then taken up more rapidly than fructose (Fig. 3). After entering the stationary phase, the cells remain viable for prolonged periods. Even after 6 weeks in the stationary phase the cells were still

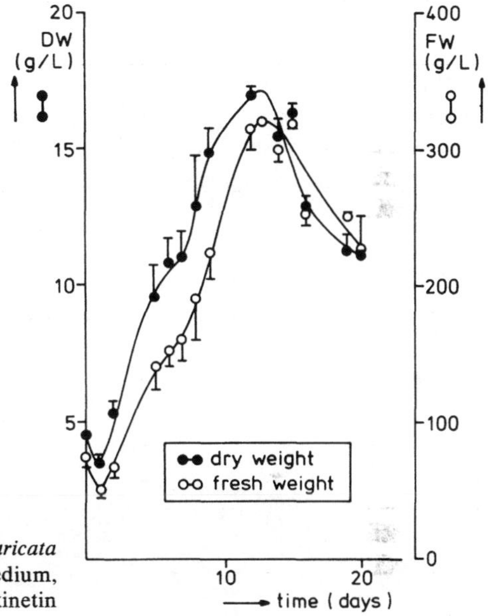

Fig. 2. Growth curve of a *Tabernaemontana divaricata* cell suspension culture on a Murashige-Skoog medium, containing 3% sucrose, 1 ppm 2,4-D and 1 ppm kinetin

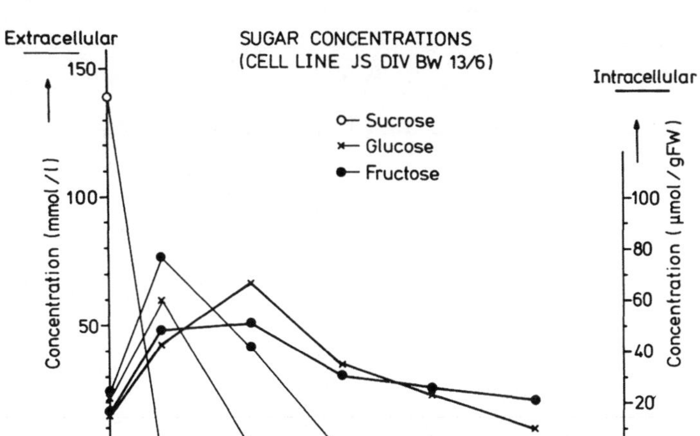

Fig. 3. Extracellular and intracellular sugar levels in a *Tabernaemontana divaricata* cell suspension culture, as determined by means of ^{1}H-NMR in aqueous solutions, using a water peak suppression method (Schripsema et al., unpublished results)

viable. By using ^{1}H-NMR with a water signal suppression method we were able to analyze the sugars in the media and the cells (Schripsema et al. unpublished results). In combination with a newly developed method to measure sugar dissimilation in a single culture by regular determination of the weight loss of a cell culture in a flask (Schripsema et al. unpublished results), we found that about 15–20% of the initial sugar was stored in the cells at the end of the growth phase. The storage sugars were mainly glucose and fructose (Fig. 3). No significant amounts of starch could be detected by determining sugars after treatment of the homogenized cells with amyloglucosidase.

By using the weight loss method the growth of a *T. divaricata* cell culture on MS medium with various carbon and nitrogen sources was studied. For the sugars, the best results were obtained with sucrose (Fig. 4). The combination of glucose and fructose gives a similar yield, which is not surprising since sucrose is usually rapidly converted into these monosaccharides by the cell cultures. Also, lactose proved to be a suitable carbon source. From the dissimilation curves the duration of a lag

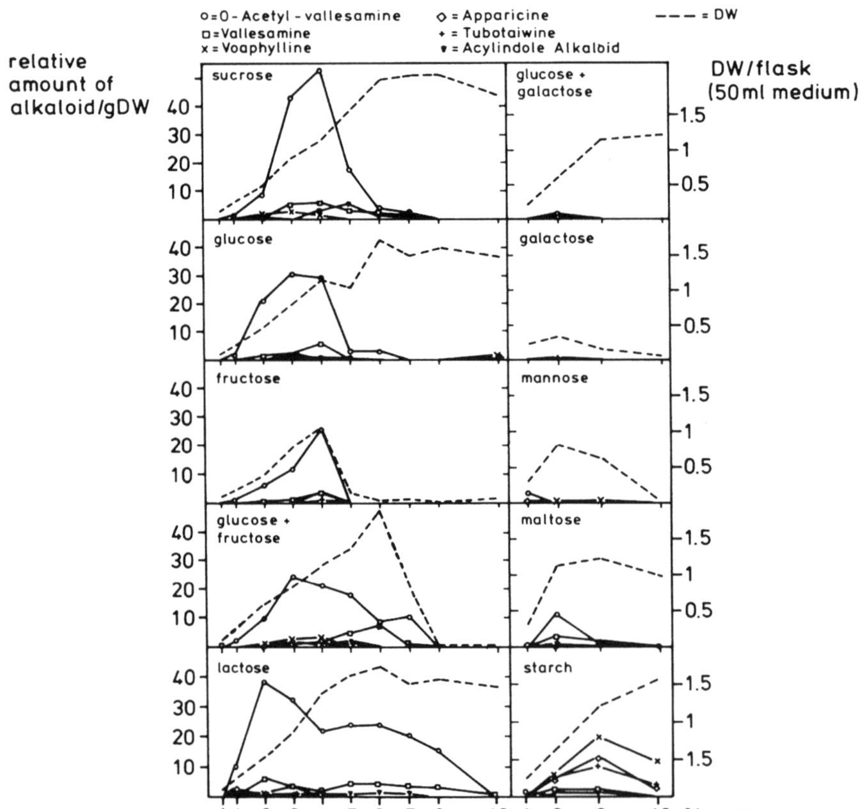

Fig. 4. Alkaloid production in *Tabernaemontana divaricata* cell suspension cultures after transferring to a medium containing various concentrations of different sugars (normal MS growth medium of this cell line contained 3% sucrose)

phase can be observed, yielding information on the initial reaction of the cell culture to the medium with the different carbon source. In fact, in most cases a clear increase of the lag phase was observed. Alkaloid content was also determined in the cells (Fig. 4). Best results again were observed in the case of sucrose. An interesting observation was that the cells can be easily grown on starch. ^{1}H-NMR spectrometry of the aqueous extract of the cells grown on starch showed the presence of considerable amounts (more than 1% dry weight) of phenylalanine and tyrosine in the cells.

Using the weight loss method, various nitrogen sources were also tested. The results are summarized in Table 2. Apparently the usual growth medium gives the best results. Only in a medium in which nitrate was replaced by aspartic acid was an increase of alkaloid production observed. Furthermore, the addition of glutamic or aspartic acid did not result in improved growth or alkaloid production. However, effects are observed on the ratio of the amounts of the various alkaloid types.

In *T. divaricata* the alkaloid production was followed over time (van der Heijden et al. 1988d). From this it was learned that the alkaloid production starts at the end of the growth phase; in fact, when the sugar in the medium is finished (Fig. 5). Similar production kinetics have been found for *Catharanthus* (e.g. Morris 1986). Interestingly, the production of one of the alkaloids, voaphylline, showed a clear maximum after which it rapidly disappears, the other alkaloids then still rising in concentration are not derived from this alkaloid. About halfway through the growth phase a small peak in the alkaloid accumulation is observed. Apparently, at different stages of the growth cycle different biosynthetic pathways are operational. It also implies that the moment of harvesting a cell culture is crucial for the optimum production of a certain alkaloid. For biosynthetic studies the use of the weight loss method is therefore of interest in order to always harvest a culture at a similar stage of the growth phase.

Table 2. Biomass yield and alkaloid production of a *Tabernaemontana divaricata* cell culture on media with various N-sources

MS medium NH_4^+	NO_3^-	Glu ac[a]	Asp ac[b]	Dry weight	Total alkaloid
+	+	–	–	100%[c]	100%[d]
–	–	60 mM	–	82	56
–	–	30 mM	–	92	61
–	–	–	60 mM	73	67
–	–	–	30 mM	86	19
+	–	30 mM	–	92	3
+	+	30 mM	–	93	2
–	+	30 mM	–	54	4
+	–	–	30 mM	105	147
+	+	–	30 mM	101	41
–	+	–	30 mM	69	0

[a] Glutamic acid.
[b] Aspartic acid.
[c] 16.6 g l^{-1} Medium.
[d] 290 $\mu g\ g^{-1}$ DW, no alkaloids in the medium.

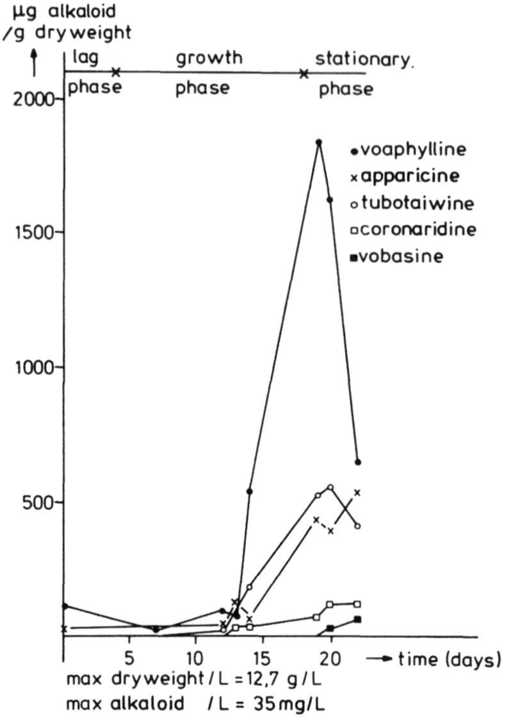

Fig. 5. Time course of alkaloid production of a *Tabernaemontana divaricata* cell suspension culture (cell line 4BW)

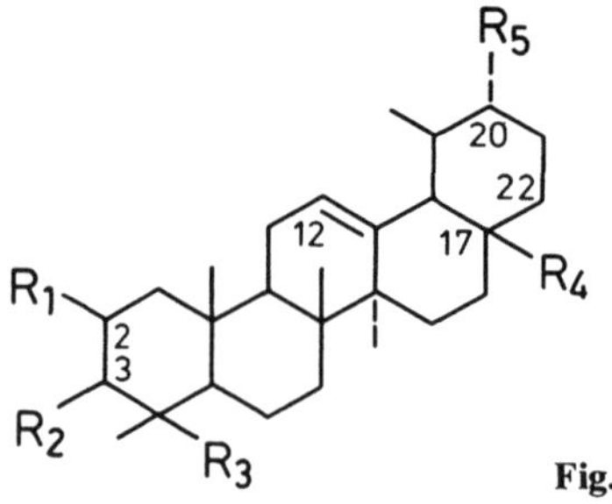

Fig. 6. Ursolic acid $R_1 = H$, $R_2 = OH$, $R_3 = R_5 = CH_3$, $R_4 = COOH$

5 Elicitation

Several authors have reported an increased alkaloid production in some *Catharanthus roseus* cell lines after treatment with fungal elicitors (Eilert et al. 1987). In the case of *Cinchona* we have not been able to detect any alkaloid production after elicitation.

Addition of various elicitors to *Tabernaemontana* cell cultures showed that the cells react by the de novo production of triterpenoids of the ursolic acid type (Fig. 6). Yields are ca. 1–2% of the dry weight (van der Heijden et al. 1988a, c). The same compounds were found to be present in different species. The elicitation causes the

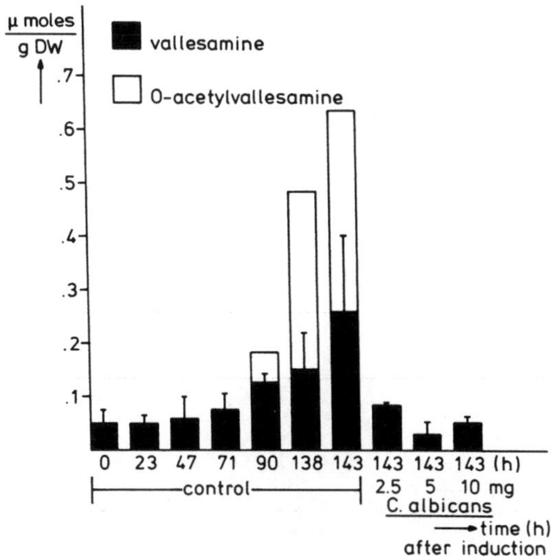

Fig. 7. Alkaloid levels in the biomass in a *Tabernaemontana divaricata* cell suspension culture after elicitation at the end of the growth phase with various amounts of a *Candida albicans* preparation. A similar result in observed for the medium

synthesis of enzymes of the triterpenoid biosynthetic pathway in the cells, which could be inhibited by cycloheximide. Also, yellow, yet unidentified compounds (probably flavonoids) are formed in the cells.

Using a *T. divaricata* cell line, which produced vallesamine and O-acetylvallesamine as major alkaloids, the influence of the elicitation on the alkaloid production was studied (van der Heijden et al. unpublished results). In Fig. 7 it can be seen that after elicitation at the end of the growth phase, growth and alkaloid production are inhibited, compared to the blank. As both triterpenoids and secologanin have mevalonic acid as a common precursor, a competition between the triterpenoid and alkaloid pathway could cause the inhibition of the alkaloid biosynthesis.

6 Biosynthesis

Using an HPLC assay for tryptophan decarboxylase (TDC) (Pennings et al. 1987, 1988) and strictosidine synthase (SSS) (Pennings et al. 1989), the activities of both enzymes could be confirmed in cell cultures of *T. divaricata* and *T. orientalis*. Quantitatively, activities in *T. orientalis* were similar to those found in *Catharanthus roseus* cells. The activity of SSS was more or less stable through a growth cycle, whereas TDC activity had a maximum in the beginning of the growth phase.

Stemmadenine is thought to be an important intermediate in the biosynthesis of a number of *Tabernaemontana* alkaloids (Fig. 8). Incubation of this alkaloid with crude enzyme preparations and various co-factors showed that conversion into vallesamine occurred (van der Heijden et al. 1988b). However, a conversion of stemmadenine into condylocarpine was more predominant. This conversion was also observed after incubation with denaturated enzyme preparations. By means

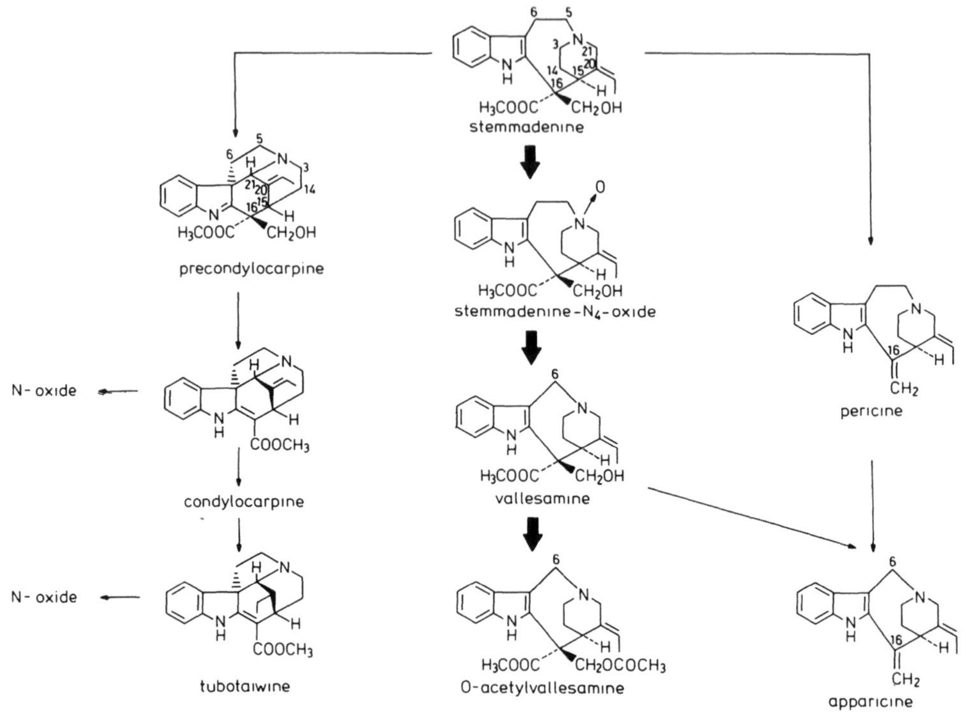

Fig. 8. Scheme of biosynthesis of some major alkaloids from *Tabernaemontana* cell cultures

of its UV spectrum, obtained by diode array detection in the HPLC assay of the alkaloids, an intermediate was tentatively identified as precondylocarpine. The rapid conversion of stemmadenine could be the reason why this alkaloid has not yet been found in the cell cultures.

7 Conclusion

Tabernaemontana cell cultures are able to produce indole alkaloids. The production remains stable for prolonged periods of subculturing. The scala of alkaloids produced is relatively simple compared to the plant. Upon elicitation the alkaloid levels are very low (increased catabolism, or decreased metabolism?), whereas other secondary pathways are induced under these conditions (e.g. triterpenoid and flavonoid pathway). This makes *Tabernaemontana* cell cultures an interesting system for further studies of the regulation of secondary metabolism.

References

Van Beek TA, Verpoorte R, Baerheim Svendsen A, Leeuwenberg AJM, Bisset NG (1984) *Tabernaemontana* L. (Apocynaceae): a review of its taxonomy, phytochemistry, ethnobotany and pharmacology. J Ethnopharmacol 10:1-156

Van Beek TA, Verpoorte R, Baerheim Svendsen A, Fokkens R (1985) Antimicrobially active alkaloids from *Tabernaemontana chippii*. J Nat Prod 48:400-423

Danieli B, Palmisano G (1986) Alkaloids from *Tabernaemontana*. In: Brossi A (ed) The alkaloids. Academic Press, New York London, pp 1-55

Eilert U, De Luca V, Constabel F, Kurz WGW (1987) Elicitor mediated induction of tryptophan decarboxylase and strictosidine synthase in cell suspension cultures of *Catharanthus roseus*. Arch Biochem Biophys 254:491-497

Van der Heijden R, Brouwer RL, Verpoorte R, Wijnsma R, van Beek TA, Harkes PAA, Baerheim Svendsen A (1986a) Indole alkaloids from *Tabernaemontana elegans* (whole plant). Phytochemistry 25:144-147

Van der Heijden R, Brouwer RL, Verpoorte R, Wijnsma R, van Beek TA, Harkes PAA, Baerheim Svendsen A (1986b) Indole alkaloids from a callus culture of *Tabernaemontana elegans*. Phytochemistry 25:843

Van der Heijden R, Wijnsma R, Verpoorte R, Harkes PAA, Baerheim Svendsen A (1986c) The influence of nutrient media, auxins and cytokinins on the initiation of alkaloid producing callus on leaf explants of eight species of *Tabernaemontana*. Fitoterapia 57:415-421

Van der Heijden R, Lamping PJ, Out PP, Wijnsma R, Verpoorte R (1987a) High-performance liquid chromatographic determination of indole alkaloids in a suspension culture of *Tabernaemontana divaricata*. J Chromatogr 396:287-295

Van der Heijden R, Hermans-Lokkerbol A, Verpoorte R, Baerheim Svendsen A (1987b) Pharmacognostical studies on *Tabernaemontana* XX. Ion-pair droplet counter-current chromatography of indole alkaloids from suspension cultures. J Chromatogr 396:410-415

Van der Heijden R, Schripsema J, Harkes PAA, Verpoorte R (1987c) Alkaloid production by plant cell cultures of *Tabernaemontana* species. In: Neijssel OM, van der Meer RR, Luyben KChAM (eds) Proc 4th Eur Congr Biotechnology 1987, vol 2. Elsevier, Amsterdam, pp 397-398

Van der Heijden R, Verheij ER, Schripsema J, Baerheim Svendsen A, Verpoorte R, Harkes PAA (1988a) Induction of triterpene biosynthesis by elicitors in suspension cultures of *Tabernaemontana* species. Plant Cell Rep 7:51-54

Van der Heijden R, van der Graaf GM, Pennings EJM, Verpoorte R (1988b) Formation and degradation of some aspidospermatan indole alkaloids in suspension cultures of *Tabernaemontana divaricata*. (submitted)

Van der Heijden R, Verpoorte R, Harkes PAA (1988c) Effects of elicitors on the production of secondary metabolites in plant cell cultures of *Cinchona, Rubia, Morinda* and *Tabernaemontana*. In: Rhodes MJC, Robins RJ (eds) Proc IAPTC-Symp, Norwich, pp 217-224

Van der Heijden R, Hermans-Lokkerbol A, de Kool LPJ, Lamping PJ, Harkes PAA, Verpoorte R (1988d) Accumulation of indole alkaloids in a suspension culture of *Tabernaemontana divaricata*. Planta Med 54:393-397

Van der Heijden R, Verpoorte R, Harkes PAA (1988e) Secondary metabolites in cell and tissue cultures of species belonging to the tribe Tabernaemontanae (Apocynaceae). In: Bajaj YPS (ed) Biotechnology in forestry and agriculture. Vol 7. Medicinal and aromatic plants, part 2. Springer, Berlin Heidelberg New York Tokyo pp 506-523

Knobloch KH, Bast G, Berlin J (1982) Medium- and light-induced formation of serpentine and anthocyanins in cell suspension cultures of *Catharanthus roseus*. Phytochemistry 21:591-594

Morris P (1986) Regulation of product synthesis in cell cultures of *Catharanthus roseus*. II. Comparison of production media. Planta Med 52:121-126

Pawelka KH, Stoeckigt (1983) Indole alkaloids from cell suspension cultures of *Tabernaemontana divaricata* and *Tabernanthe iboga*. Plant Cell Rep 2:105-107

Pennings EJM, Hegger I, van der Heijden R, Duine JA, Verpoorte R (1987) Assay of tryptophan decarboxylase from plant cell cultures. Anal Biochem 165:133-136

Pennings EJM, van der Heijden R, van den Bosch RA, van der Leer T, Verpoorte R (1988) Rapid assay of tryptophan decarboxylase from plant cell cultures. In: Rhodes MJC, Robins RJ (eds) Proc IAPTC-Symp, Norwich, pp 79-92

Pennings EJM, van den Bosch RA, van der Heijden R, Stevens LH, Duine JA, Verpoorte R (1989) Assay of strictosidine synthase from plant cell cultures by HPLC. Anal Biochem 176:412–415

Verpoorte R (1987) Antimicrobially active alkaloids. In: Kramer A, Weuffen W, Krasilnikov AP, Groeschel D, Bulke E, Rehn D (eds) Handbuch der Antiseptik, vol 2, pt 3. Volk und Gesundheit, Berlin, pp 385–422

Zenk MH, El-Shagi H, Arens H, Stoeckigt J, Weiler EW, Deus B (1977) Formation of the indole alkaloids serpentine and ajmalicine in cell suspension cultures of *Catharanthus roseus*. In: Barz W, Reinhard W, Zenk MH (eds) Plant tissue culture and its biotechnological application. Springer, Berlin Heidelberg New York, pp 27–43

Distribution of Catharanthine, Vindoline and 3′,4′-Anhydrovinblastine in the Aerial Parts of some *Catharanthus roseus* Plants and the Significance Thereof in Relation to Alkaloid Production in Cultured Cells*

J. Balsevich and G. Bishop[1]

1 Introduction

Vinblastine (VLB) and vincristine (VCR) are two high cost anti-cancer drugs derived from *Catharanthus roseus*. For more than 10 years a number of laboratories had focussed on the use of tissue culture technology to develop and screen numerous cell suspensions in search of a high VLB/VCR producer. Despite the use of different explants, media alterations, and growth conditions this search was unsuccessful. Thus, while the production and accumulation of many indole alkaloids, including catharanthine (Kurz et al. 1985), one of the two "monomeric" precursors of VLB and VCR had been reproducibly observed in cultured *Catharanthus* cells; the reproducible production and accumulation of vindoline, the other "monomeric" precursor, had not. This lack of vindoline production likely accounted for the lack of production of bisindoles such as VLB and VCR and stimulated some interest in studying the regulation of these alkaloids' biosynthesis. Earlier, Constabel et al. (1982) noted that production of vindoline required a certain degree of tissue differentiation as evidenced by the renewed presence of vindoline in shoots regenerated from a callus. More recently, DeLuca and Cutler (1987) noted that one of the enzymes involved in the late stages of vindoline biosynthesis was associated with thylakoid membranes while another was activated by light (DeLuca et al. 1986, 1988). These results prompted us to develop further studies with plants in the hope of obtaining additional information pertinent to the development of tissue culture systems for the production of VLB, VCR, and vindoline.

The present study was initiated subsequent to the discovery of a flowerless plant which exhibited different alkaloid profiles as compared to "normal" *Catharanthus* plants, and as a consequence of the observation that normal plants, in contrast to the impression afforded by earlier literature, contain significant quantities of bisindoles, consisting predominantly of one, 3′,4′-anhydrovinblastine (AVLB).

Earlier, Westekemper et al. (1980) and Deus-Neumann et al. (1987) had reported the development of radioimmunoassays for vindoline and catharanthine respectively and determined their distributions in various plants. In this study we sought to extend these results by determining the distribution of catharanthine, vindoline, and AVLB in our flowerless plant as well as normal plants, thus relating

[1]Plant Biotechnology Institute, National Research Council of Canada, Saskatoon, Saskatchewan S7N OW9, Canada
*NRCC No. 29595

Primary and Secondary Metabolism of Plant Cell Cultures, ed. by W.G.W. Kurz

their accumulation to the developmental stage of the various leaves in the mature plants, as well as comparing the distributions between the two plants.

AVLB is a bisindole which is closely related, both structurally and biosynthetically, to VLB and VCR. Catharanthine and vindoline are precursors of AVLB. These three alkaloids are the major alkaloids of the aerial portions of *C. roseus* and are found mainly in the leaves.

2 Materials and Methods

For the study, the normal plants used were *C. roseus* var. Little Bright Eye (LBE). These plants were initiated from commercially available seed and were approximately 1 year of age. The flowerless plant (FP) was discovered in a field plot of normal plants in the summer of 1986 and was propagated by cuttings. The plants used for the study were of the same approximate size as the LBE variety. The morphology of the flowerless mutant was similar to the LBE variety except for its floral apparatus which appeared as vegetative structures which consisted of numerous sepals with the occasional formation of apparent pistils but no further floral development.

The alkaloid quantitation was performed by obtaining batches of 10 to 40 leaves from 2 to 10 plants with each batch representing leaves of a similar age. Thus, leaf # 1 represented the very small young leaves at the top of a shoot, while leaf #9 represented the leaves obtained by counting down the stem from the top of the shoot to the ninth pair of leaves. Alkaloids were obtained by extracting the leaves with acidified methanol followed by concentration of the extract, removal of pigments, basification, and extraction of alkaloids into ethyl acetate. Catharanthine, vindoline, and AVLB were isolated from the alkaloid mixture by thin-layer chromatography on silica gel and quantified by ultraviolet spectrophotometry.

3 Results

The results of the measurements are listed in Table 1a,b and Fig. 1a, b and c. In the LBE variety (Fig. 1a) the youngest leaves (# 1) contained the highest concentration of catharanthine and the lowest concentration of AVLB, whereas in the older leaves (#5–9) the lowest catharanthine and highest AVLB concentrations were noted. Vindoline appeared in moderate to high concentrations in leaves of all ages with the mature leaves (#5–7) having the highest concentrations. These overall profiles were relatively stable over a 3-month period.

The FP variety exhibited profiles which were different from the LBE variety but which also varied with time (Fig. 1b,c). In the initial sampling (Fig. 1b) the main profile differences between it and the LBE variety were exhibited by leaves # 1–3 where the concentrations of vindoline and AVLB were up to 12 times greater than levels observed in the LBE variety. In contrast, the older leaves (#5–9) of both varieties exhibited similar profiles.

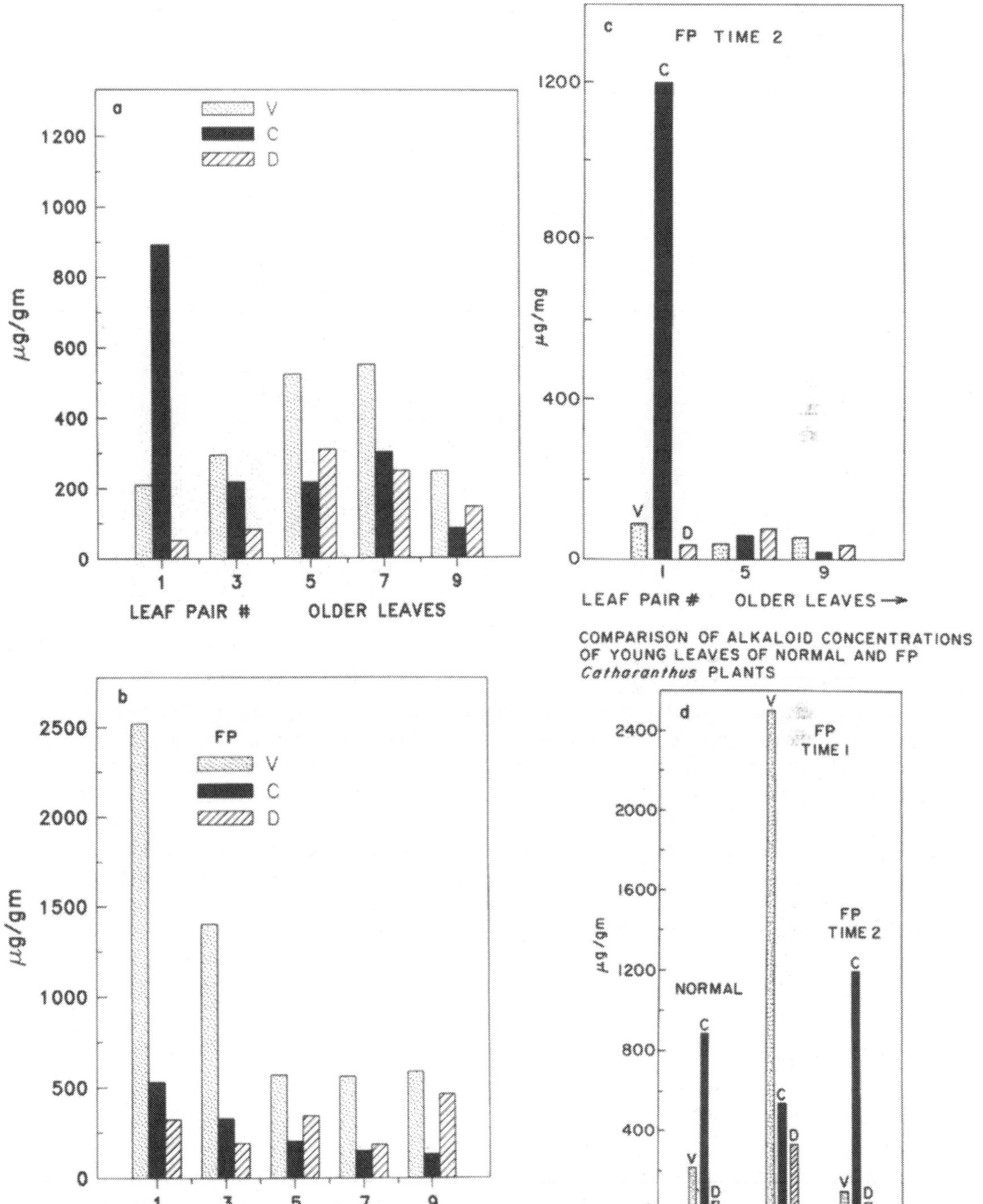

Fig. 1a-d. Concentrations of catharanthine (*C*), vindoline (*V*), and AVLB (*D*) in leaves of various ages in **a** LBE variety; **b** FP variety, initial sampling; **c** FP variety, second sampling; **d** LBE, FP initial sampling, FP second sampling; leaf one only comparison

Table 1. Distribution of alkaloids on a per leaf basis in the (**a**) LBE and (**b**) FP varieties

a Normal

	Leaf No.	Average leaf Wt. (mg)	Amount of alkaloids per leaf		
			Catharanthine (μg)	Vindoline (μg)	AVLB (μg)
	1	23	20	5	1
	3	223	49	66	19
	5	267	58	140	92
Older	7	265	80	146	66
leaves	9	238	20	59	34

b FP (Initial sampling)

	Leaf No.	Average leaf Wt. (mg)	Amount of alkaloids per leaf		
			Catharanthine (μg)	Vindoline (μg)	AVLB (μg)
	1	18	10	45	6
	3	200	60	280	38
	5	253	51	144	87
Older	7	227	34	126	35
leaves	9	225	29	130	104

In a second sampling of the FP variety taken 7 weeks after the initial sampling, a surprising set of profiles emerged (Fig. 1c). The concentration of vindoline and AVLB had dropped to extremely low levels in leaves of all ages. Catharanthine concentrations followed the general trend observed earlier with the youngest leaves (#1) containing the highest concentrations and the older leaves (#5–9) having much lower levels. This dynamic behavior in vindoline production and accumulation was noted in subsequent samplings, however, the periodicity and factors influencing profile changes have not yet been determined. Nonetheless some very interesting profile changes were noted among the LBE and FP varieties, particularly when the profiles of the youngest leaves were compared as illustrated in Fig. 1d.

Table 1 lists the average weights of the various leaves and the amounts of alkaloids present on a per leaf basis in both varieties.

4 Discussion and Conclusions

The fact that the results obtained with catharanthine were similar to those reported by Deus-Neumann et al. (1987) using different plants of different ages grown in different environments suggests that production, accumulation, and distribution of catharanthine is quite strictly controlled. Similarly, the results obtained with vindoline were analogous to those reported by Westekemper et al. (1980) except in the case of the youngest leaves where they observed only very low concentrations in contrast to the moderate levels observed here. This discrepancy might have been due in part to the lack of specificity of the assay method which we employed and which would not have differentiated between vindoline and such closely related derivatives as desacetoxyvindoline or vindorosine. Nonetheless, the similarities in the general trends suggested that vindoline production, accumulation, the distribution is normally tightly controlled, although the discovery of the FP variety and its extreme high/low accumulation pattern indicated that normal regulatory mechanisms can be altered. There were no other studies on the distribution of AVLB with which to compare, however, we would expect the pattern observed with the LBE variety to be typical.

Regarding tissue culture strategies, the implications of this study are that leaf and shoot cultures for production of vindoline and catharantine should be further examined. In particular, it would be interesting to note if alkaloid patterns observed in the young rapidly growing leaves of the plant could be reproduced in rapidly growing shoot cultures. If this were the case, then one would expect some interesting results from shoot cultures derived from a plant such as the FP variety where the young leaves were able to accumulate vindoline in such high concentrations.

References

Constabel F, Gaudet-LaPrairie P, Kurz WGW, Kutney JP (1982) Alkaloid production in *Catharanthus roseus* cell cultures. XII. Biosynthetic capacity of callus from original explants and regenerated shoots. Plant Cell Rep 1:139–142

DeLuca V, Cutler AJ (1987) Subcellular localization of enzymes involved in indole alkaloid biosynthesis in *Catharanthus roseus*. Plant Physiol 85:1099–1102

DeLuca V, Balsevich J, Tyler RT, Eilert U, Panchuk BD, Kurz WGW (1986) Biosynthesis of indole alkaloids: developmental regulation of the biosynthetic pathway from tabersonine to vindoline. J Plant Physiol 125:147–156

DeLuca V, Alvarez Fernandez J, Campbell D, Kurz WGW (1987) Developmental regulation of enzymes of indole alkaloid biosynthesis in *Catharanthus roseus*. Plant Physiol 86:447–450

DeLuca V, Alvarez Fernandez J, Campbell D, Kurz WGW (1988) Developmental regulation of enzymes of indole alkaloid biosynthesis in *Catharanthus roseus*. Plant Physiol 86:447–450

Deus-Neumann B, Stöckigt J, Zenk MH (1987) Radioimmunoassay for the quantitative determination of catharanthine. Planta Med 53:184–188

Kurz WGW, Chatson KB, Constabel F (1985) Biosynthesis and accumulation of indole alkaloids in cell suspension cultures of *Catharanthus roseus* cultivars. In: Neumann KH, Barz W, Reinhard E (eds) Primary and secondary metabolism of plant cell cultures. Springer, Berlin Heidelberg New York Tokyo, pp 143–153

Westekemper P, Wieczorek U, Gueritte F, Langlois N, Langlois Y, Potier P, Zenk MH (1980) Radioimmunoassay for the determination of the indole alkaloid vindoline in *Catharanthus*. Planta Med 39:24–37

Regulation of Vindoline Biosynthesis in *Catharanthus roseus*: Molecular Cloning of the First and Last Steps in Biosynthesis*

V. De Luca[1], N. Brisson[2], J. Balsevich[1], and W.G.W. Kurz[1]

1 Introduction

A large body of research has concerned itself with the production of commercially important phytochemicals by using plant cells in culture. These studies have identified several key areas where basic knowledge is generally limiting exploitation [1,2]. One fundamental problem occurs when non-differentiated tissue cultures are unable to produce and to accumulate complex chemicals originally found in specialized tissues or organs in the intact plant [3,4]. It is this need for specialized organs or tissues for metabolite production that has led to the development of rapidly growing and phytochemical producing organized hairy root cultures following genetic transformation of plants with *Agrobacterium rhizogenes* [5, and references cited therein]. Such cultures offer the possibility of producing root-derived phytochemicals but not those occurring in other plant tissues [5]. These technical innovations, however, fail to address the fundamental regulatory controls that maintain the specific developmental states which induce the production and accumulation of a given phytochemical.

In this report, we will briefly describe the recent work carried out in our laboratory on the *Catharanthus roseus* model system. The majority of earlier work on this system has been carried out in tissue cultures with the hope of capitalizing on the totipotent nature of plant cells to produce the commercially important dimeric indole alkaloids, vinblastine and vincristine [2,3]. The discovery of catharanthine producing cell lines has demonstrated the biosynthetic potential of *Catharanthus* cell cultures, but this has not been reproduced with vindoline or the derivative dimeric indole alkaloids [6–8]. The capability of regenerated shoots from cell cultures [9] to produce vindoline as well as dimers has demonstrated that this biosynthetic potential is present but silent in culture. Our studies on the regulation of vindoline biosynthesis have therefore been performed in the intact plant.

[1]Plant Biotechnology Institute, National Research Council of Canada, Saskatoon, Saskatchewan, S7N OW9, Canada
[2]Department of Biochemistry, University of Montréal, P.O. Box 6128, Station "A", Montréal, Québec, H3C 3J7, Canada
*NRCC No. 29490

Primary and Secondary Metabolism of Plant Cell Cultures, ed. by W.G.W. Kurz

2 Patterns of Indole Alkaloid Accumulation

The biosynthesis of vindoline from tabersonine occurs via three hydroxylations, two methylations and one acetylation step. Two alternative pathways have been proposed for the enzymatic conversion of tabersonine to vindoline in the intact plant [10–13]. Studies in developing seedlings have demonstrated that in the absence of light, tabersonine as well as four of five vindoline biosynthetic pathway intermediates accumulate (Table 1), whereas upon illumination tabersonine as well as intermediates are largely transformed to vindoline. In contrast, the level of catharanthine, which is already present in etiolated seedlings (Table 1), does not change significantly after light treatment. When the distribution of vindoline biosynthetic pathway intermediates was examined in different etiolated plant tissues (Table 2), it was found that those beyond 16-hydroxytabersonine were

Table 1. Accumulation of vindoline biosynthetic pathway intermediates in etiolated seedlings and their conversion to vindoline by light treatment[a]

	Amount of alkaloid ($mg\ kg^{-1}$ of seedling) fresh weight	
Alkaloid	Dark-grown seedlings	Light-grown seedlings
Tabersonine	40	13
16-Methoxytabersonine	15	10
16-Hydroxytabersonine	trace	–
Desacetoxyvindoline	10	trace
Vindoline	5	54
Deacetylvindoline	4	trace
Catharathine	31	37

[a]Light-grown seedlings were germinated (4 days) in the dark and then transferred to a lighted environment (4 days). Dark-grown seedlings were germinated in the dark (4 days) and maintained in the dark (4 days). This table is reproduced from [12].

Table 2. Distribution of vindoline biosynthetic pathway intermediates in developing seedlings

	Seedlings		
Aspidosperma alkaloids	Roots	Hypocotyl	Cotyledons
Tabersonine	+	+	++
16-Hydroxytabersonine	+	+	++
16-Methoxytabersonine	–	+	++
16-Methoxy-2,3-Dihydro-[a] -3-hydroxytabersonine	–	–	–
Desacetoxyvindoline	–	+	++
Deacetylvindoline	–	+	++
Vindoline	–	+	++

[a]This vindoline pathway intermediate was never detected in any of the seedling parts.

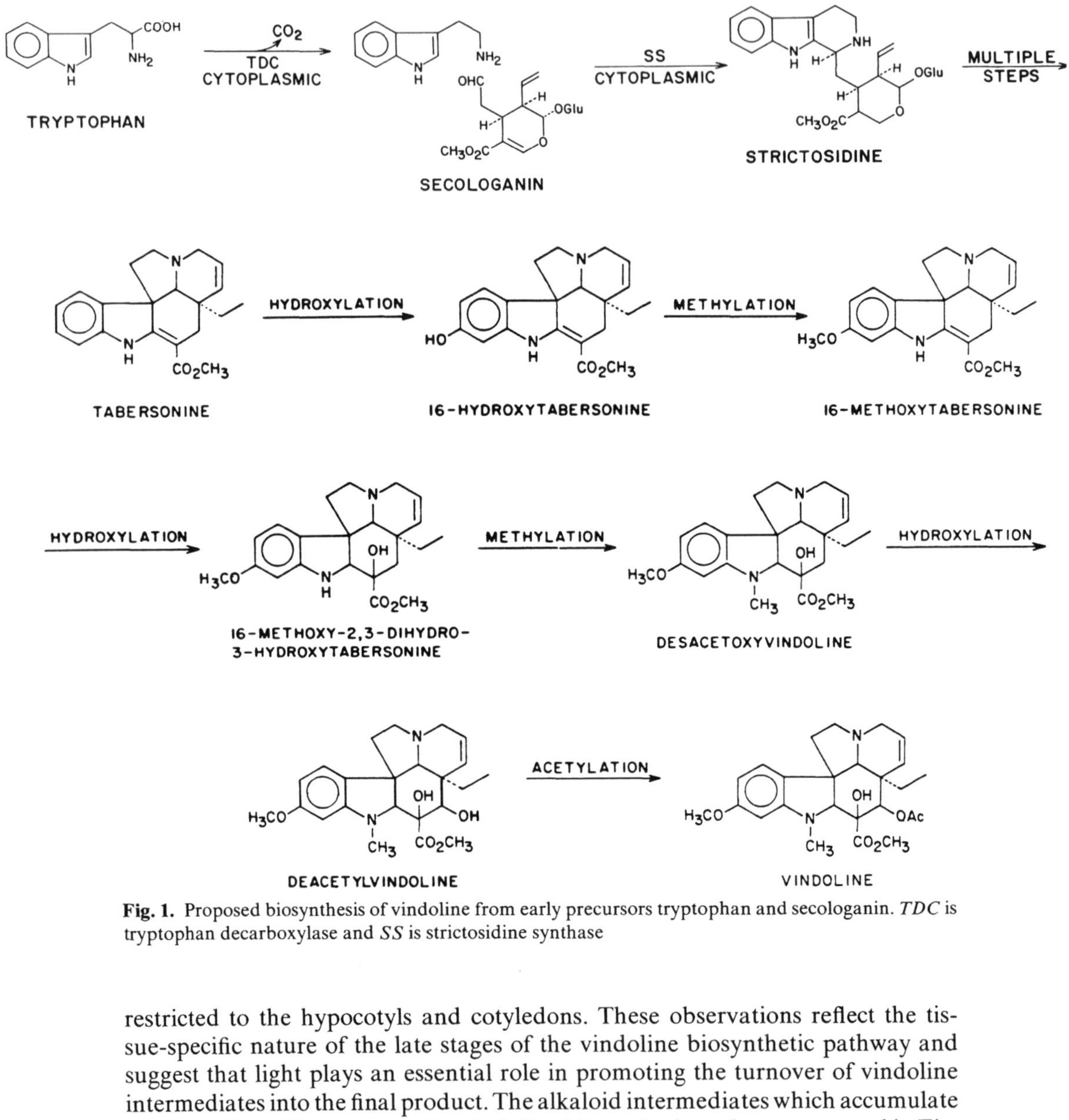

Fig. 1. Proposed biosynthesis of vindoline from early precursors tryptophan and secologanin. *TDC* is tryptophan decarboxylase and *SS* is strictosidine synthase

restricted to the hypocotyls and cotyledons. These observations reflect the tissue-specific nature of the late stages of the vindoline biosynthetic pathway and suggest that light plays an essential role in promoting the turnover of vindoline intermediates into the final product. The alkaloid intermediates which accumulate in etiolated seedlings suggest the vindoline biosynthetic pathway proposed in Fig. 1. The order of biosynthesis from tabersonine appears to involve hydroxylation, O-methylation, hydroxylation, N-methylation, hydroxylation and O-acetylation, and is in contrast to the alternative pathway proposed by Fahn et al. [10, 11] which involves three hydroxylation reactions prior to further substitution.

3 Enzymes of Indole Alkaloid Biosynthesis

We have developed novel enzyme assays for the last three steps of vindoline biosynthesis based on the pathway proposed in Fig. 1. We have discovered an N-methyltransferase which uses S-adenosyl-L methionine as a methyl group donor, and which specifically N-methylates advanced monoterpenoid indole alkaloids (Fig. 2A) [14]. This enzyme showed a high degree of specificity for the N-indole of 2,3-dihydro-3-hydroxytabersonine but not N-desmethyl-deacetyl-vindoline which indicates that hydroxylation at least at position 16 may occur after N-methylation. This biochemical evidence provides support for the pathway proposed in Fig. 1.

The second to last step in vindoline biosynthesis is catalyzed by a novel 2-oxoglutarate-dependent dioxygenase which we have recently isolated (De Luca et al., unpublished) (Fig. 2B). The enzyme required desacetoxyvindorosine, 2-oxoglutarate, ascorbate, ferrous ions and molecular oxygen for activity. Ascorbate and catalase stimulated enzyme activity and the presence of catalase reduced the background levels of activity. Substrate specificity studies indicated that this enzyme catalyzes the second to last reaction in vindoline biosynthesis (De Luca et al., unpublished).

The last step in vindoline biosynthesis is catalyzed by an O-acetyltransferase which transfers acetate from acetyl-CoA to deacetylvindoline (Fig. 2C). The catalytic properties of this enzyme have been reviewed previously [2]. We have

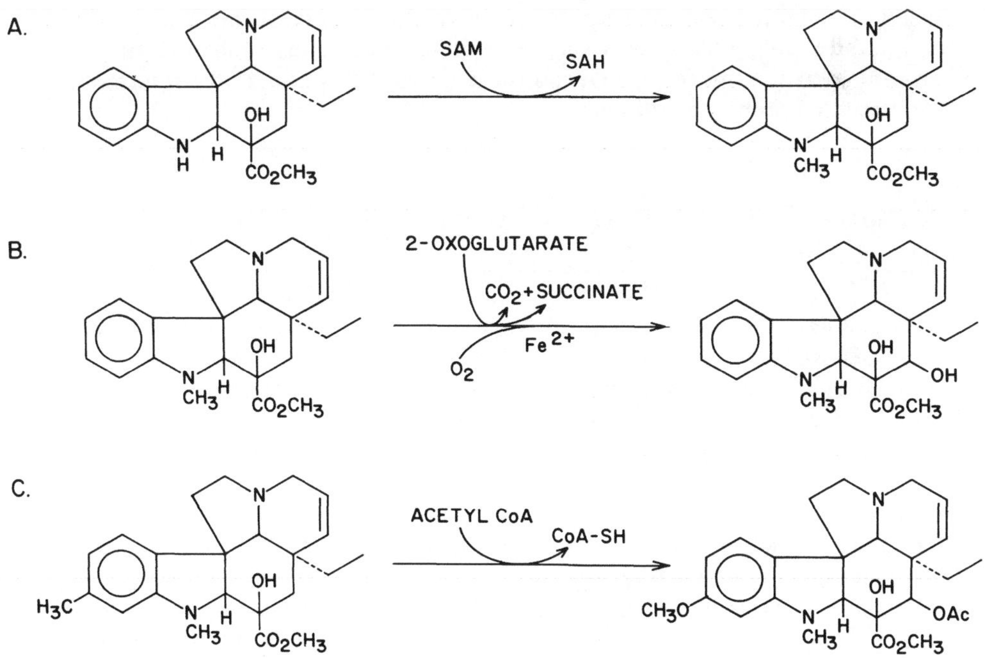

Fig. 2. Enzyme reactions which catalyze the last three steps in vindoline biosynthesis. The substrates used for enzyme assays were the best ones available

recently purified this protein to homogeneity using conventional column chromatography as well as affinity chromatography on coenzyme A-Sepharose (Powers et al., unpublished). The purified protein is a dimer with an M_r of 44 000 (gel filtration) and is composed of unequal subunits with an M_r of 30 000 and 15 000 (SDS-Page).

4 Developmental Regulation of Enzyme Activities Involved in Vindoline Biosynthesis

Developing *Catharanthus* seedlings were analyzed for appearance of tryptophan decarboxylase (TDC), strictosidine synthase (SS), N-methyltransferase (NMT) and O-acetyltransferase (DAT) enzyme activities. SS enzyme activity appeared early after germination and the enzyme was present throughout seedling growth. TDC appeared for a brief 48-h period (maxima at day 5) which coincided with the period of accumulation of tabersonine in etiolated seedlings [12, 13]. Both TDC and SS were present in all tissues of the seedling. NMT and DAT enzyme activities were induced after TDC and SS had peaked and the former activities could only be found in hypocotyls and cotyledons. TDC, SS and NMT did not require light for induction, whereas DAT enzyme activity was increased approximately ten-fold after light treatment of dark-grown seedlings. These findings clearly demonstrate that the product accumulation observed in seedlings under different lighting regimes (Tables 1 and 2) is strictly dependent on the presence or absence of specific enzyme activities involved in vindoline biosynthesis. Light seems to play a crucial role in the expression of the last step in vindoline biosynthesis and induction of this enzyme activity appears to be responsible for turnover of alkaloid intermediates into the final product, vindoline.

5 Biochemical Localization of Enzymes of Indole Alkaloid Biosynthesis

The cellular compartmentation of TDC, SS, NMT, α-ketoglurate-dependent dioxygenase and DAT was investigated. NMT enzyme activity was associated with broken and intact chloroplasts and studies with isolated chloroplast fractions indicated that NMT enzyme activity may be associated with chloroplast thylakoids [14]. The compartmentation of the other four biosynthetic activities appeared to be cytoplasmic and suggests that an advanced indole alkaloid precursor is transported from the cytoplasm to the chloroplast for N-methylation to occur. The methylated product is then transported out of the chloroplast to the cytoplasm where the final two steps in vindoline biosynthesis may occur. The final product, vindoline, would presumably be transported across the tonoplast membrane and would accumulate in the vacuole [16]. The reasons for this complex compartmentation are not clear, however, one or more of these factors may be responsible for lack of production of vindoline in cell suspension cultures.

6 Expression of Indole Alkaloid Biosynthesis Pathways in Plant Tissue Cultures

Plant cell suspension cultures have been developed in this laboratory which produce monoterpenoid indole alkaloids when cells are transferred to an alkaloid production medium [3] or when they are treated with fungal elicitors [17, 18]. Photoautotrophic [19] and hormone autotrophic [20] (habituated or tumorous) cell suspension cultures have been developed, with only the hormone autotrophs consistently producing and accumulating alkaloids [20]. None of these cultures, however, accumulated vindoline. When each of these cell suspension culture types were assayed for TDC, SS, NMT and DAT enzyme activities, all contained constant or inducible levels of TDC and SS, but none expressed NMT or DAT enzyme activities. These observations indicated that lack of vindoline accumulation was directly related to missing late stage biosynthetic enzymes.

7 cDNA Cloning of TDC and DAT from Developing Seedlings

It has been proposed that a capacity for both biosynthesis and storage are prerequisites for the production of secondary metabolites in plants [4]. Our work with the *Catharanthus* system has indicated that lack of expression of DAT enzyme activities in plant tissue cultures may be responsible for vindoline negative cultures.

Studies with developing seedlings have demonstrated that unlike the enzyme in tissue cultures [1, 18], TDC behaves as a classical regulatory enzyme, since its appearance in seedlings, unlike SS, correlated strictly with the biosynthesis and accumulation of indole alkaloids [13, 21]. These observations, together with the fact that vacuoles isolated from cell suspension cultures are capable of specific uptake and storage of vindoline [14], suggest that induction of DAT biosynthetic activity may result in vindoline biosynthesis and accumulation.

Modification of tissue-specific gene expression by genetic engineering requires precise knowledge of the mechanisms controlling the regulation of this gene. To gain a better understanding of DAT and TDC gene regulation, DAT and TDC cDNA clones were isolated. A lambda ZAP cDNA expression library was made using poly(A) $^{+}$RNAs isolated from 6-day-old *Catharanthus* seedlings. The library was screened with monospecific anti-TDC polyclonal antibodies (DeLuca, unpublished) and 27 clones were isolated and further characterized. The screening of the library was repeated using DAT monospecific anti-DAT polyclonal antibodies (DeLuca, unpublished) and 13 clones were isolated. The nucleotide sequence of a full length TDC-cDNA clone was determined [22] and the amino acid sequence of the protein was deduced using the largest open reading frame present in the cDNA sequence clone. This reading frame corresponded to a protein of 55.5 kD, which is the approximate size estimated for TDC (54 kD) [23]. Comparison of the deduced amino acid sequence with the sequence of the L-dopa-decarboxylase from *Drosophila melanogaster* showed that 40% of the amino acid residues are identical in the two sequences [24, 25]. The homology extended throughout the protein and was not restricted to a particular portion of the structure. The two decarboxylases showed absolute conservation of the pyridoxal-phosphate binding site and surrounding amino acids. This high degree of homology between two aromatic amino acid decarboxylases from widely divergent species indicated that the corresponding genes have a common ancestor. Determination of the DAT cDNA clone nucleotide sequence is in progress and it is hoped that comparison of the deduced amino acid sequence with that of a partial DAT protein sequence will identify this as the correct coding sequence.

8 Future Prospects

The capability to selectively clone genes whose expression is suppressed in culture systems has allowed the study of developmentally regulated and tissue-specific expression of these genes and their related biosynthetic pathways. In *Catharanthus roseus,* the time course of transcript production for TDC and DAT, the first and last steps in vindoline biosynthesis, will help to understand the controlled production of monoterpenoid indole alkaloids. The two cDNA clones will be used to screen a total genomic library in order to isolate the individual genes and to identify regulatory sequences which control tissue-specific gene expression. The DAT cDNA clone will be attached to a constitutive promotor and this construct will be transferred to *Catharanthus* cell suspension cultures using an *Agrobacterium*-mediated transformation system. Transformed cultures will be monitored for DAT biosynthetic activities as well as vindoline biosynthesis and accumulation. These studies will demonstrate the feasibility of using genetic engineering technology to allow constitutive expression and accumulation of vindoline in *Catharanthus* plant cell suspension cultures.

References

1. Berlin J (1984) Plant cell cultures – a future source of natural products? Endeavour N S 8:5–8
2. DeLuca V, Kurz WGW (1988) Monoterpene indole alkaloids (*Catharanthus* alkaloids). In: Constabel F, Vasil IK (eds) Cell culture and somatic cell genetics of plants, vol 5. Academic Press, New York London, pp 385–401
3. Kurz WGW, Constabel F (1985) Aspects affecting biosynthesis and biotransformation of secondary metabolites in plant cell cultures. CRC Crit Rev Biotechnol 2(2):105–118
4. Wink M (1987) Physiology of the accumulation of secondary metabolites with special reference to alkaloids. In: Constabel F, Vasil IK (eds) Cell culture and somatic cell genetics of plants, volume 4. Academic Press, New York London, pp 17–42
5. Hamill JD, Parr AJ, Rhodes MJC, Robins RJ, Walton NJ (1987) New routes to plant secondary products Bio/Technology 5:800–804
6. Kurz WGW, Chatson KB, Constabel F, Kutney JP, Choi LSL, Kolodziejczyk P, Sleigh SK, Stuart KL, Worth BR (1981) Alkaloid production in *Catharanthus roseus* cell cultures. VIII. Characterization of the PRL #200 cell line. Planta Med 42:22–31
7. Scott AI, Mizukami H, Hirato T, Lee SL (1980) Formation of catharanthine, akuammicine and vindoline in *Catharanthus roseus* suspension cells. Phytochemistry 19:488–489
8. Stöckigt J, Soll HJ (1980) Indole alkaloids from cell suspension cultures of *Catharanthus roseus* and *C. ovalis*. Planta Med 40:22–30

9. Constabel F, Gaudet-LaPrairie P, Kurz WGW, Kutney JP (1982) Alkaloid production in *Catharanthus roseus* cell cultures. XII. Biosynthetic capacity of callus from original explants and regenerated shoots. Plant Cell Rep 1:139–142
10. Fahn W, Grundlach H, Deus-Neumann B, Stöckigt J (1985a) Late enzymes in vindoline biosynthesis. Acetyl-CoA:17-O-deacetylvindoline 17-O-acetyl-transferase. Plant Cell Rep 4:333–336
11. Fahn E, Lausermair E, Deus-Neumann B, Stöckigt J (1985b) Late enzymes in vindoline biosynthesis. S-adenosyl-L-methionine: 11-O-demethyl-17-O-deacetylvindoline 11-O-methyltransferase, and unspecific acetylesterase. Plant Cell Rep 4:337–340
12. Balsevich J, DeLuca V, Kurz WGW (1986) Altered alkaloid pattern in dark grown seedlings of *Catharanthus roseus*. The isolation and characterization of 4 desacetoxyvindoline: a novel indole alkaloid and proposed precursor of vindoline. Heterocycles 24:2415–2421
13. DeLuca V, Balsevich J, Tyler RT, Eilert U, Panchuk BD, Kurz WGW (1986) Biosynthesis of indole alkaloids: developmental regulation of the biosynthetic pathway from tabersonine to vindoline in *Catharanthus roseus*. J Plant Physiol 125:147–156
14. DeLuca V, Balsevich J, Tyler RT, Kurz WGW (1987) Characterization of a novel N-methyltransferase from *Catharanthus roseus* plants. Plant Cell Rep 6:458–461
15. DeLuca V, Cutler A (1987) Subcellular localization of enzymes involved in indole alkaloid biosynthesis in *Catharanthus roseus*. Plant Physiol 85:1099–1102
16. Deus-Neumann B, Zenk MH (1984) A highly selective alkaloid uptake system in vacuoles of higher plants. Planta 162:250–260
17. Eilert U, Constabel F, Kurz WGW (1986) Elicitor stimulation of monoterpene indole alkaloid formation in suspension cultures of *Catharanthus roseus*. J Plant Physiol 126:11–22
18. Eilert U, DeLuca V, Constabel F, Kurz WGW (1987) Elicitor-mediated induction of tryptophan decarboxylase and strictosidine synthase activities in cell suspension cultures of *Catharanthus roseus*. Arch Biochem Biophys 254:491–497
19. Tyler RT, Kurz WGW, Panchuk BD (1986) Photoautotropic cell suspension cultures of periwinkle *Catharanthus roseus* (L.) G. Don: transition from heterotrophic to photoautotropic growth. Plant Cell Rep 5:195–198
20. Eilert U, DeLuca V, Kurz WGW, Constabel F (1987) Alkaloid formation by habituated and tumorous cell suspension cultures of *Catharanthus roseus*. Plant Cell Rep 6:271–274
21. DeLuca V, Alvarez Fernandez F, Campbell D, Kurz WGW (1988) Developmental regulation of enzymes of indole alkaloid biosynthesis in *Catharanthus roseus*. Plant Physiol 86:447–450
22. Sanger F, Nicklen S, Coulson AR (1977) DNA sequencing with chain terminating inhibitors. Proc Natl Acad Sci USA 74:5463–5467
23. Noé W, Mollenschott C, Berlin J (1984) Tryptophan decarboxylase from *Catharanthus roseus* cell suspension cultures: purification, molecular and kinetic data of the homogeneous protein. Plant Mol Biol 3:281–288
24. Eveleth DD, Gietz RD, Spencer CA, Nargang FE, Hodgetts RB, Marsh JL (1986) Sequence and structure of the dopa decarboxylase gene of *Drosophila*: evidence for novel RNA splicing variants. EMBO J 5:2663–2672
25. Morgan BA, Johnson WA, Hirsh J (1986) Regulated splicing produces different forms of dopa decarboxylase in the central nervous system and hypoderm of *Drosophila melanogaster*. EMBO J 5:3335–3342

Regulation of Capsaicin Synthesis in Cultured Cells of *Capsicum frutescens*

M.M. Yeoman, M.A. Holden, R.D. Hall, P.R. Holden, and S.S. Holland[1]

1 Introduction

Capsaicin belongs to a group of phenolic compounds the capsaicinoids. These molecules, which are the end products of metabolism, are responsible for the hot taste in the mature fruits of the chilli pepper, *Capsicum frutescens* (Yeoman et al. 1980). They have a modest commercial value and at today's prices one kilogram of the oleoresin would be worth about £450. Apart from the commercial implications there are certain features of capsaicin, its synthesis and accumulation, which form an attractive basis for those wishing to work with a system to explore the regulation of secondary product biosynthesis in general. For example, capsaicin, the predominant capsaicinoid in fruits, constitutes about 70% of the total with dihydrocapsaicin at about 28%. These capsaicinoids are only synthesized and accumulated in the maturing fruit at one particular stage of development, and have not been detected in other parts of the plant (Holden et al. 1987a). Indeed, all attempts to demonstrate capsaicin synthesis in leaves, stems or roots have so far failed (Holden et al. 1988a,b). From this it can be seen that a highly focussed approach is possible as the site and timing of capsaicin synthesis are known and the product is stable.

2 Studies on Attached Fruits

It can be seen from Fig. 1 that capsaicin is not present in fruits at an early stage of development. Unfortunately, however, there are no clear correlations between fruit length, fresh weight, colour and the onset of capsaicin synthesis (Hall et al. 1987). This imposed a difficulty in advancing the research until it was discovered that a loose correlation did exist between the age of the fruit and the start of capsaicin synthesis. Therefore, it was possible to select and mark populations of developing fruits and be reasonably sure of the point at which capsaicin began to be synthesized and accumulate. This was the procedure used throughout this investigation into capsaicin metabolism in chilli pepper fruits.

If radioactive precursors of the capsaicin biosynthetic pathway (Leete and Louden 1968), phenylalanine and valine, are injected into the loculus of the fruit

[1]Department of Botany, University of Edinburgh, Edinburgh EH9 3JH, Scotland, UK

Primary and Secondary Metabolism of Plant Cell Cultures, ed. by W.G.W. Kurz

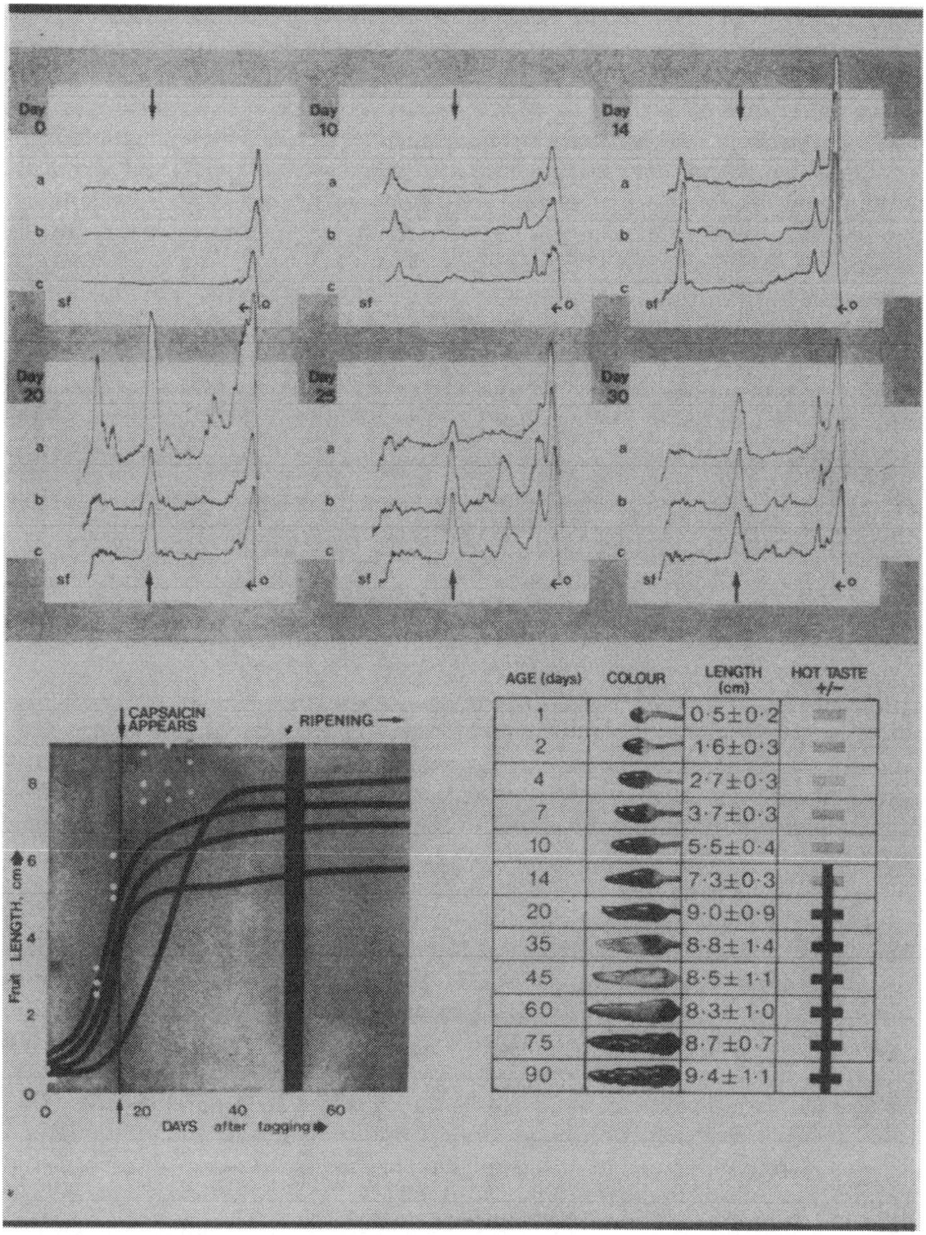

AGE (days)	COLOUR	LENGTH (cm)	HOT TASTE +/−
1		0·5±0·2	
2		1·6±0·3	
4		2·7±0·3	
7		3·7±0·3	
10		5·5±0·4	
14		7·3±0·3	
20		9·0±0·9	
35		8·8±1·4	
45		8·5±1·1	
60		8·3±1·0	
75		8·7±0·7	
90		9·4±1·1	

Fig. 1. Time course of capsaicin production in fruits of chilli: pepper, *Capsicum* frutescens. The diagram is compiled from optical density scans of autoradiographs of extracts of *Capsicum* fruits taken at 0, 10, 14, 20, 25 and 30 days after injection with the radioactive precursors. Separation of capsaicin (*arrows*) was achieved on TLC plates of SILICAGEL + 1% $AgNO_3$ + 1% HBO_3 developed in 50:50 chloroform: ethyl acetate

then they will only be incorporated into capsaicin at one particular stage of development. This suggests either that the pathway is incomplete or that precursors and the enzymes of the pathway are in different cellular compartments not accessible to the radiolabel. Of course, whatever the developmental state the greater part of the radioactivity is incorporated into protein, wall components including lignin, and into cell membranes (Lindsey 1986, Hall and Yeoman, unpublished results). However, the proportions of the added precursors incorporated into cellular components decrease as the fruit matures and begins to accumulate capsaicin. Even at the height of capsaicin synthesis the amount of radioactivity which finds its way into capsaicin is still only a small proportion of the total amount added (Hall et al. 1987). A detailed analysis of the incorporation of radioactivity into different components during capsaicin synthesis provides further important information about the process. For example, radioactivity appears in capsaicin less than 5 min after feeding ^{14}C phenylalanine to the attached fruit. Also, it is not possible using a single fruit and HPLC to detect the free phenolics which constitute the pathway, only cinnamic acid can occasionally be shown to be present (Hall et al. 1987). However, ^{14}C phenylalanine fed to the fruit at this time is followed by the appearance of radioactivity in all of the intermediates as shown by the radioactive spots on an autoradiograph of a two-dimensional TLC (Holden et al. 1987a). This shows the high specific activity of these intermediates, suggesting rapid turnover in a small metabolic pool. On the other hand, free esters and glycosides of these phenolics are detectable in significant amounts using both HPLC and TLC and if the wall-bound phenolics are examined, it can be seen that they are also present in significant amounts (Lindsey 1986). However, these wall-bound intermediates are probably not available for the further synthesis of capsaicin because experiments in which labelled wall fragments were fed to cultured cells did not provide a source of intermediates for capsaicin synthesis (Holden et al. 1987b).

In a system where there is intense competition for intermediates between protein, wall, membranes and capsaicin synthesis any treatment which alters the balance between these components tends to increase the flow of metabolites down the phenylpropanoid pathway (Yeoman 1987). Here, the importance of phenylalanine ammonia-lyase (PAL), the first enzyme of the limiting part of the pathway, would appear to be paramount as it is the enzyme which controls the metabolite flow into the phenylpropanoid pathway. Extensive studies with fruits at various stages of development have failed to show any significant differences in PAL activity associated with the onset of capsaicin synthesis and accumulation (Table 1). From this it may be presumed that the process is not controlled at the top of the pathway but that some control must be exerted later in the biosynthetic sequence. Very preliminary studies on capsaicin synthase, the ultimate enzyme in the sequence in developing fruits, have shown, perhaps surprisingly, that enzyme activity can be detected long before capsaicin is synthesized and therefore the role of compartmentation may be an important issue. However, the fact that capsaicin synthase cannot be detected elsewhere in the plant suggests that the synthesis of the enzyme must be developmentally regulated in the fruit but that the onset of capsaicin synthesis may be induced through the provision of substrates for the enzyme, i.e. substrate limitation. It is conceivable, indeed likely, that the dramatic

Table 1. Phenylalanine ammonia-lyase activity and capsaicin content of *Capsicum frutescens* at different stages of fruit ripening

	PAL activity in comparison with fruit during capsaicin synthesis	Capsaicin in comparison with fruit during active synthesis
	(%)	(%)
Fruit during synthesis	100	100
Fruit pre-capsaicin synthesis	100	0
Fruit post-capsaicin synthesis	93	113

changes that take place in the walls of the pericarp cells during ripening (Gross et al. 1986) would provide large amounts of precursors for the synthesis of capsaicin and that the availability of these substances would remove substrate limitation. Also, as it has been shown that cell wall fragments may display biological activity (Albersheim and Darvill 1985; Eilert 1987), it is possible that components within the walls once freed from the binding matrix could elicit changes leading to the onset of capsaicin synthesis.

How then can we use the information obtained from studies on capsaicin synthesis in the fruit to increase capsaicin yield in cell cultures? Before directly addressing the point it is necessary to set out the background to the cell culture research.

3 Studies on Cells in Culture

Studies on cell cultures offer clear advantages over investigations with attached fruits primarily because of their ease of experimental manipulation, for example in studies involving cell immobilization, nutrient limitation and elicitation. An important fact that emerges from these investigations on chilli pepper cells in culture is that without exception when ^{14}C phenylalanine is fed to callus, suspension cultures or immobilized cells, radioactivity can be detected in capsaicin within minutes (cf. fruit). This is particularly surprising when it is pointed out that many of these cultures produce vanishingly small amounts of capsaicin during the radioactive feeding experiments. However, these results provide unequivocal evidence for the existence and operation of the complete pathway in cell cultures, although low yields are highly suggestive of the inefficient conversion or lack of precursors in these cells (Hall et al. 1987). This is of course a common phenomenon in many cell systems producing, or with the potential to produce, a designated product (Yeoman et al. 1982a). It is the solution to this problem that we have sought over the last 10 years and the approach employed is described below.

4 Manipulation of Cultures

The relationship between differentiation and product yield has dominated our thinking for some time (Yeoman et al. 1980, 1982b; Lindsey and Yeoman 1985). It is clear that in most cases, but not all, the secondary product, or products, characteristic of the parent plant tend to be associated with the down-turn in culture growth and also with biochemical and structural differentiation, or both (Lindsey and Yeoman 1985, Collin 1987). There are, however, exceptions where the product is produced during growth and is closely related to growth, e.g. Betacyanin in *Beta vulgaris*. As described earlier the product is accumulated in the fruit, a highly differentiated organ, around the down-turn of fruit growth and therefore the logical approach with the cultures was to place the cells under conditions which would promote differentiation and slow down growth. It was this approach which led to the development of a series of immobilized cell systems from the "flat-bed" (Yeoman et al. 1978; Lindsey and Yeoman 1983) through alginate and agarose entrapment to the polyurethane systems used in our laboratory today (Lindsey et al. 1983; Lindsey and Yeoman 1987). These systems were not only successful in raising the yield of capsaicin from cultured cells but have also provided excellent systems for exploring the regulation of the biosynthetic pathway leading to capsaicin and it is this aspect which will dominate the remainder of this chapter rather than the commercial aspects (Yeoman and Lindsey 1985; Yeoman 1987).

5 Measurement of Flux in the Biosynthetic Pathway

A close examination of the composition of immobilized cells before, and during the period when capsaicin is accumulated, shows that the situation with respect to free phenols, phenols as esters or glycosides and phenols bound to the wall is remarkably similar to that previously encountered with the developing fruit. Similarly, after brief periods of feeding with ^{14}C phenylalanine all of the intermediates of the phenylpropanoid pathway can be seen on autoradiographs as clear spots (Hall et al. 1987). There are in addition to these known phenolics others which have been detected by HPLC which have so far not been identified. Also, as in the fruit, free phenolics as esters and glycosides, and bound in the cell wall, are detectable at low levels by TLC and HPLC.

When relatively undifferentiated cells in suspension and callus are compared with immobilized cells, some differences in the free phenolics are observed, namely more unknown compounds are present in the immobilized cells (Fig. 2). All of this evidence presents a picture in which the free phenolic pools are very small and rates of turnover are high, as in the developing fruit.

Serious attempts to measure the rates of flux down the biosynthetic pathway to capsaicin have been hampered by the difficulty of measuring the amounts of the intermediates at any one time and hence their specific activities and how these change following the addition of the radioactive precursor. However, this point will be discussed later.

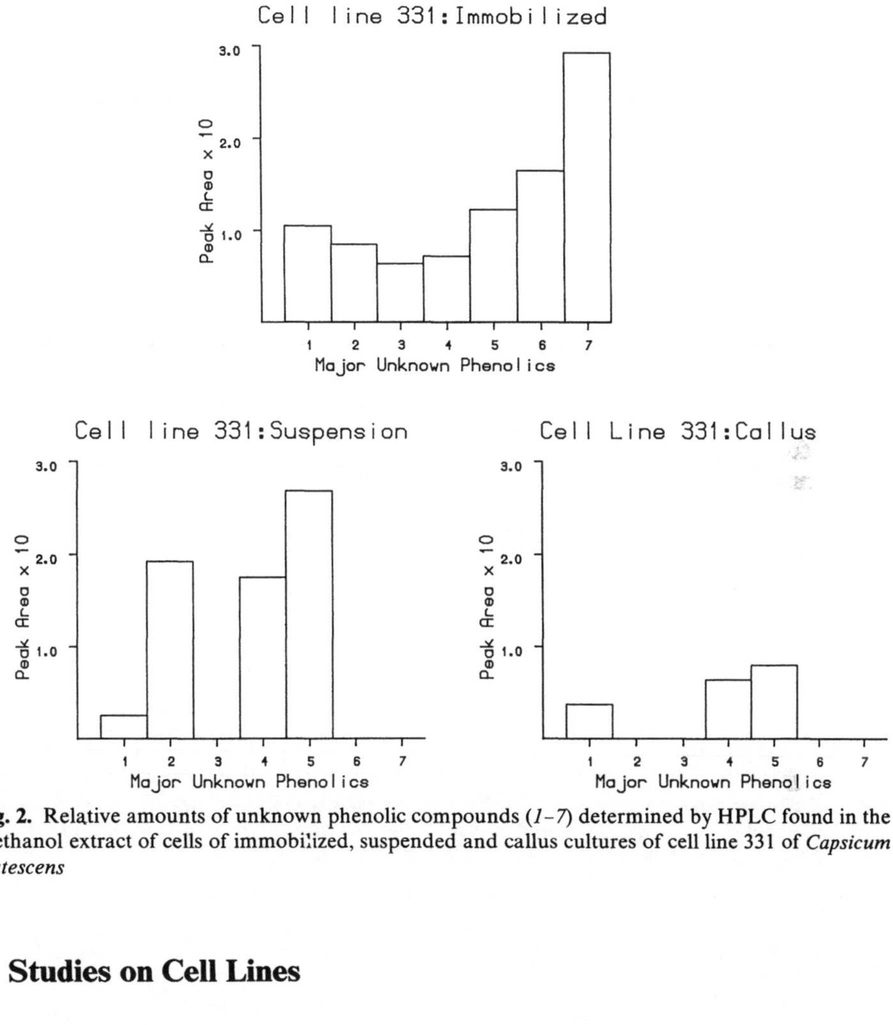

Fig. 2. Relative amounts of unknown phenolic compounds (*1–7*) determined by HPLC found in the methanol extract of cells of immobilized, suspended and callus cultures of cell line 331 of *Capsicum frutescens*

6 Studies on Cell Lines

An interesting and perhaps important piece of information has been obtained from a study of the free phenolic component of cell lines raised from single cells. These cell lines show different morphological and growth characteristics as well as differing abilities to accumulate capsaicin (Lindsey and Yeoman 1985). However, when a close comparative study was made of the free phenolics present, it was observed that some of the cell lines not accumulating capsaicin produced certain other pathway intermediates, for example one line accumulated vanillin, while another accumulated ferulic acid (Fig. 3). Clearly, there are changes in the metabolic pattern of these selected cell lines which demonstrate that the pathway can be altered and remain altered. This evidence, although fragmentary, also supports the view that it is possible to select cell lines in which the regulatory steps may be different from those of other cell lines and in the plant. These cell lines with altered metabolism provide useful tools for the study of phenylpropanoid metabolism.

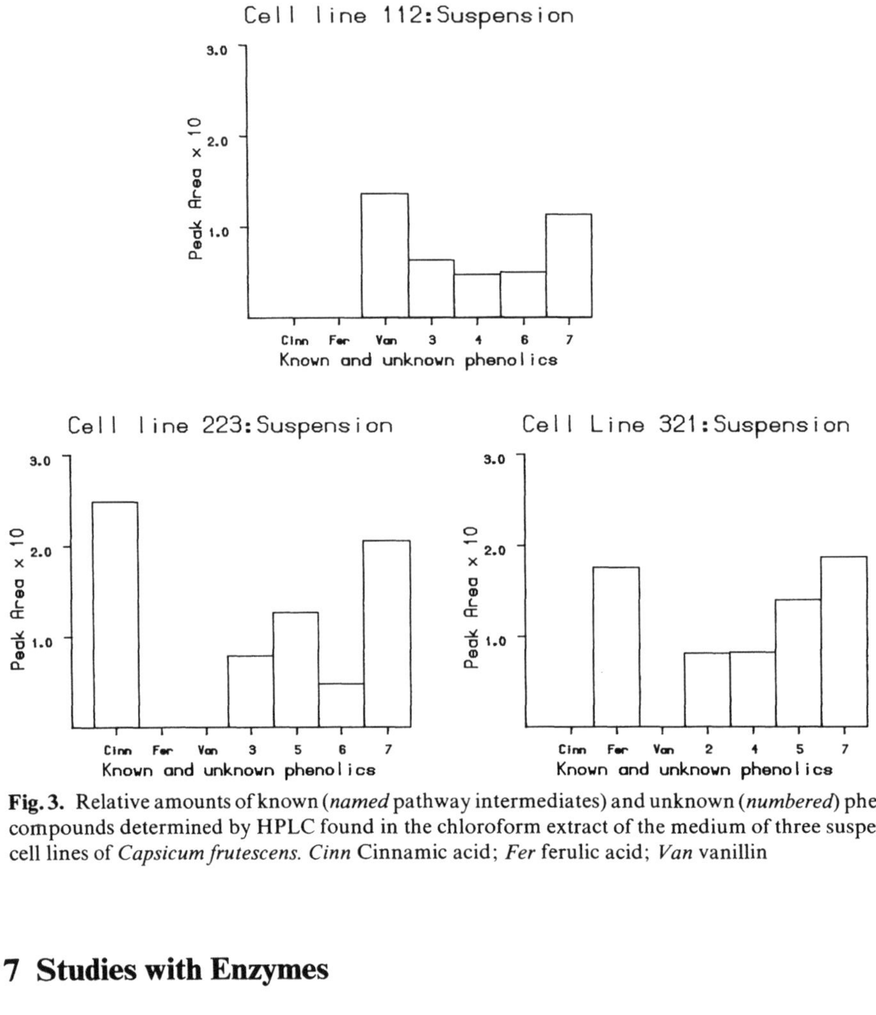

Fig. 3. Relative amounts of known (*named* pathway intermediates) and unknown (*numbered*) phenolic compounds determined by HPLC found in the chloroform extract of the medium of three suspended cell lines of *Capsicum frutescens*. *Cinn* Cinnamic acid; *Fer* ferulic acid; *Van* vanillin

7 Studies with Enzymes

It was shown previously that PAL activity remained constant during fruit development and did not change at the onset of capsaicin synthesis. In a series of experiments where PAL activity was measured in immobilized cells synthesizing capsaicin and in suspended cells in which capsaicin synthesis is lower, no difference was detected in the activity of the enzyme or in the second enzyme in the pathway cinnamate-4-hydroxylase (Table 2). A similar result was obtained when a comparison was made between immobilized cells on growth medium and under nutrient limitation on a production medium. The general conclusion here is the same as that reached already with fruits, i.e.the first and second enzymic steps in the phenylpropanoid pathway are not limiting capsaicin production under these conditions.

Preliminary studies on capsaicin synthase in cultured cells show the enzyme to be present in suspended cells not actively synthesizing capsaicin and also show the presence of an enzyme oxidatively deaminating vanillylamine to vanillin.

Table 2. A time course experiment to show the effect of immobilization on the activities of phenylalanine ammonia-lyase (PAL) and cinnamate 4-hydroxylase (C4-H) and on the number of labelled phenolics produced upon feeding [^{14}C]phenylalanine[a]

Time after subculture (h)	Suspended cells			Immobilized Cells		
	SA (μKat kg^{-1} protein)		No. labelled phenolics	SA (μKat kg^{-1} protein)		No. labelled phenolics
	PAL	C4-H		PAL	C4-H	
0	1.00 ± 0.19	nd	12	0.99 ± 0.15	nd	15
13	39.69 ± 5.24	nd	12	28.49 ± 3.87	nd	19
26	10.03 ± 3.29	8.44 ± 5.25	13	8.78 ± 1.77	7.36 ± 2.55	16

[a] Enzyme activity results represent the mean of three replicates ± SE. SA = Specific activity.

8 The Fate of Added Non-Radioactive Precursors

A major breakthrough in raising the concentration of capsaicin in cultures of *Capsicum* was the use of iso-capric acid as a precursor. As can be seen from the pathway (Yeoman et al. 1980) iso-capric acid is very close to the end of the biosynthetic pathway and gives rise to the product (8-methylnonenoic acid) which condenses with vanillylamine to produce capsaicin (Fig. 4). Using 5 mM iso-capric acid quite spectacular results were obtained with the cell lines in use at the time (Lindsey and Yeoman 1984a,b). However, it was also discovered that all cell lines do not respond as effectively to the addition of iso-capric acid but do react positively to the addition of intermediates of the phenylpropanoid pathway. The general conclusion was that the closer the added precursor is to the end of the pathway, the higher the yield of capsaicin, so that ferulic acid was much more efficient than cinnamic acid or phenylalanine. It must also be pointed out, however, that cultured cells of *Capsicum* do not retain their viability after prolonged exposure to relatively modest concentrations of these precursors, e.g. 1–5 mM. It was also observed when feeding phenylpropanoids that the major conversions were to the immediate precursor and product, for example cinnamic acid would be converted to phenylalanine and coumaric acid and ferulic acid would be converted to vanillin and caffeic acid. Feeding experiments with vanillylamine have been much less successful as the molecule is not easily taken up by the cells and is toxic at low concentrations.

9 Effects of Added Capsaicin

At an early stage in this study it was noted that almost all of the capsaicin produced appears in the culture medium with only traces remaining within the cells (Lindsey and Yeoman 1984b). It was also observed that if the culture medium around the cells was replaced periodically, then the yield of capsaicin was markedly elevated. Lindsey (1986) showed convincingly that the addition of capsaicin to immobilized

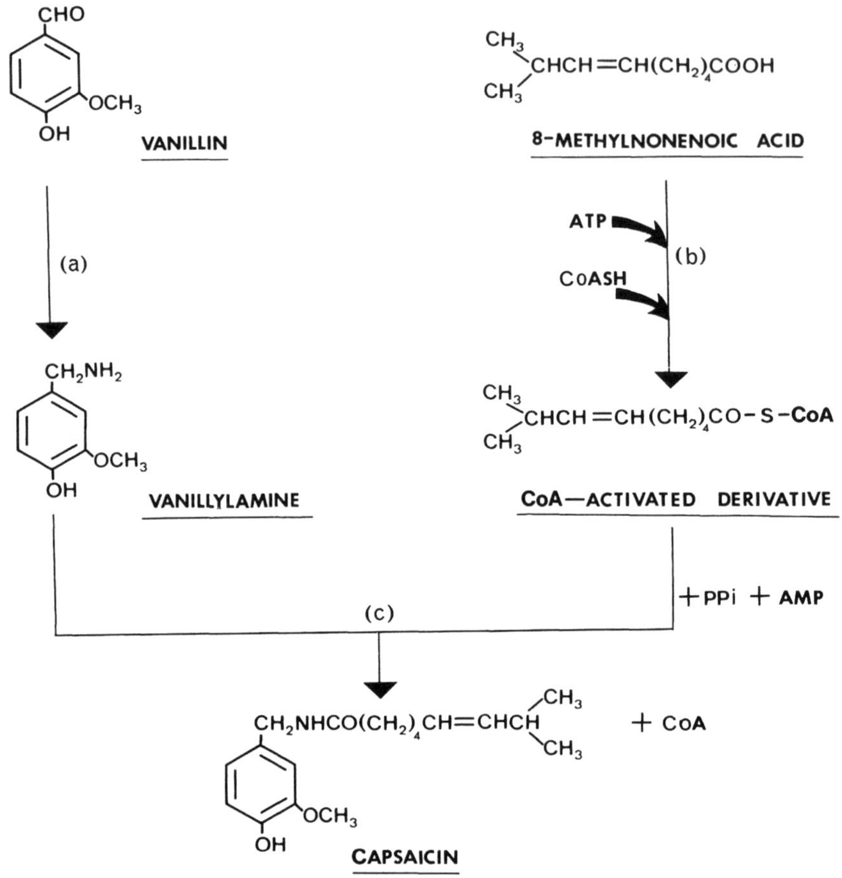

Fig. 4. Proposed terminal steps in the biosynthetic pathway. *a* Transaminase; *b* acyl-CoA synthase; *c* acyltransferase

cells synthesizing capsaicin reduced the extent of incorporation of ^{14}C phenylalanine into capsaicin. Subsequently, Mavituna (personal communication) has demonstrated that the continuous removal of capsaicin into vegetable oil improves capsaicin yield from immobilized cells in a bioreactor system. Clearly, from this evidence capsaicin can inhibit its own synthesis by a negative feedback mechanism. Further, experiments with crude preparations of capsaicin synthase (Holland, unpublished results) have shown that this enzyme is strongly inhibited, as might be expected, by capsaicin in the reaction mixture: so far, there is no evidence that capsaicin is metabolized by cultured cells or in crude cell homogenates. An experimental approach has been described (Newsholme and Start 1973) in which the non-equilibrium enzymes in a pathway can be identified by feeding a radioactive precursor down the pathway in the presence of the end product. As such non-equilibrium enzymes are likely to provide regulatory points in a pathway, the results of such experiments can supply useful information which may lead to a better understanding of how the pathway is regulated. Clearly, the ideal situation

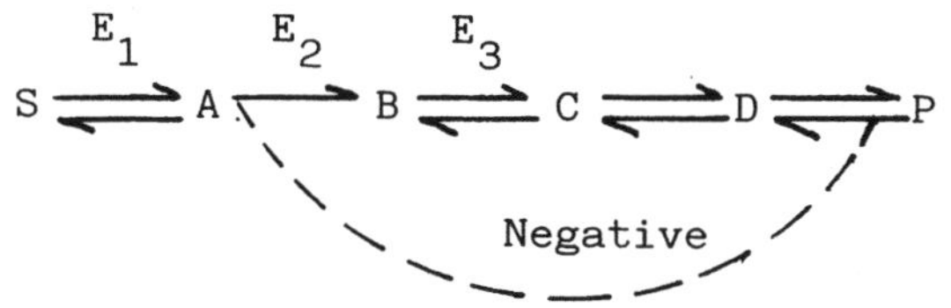

Fig. 5. Diagram showing a simple example of negative feedback regulation. *S* substrate; *P* product; enzyme converting $A \rightarrow B$ is non-equilibrium so that the activity of this enzyme limits overall flux from *S* to *P* (Newsholme and Start 1973)

(see Fig. 5) is one in which the pathway is linear without significant side reactions, the end of the phenylpropanoid pathway only approximates to this ideal situation and, therefore, interpretation of the results obtained must be tempered accordingly. In such a study Hall, Holden and Yeoman (unpublished results) have shown that in a complex time course experiment in which ^{14}C phenylalanine was fed to immobilized chilli pepper cells accumulating capsaicin, the pattern of incorporation of the precursor into the intermediates of the pathway was modified significantly in the presence of capsaicin. From these results it is evident that, firstly, incorporation of radioactivity into capsaicin is strongly depressed and, secondly, that there are significant differences between the extent of incorporation into coumaric acid, caffeic acid and ferulic acid with the largest effect on caffeic acid, the substrate for the enzyme catalyzing its conversion to ferulic acid. However, there is also an effect on the labelling pattern of ferulic acid which is enzymically converted to vanillin. While these differences are clear, they are difficult to interpret with any certainty but at least are consistent with the pathway being controlled at the end of the biosynthetic sequence rather than at the beginning or in the middle (Fig. 6).

10 Studies with Biotic Elicitors

It is a well-known but poorly understood phenomenon that preparations of fungal mycelia or spores can elicit changes in the secondary metabolism of plants and cell cultures derived from plants (Funk et al. 1987; Hahlbrock et al. 1981; Holden et al. 1988a,c). This approach has been used as a further means to probe the biosynthetic pathway leading to capsaicin. It has been shown that the addition of a fungal elicitor increases the amount of ^{14}C phenylalanine incorporated into capsaicin, increases the yield of capsaicin and brings about a rise in PAL activity (Holden et al. 1988a,b). There is also an associated increase in the activity of the phenylpropanoid pathway leading to the deposition of polyphenolics and lignin into the cell wall in what perhaps is a defence reaction. Here, it would appear is an example where increasing the flow of metabolites down the phenylpropanoid pathway from phenylalanine not only increases wall building but also provides additional precursors, not required for wall synthesis, to the production of capsaicin. This raises the possibility of a composite strategy for capsaicin production involving immobilization, precursor feeding, product removal and elicitation.

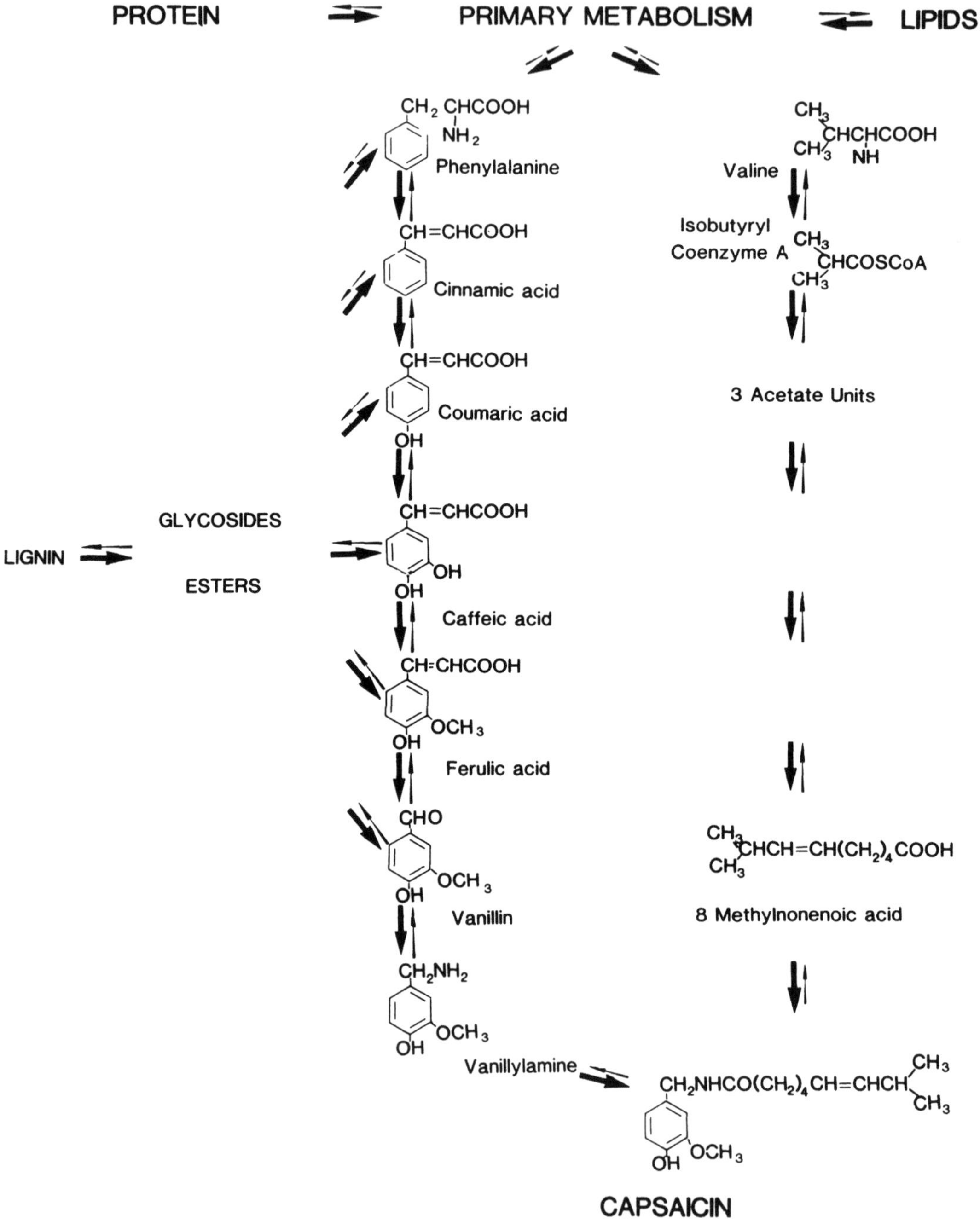

Fig. 6. The capsaicin biosynthetic pathway under idealized conditions of product synthesis. The *bold arrows* represent the situation resulting in optimized capsaicin synthesis. The *fine arrows* represent the situation resulting in low capsaicin synthesis

11 Concluding Remarks

Evidence has been presented which is consistent with a situation in which the regulation of capsaicin synthesis and accumulation in cell cultures occurs towards the end of the biosynthetic pathway but that, as in the fruit, the provision of precursors will determine the product yield (Fig. 6). Labelling experiments with ^{14}C phenylalanine in both the fruit and in all of the cell cultures examined so far have shown that the biosynthetic pathway leading to capsaicin is operative, however, capsaicin does not always accumulate. As it is known that capsaicin is not degraded by pepper cells either in vivo or in vitro and apparently cannot be esterified or glycosylated or bound into the cell wall, failure to accumulate capsaicin must be due to substrate limitation, or compartmentation of pathway enzymes and substrates, or the presence of a major regulatory step towards the end of the pathway or combinations of these. The fact that precursor feeding is most effective in slow-growing, aggregated cultures suggests that changes intrinsic to this kinetic state either relieve compartmentation or stimulate a change in the regulatory step(s) in the pathway, which may or may not result in enhanced activity of these regulatory enzymes. These tentative conclusions may be tested further (1) by comparing the activities of the non-equilibrium enzymes in cell lines accumulating different amounts of capsaicin or in immobilized cells before and after induction; (2) by genetic manipulation of the pathway. Of course, the modification of the pathway by the amplification of the non-equilibrium (regulatory) enzymes could produce quite unexpected results. Pathway regulation is complex and is the result of an evolutionary process over several million years. Indeed, it is possible that each pathway has the means to compensate for changes in enzyme levels, which occur by genetic change, in order to protect the metabolic balance. This could make it difficult to genetically engineer biosynthetic pathways in higher plants. However, before answers can be provided to these questions, the critical experiments must be performed to test the possibility of producing, by genetic manipulation, cells "dedicated" to the synthesis and accumulation of a designated metabolite. Current advances in molecular biological techniques make this approach well-suited to the exploitation of plant cell cultures in the production of valuable secondary products.

Acknowledgements. The authors wish to thank Albright and Wilson Ltd., Glaxo Group Research Ltd., the S.E.R.C. for financial support, and Mrs. E. Raeburn and Mrs. J. Summers for typing the manuscript.

References

Albersheim P, Darvill AG (1985) Oligosaccharides. Sci Am 253:44–50

Collin HA (1987) Determinants of yield of secondary products in plant tissue cultures. Adv Bot Res 113:146–187

Eilert U (1987) Elicitation: methodology and aspects of application. In: Constabel F, Vasil IK (eds) Cell culture and somatic cell genetics of plants, vol 4. Cell culture in phytochemistry. Academic Press, New York London, pp 153–196

Funk C, Gugler K, Brodelius P (1987) Increased secondary product formation in plant cell suspension cultures after treatment with a yeast carbohydrate preparation (elicitor). Phytochemistry 26:401–405

Gross KC, Watada AE, Kang MS, Kim SD, Lee SW (1986) Biochemical changes associated with the ripening of hot pepper fruit. Physiol Plant 66:31–36

Hahlbrock K, Lamb CL, Purwin C, Ebel J, Fautz E, Schafer E (1981) Rapid response of suspension-cultured parsley cells to elicitors from *Phytophthora megasperma* var. sogae. Plant Physiol 67:768–777

Hall RD, Holden MA, Yeoman MM (1987) The accumulation of phenylpropanoid and capsaicinoid compounds in cell cultures and whole fruit of the chilli pepper *Capsicum frutescens* Mill. Plant Cell Tissue Org Cult 8:163–176

Holden MA, Hall RD, Lindsey K, Yeoman MM (1987a) Capsaicin biosynthesis in cell cultures of *Capsicum frutescens*. In: Webb C, Mavituna F, Faria JJ (eds) Process possibilities for plant and animal cell cultures. Horwood, Chichester, pp 45–63

Holden MA, Hall RD, Yeoman MM (1987b). Incorporation of cell-wall bound cinnamic acid derivatives into capsaicin in cultured cells of the chilli pepper *Capsicum frutescens*. In: Webb C, Mavituna F, Faria JJ (eds) Process possibilities for plant and animal cell cultures. Horwood, Chichester, pp 297–300

Holden MA, Holden PR, Yeoman MM (1988a) The effects of fungal elicitation on secondary metabolism in cell cultures of *Capsicum frutescens*. In: Robins RJ, Rhodes MJC (eds) Manipulating secondary metabolism in culture. CUP (in press)

Holden MA, Holden PR, Yeoman MM (1988b) Elicitation of secondary product formation in *Capsicum frutescens* cultures. In: Robins RJ, Rhodes MJC (eds) Manipulating secondary metabolism in culture. CUP

Holden MA, Holden PR, Yeoman MM (1988c) Altered phenylpropanoid metabolism in cultures of *Capsicum frutescens* following elicitation. In: Boccon-Gibod J (ed) Cell culture techniques applied on plant production and plant breeding. ENTTAH, Angers, Fr (in press)

Leete E, Louden LCL (1968) Biosynthesis of capsaicin and dihydrocapsaicin in *Capsicum frutescens*. J. Am Chem Soc 90:6837–6841

Lindsey K (1986) Incorporation of ^{14}C Phenylalanine and ^{14}C cinnamic acid into capsaicin in cultured cells of *Capsicum frutescens*. Phytochemistry 25:2793–2801

Lindsey K, Yeoman MM (1983) Novel experimental systems for studying the production of secondary plant metabolites by plant tissue culture. In: Mantell SH, Smith H (eds) Plant biotechnology. Soc Exp Biol Sem Ser, vol 18. Univ Press, Cambridge, pp 39–66

Lindsey K, Yeoman MM (1984a) The synthetic potential of immobilised cells of *Capsicum frutescens* Mill. cv. Annuum. Planta 162:495–501

Lindsey K, Yeoman MM (1984b) The viability and biosynthetic activity of cells of *Capsicum frutescens* Mill. cv. Annuum immobilised in reticulate polyurethane. J Exp Bot 35:1684–96

Lindsey K, Yeoman MM (1985) Immobilised plant cell culture systems. In: Neumann KH (ed) Primary and secondary metabolism of plant cell cultures. Springer, Berlin Heidelberg New York Tokyo, pp 304–315

Lindsey K, Yeoman MM (1987) Techniques for the immobilisation of plant cells. In: Mosbach K (ed) Immobilised enzymes and cells: methods in enzymology, vol 135. Academic Press, New York London, pp 410–421

Lindsey K, Yeoman MM, Black G, Mavituna F (1983) A novel method for the immobilisation and culture of plant cells. FEBS Lett 155:143–149

Newsholme EA, Start C (1973) Regulation in metabolism. John Wiley & Sons, Chichester New York, Brisbane, Toronto

Yeoman MM (1987) Bypassing the plant. Ann Bot 60 (Suppl):157–174

Yeoman MM, Lindsey K (1985) The scientific and commercial potential of immobilised plant cells. In: Advances in fermentation 85. Turret-Wheatland, UK, pp 87–92

Yeoman MM, Fagandini DAA, Childs AF (1978) Callus culture in nutrient flow. Br Pat BP 32185/78

Yeoman MM, Miedzybrodzka MB, Lindsey K, McLauchlan WR (1980) The synthetic potential of cultured plant cells. In: Sala F, Parisi B, Cella R, Ciferri O (eds) Plant cell cultures: results and perspectives. Elseveir-North Holland, pp 327–343

Yeoman MM, Lindsey K, Miedzybrodzka MB, McLauchlan WR (1982a) Accumulation of secondary products as a facet of differentiation in plant cell and tissue cultures. In: Yeoman MM, Truman DES (eds) Differentiation in vitro. Br Soc Cell Biol Symp vol 4. Cambridge University Press, Cambridge and London, pp 65–82

Yeoman MM, Lindsey K, Hall R (1982b) Differentiation as a prerequisite for the production of secondary metabolites. In: Proc Plant cell culture conf. Oyez, London, pp 1–7

The Formation and Regulation of 1-Methoxycanthin-6-one Production in *Ailanthus altissima* Cell Suspension Cultures

M. F. Roberts[1,4], C. A. Hay[2], L.A. Anderson[3], and J. D. Phillipson[1]

1 Introduction

Ailanthus altissima (Mill) Swingle. (syn. *A. glandulosa* Desf.), a deciduous tree of Chinese origin (the Tree of Heaven), has been introduced into North America and Europe where it has a long history of use in traditional medicine. *A. altissima* is noted as an antibacterial, anthelmintic, amoebicide and insecticide [1,2,3,4]. In recent years it has been used as an ornamental tree in urban areas throughout Europe and North America [4]. Phytochemical investigations of *Ailanthus altissima* have demonstrated compounds with biological activities. The most notable are quassinoids, of interest as potential new anticancer [5] and antimalarial agents [6]. *Ailanthus* also contains two categories of indole alkaloids: the non-iridoid β-carbolines and the canthin-6-ones. There are, however, conflicting reports on the anticancer activity of these latter alkaloids [7,8]; canthin-6-one has some antimicrobial activity [9] and the canthin-6-ones are strongly inhibitory to c-AMP phosphodiesterase [10].

Although there is one report of the isolation of the quassinoid, ailanthone [11], cell cultures of *A. altissima* are remarkable for their high yields of canthin-6-ones (100-fold greater than the plant), and cell suspension cultures developed in our laboratories produce 1-methoxycanthin-6-one (7) as a major constituent with canthin-6-one (5) and 1-hydroxycanthin-6-one (6) as minor constituents (Fig. 1). These cultures were used as a model system for the production of alkaloids [12–15].

2 Development of *A. altissima* Cell Cultures High Yielding in 1-Methoxycanthin-6-one

Callus tissue and cell suspension cultures of *A. altissima* were developed as detailed in [12 and 13]. Examined microscopically, these cultures were found to consist of undifferentiated cells, either as single cells or in small aggregates of two to eight cells. Staining with Dragendorff's reagent [16] demonstrated the presence of both

[1]Department of Pharmacognosy, The School of Pharmacy, London University, 29/39 Brunswick Square, London, WC1N 1AX, England
[2]Department of Botany, University of Toronto, Toronto, Ontario, M5S 1A1, Canada
[3]Department of Health and Social Security, Market Towers, 1 Nine Elms Lane, London, SW8 5NQ, England
[4]To whom correspondence should be addressed

Primary and Secondary Metabolism of Plant Cell Cultures, ed. by W.G.W. Kurz

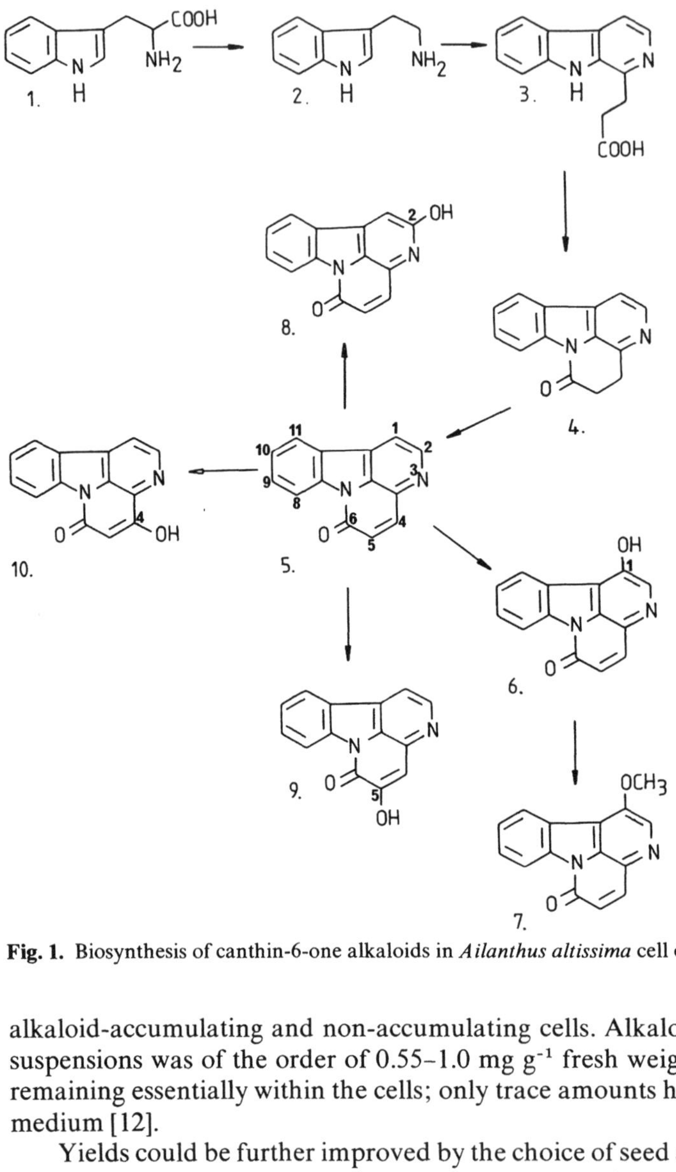

Fig. 1. Biosynthesis of canthin-6-one alkaloids in *Ailanthus altissima* cell cultures

alkaloid-accumulating and non-accumulating cells. Alkaloid production by cell suspensions was of the order of 0.55–1.0 mg g^{-1} fresh weight, with the alkaloids remaining essentially within the cells; only trace amounts have been found in the medium [12].

Yields could be further improved by the choice of seed and basal medium for the development of callus and suspension cultures [13].

3 The Biosynthesis and Regulation of 1-Methoxycanthin-6-one Production

Until recently, little was known of the biosynthesis of the canthin-6-one alkaloids. They were thought to be derived either from a β-carboline intermediate with the addition of a C-2 subunit derived from acetate or pyruvate or from tryptamine with the incorporation of a C-4 unit [17]. It was suggested by Crespi-Perellino et al. [18] that canthin-6-one may be derived from tryptophan via β-carboline-1-propionic acid (3) and either 4,5-dihydrocanthin-6-one (4) or β-carbolyl-1-acrylic acid. Their large-scale experiments, utilizing (DL)-[methylene ^{14}C]-tryptophan and (L)-[methylene ^{14}C]-tryptophan, demonstrated the incorporation of ^{14}C from tryptophan as follows: When the canthin-6-one alkaloids were isolated, 70% of the ^{14}C-label occurred in canthin-6-one (5) which was 70% of the total alkaloid fraction. The rest of the ^{14}C-label was distributed among 4-hydroxycanthin-6-one (10), 5-hydroxycanthin-6-one (9), 4,5-dihydrocanthin-6-one (4), canthin-6-one-3*N*-oxide and 1-methoxycanthin-6-one-3*N*-oxide (Fig. 1). The existence and label in 4,5-dihydrocanthin-6-one indicate that β-carbolyl-1-acrylic acid was not involved in the formation of canthin-6-one alkaloids.

In contrast to the cell cultures developed by Crespi-Perellino et al. [18,19], *A. altissima* cell cultures developed in our laboratories produced 1-methoxycanthin-6-one as the major constituent and 1-hydroxycanthin-6-one and canthin-6-one as minor constituents.

Studies of the growth cycle of *A. altissima* cell suspensions (Fig. 2) [13] demonstrated that the production of canthin-6-one commences around day 16 and reaches a maximum around day 24. The level of 1-methoxycanthin-6-one, on the other hand, remains constant until day 22 when rapid production of this latter alkaloid occurs until day 28, at which stage maximum accumulation is reached. The yields of 1-methoxycanthin-6-one (62.5 mg l^{-1}) are always significantly greater than those of the parent alkaloid canthin-6-one (20 mg l^{-1}). The level of isolatable

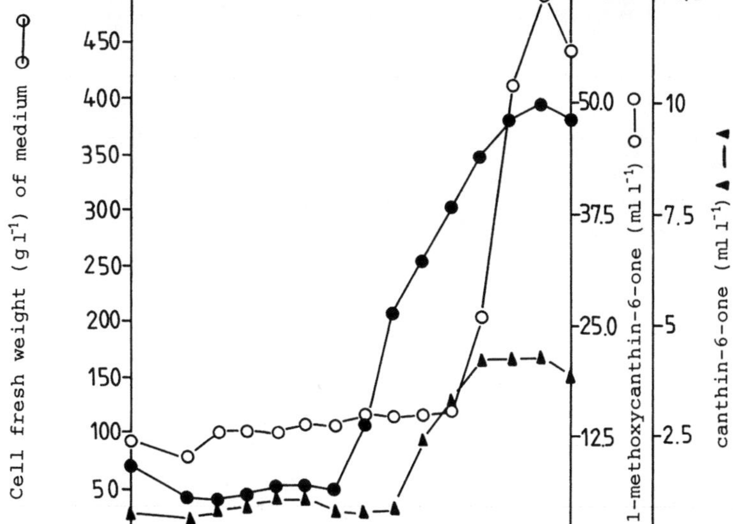

Fig. 2. The accumulation of canthin-6-one and 1-methoxycanthin-6-one in *Ailanthus altissima* cell suspension cultures. Details of isolation and analysis of alkaloids are given in (12, 14)

1-hydroxycanthin-6-one remained low throughout the growth cycle suggesting that the hydroxylase required for the conversion of canthin-6-one to 1-hydroxycanthin-6-one has a regulatory effect on the conversion of canthin-6-one to 1-methoxycanthin-6-one. The methyltransferase required for the conversion of 1-hydroxycanthin-6-one to 1-methoxycanthin-6-one would appear to be operative throughout most of the growth cycle.

4 The Utilization of Tryptophan and Tryptamine in the Production of Canthin-6-one Alkaloids

4.1 L-Tryptophan

L-tryptophan, fed to suspension cultures, was rapidly taken up by the cells (80% in 60–120 min) and enhancement of alkaloid yield was observed [14].

In time course studies (Fig. 3) L-[methylene^{14}C]-tryptophan, fed to suspension cultures at day 0 of the growth cycle, was shown to be well incorporated into the canthin-6-one alkaloids which were found to be labelled after 24 h. Compared with the continuous accumulation of canthin-6-one (Fig. 2), the label in the alkaloids fluctuated during the time course with an increase at day 16 (Fig. 3) in the label of canthin-6-one corresponding with the onset of exponential growth. Prior to the onset of exponential growth, the label found in canthin-6-one and 1-methoxycanthin-6-one was similar and around 3–4% of the total label found in the cells. With the onset of exponential growth, whilst rapid increase in the ^{14}C label in canthin-6-one occurs, 1-methoxycanthin-6-one shows an apparent loss of ^{14}C

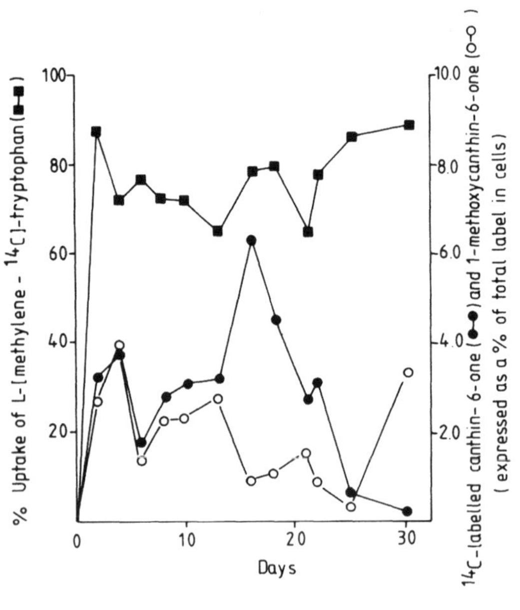

Fig. 3. The incorporation of radiolabel from L-[methylene ^{14}C]-tryptophan into canthin-6-one and 1-methoxycanthin-6-one by *Ailanthus altissima* cell suspension cultures. Cell suspensions fed L-[methylene ^{14}C]-tryptophan 1 μCi 40 ml^{-1} (specific activity: 50 μCi mmol^{-1}) to cell cultures 3 g 40 ml^{-1}) at day 0. Details of extraction and analysis of alkaloids are given in (12, 14)

label. Similarly, ^{14}C label is apparently lost from canthin-6-one (drops below 3%) after day 16 of the growth cycle. Compared with the continuous accumulation of the canthin-6-one alkaloids shown in Fig. 2, the results from feeding L-[methylene ^{14}C]-tryptophan to the cell suspensions of *A. altissima* suggest a complex turnover of alkaloids during the growth cycle. This pattern is moderated by changes in metabolic pools of tryptophan and variations in activity of the key enzymes during the growth cycle. Experiments in which L-tryptophan or L-[methylene ^{14}C]-tryptophan were fed to cultures on day 0, 7, 14, or 21 with cells harvested at day 30 shed light on the effects of L-tryptophan pools and/or enzyme activity on canthin-6-one formation.

Feeding large quantities (250 and 500 mg l^{-1} L-tryptophan) increased alkaloid production in the order of 67–76% [13], with the best results obtained if unlabelled L-tryptophan was fed during the lag phase (0 to 7 days) (Fig. 4) [14]. Parallel experiments feeding L-[methylene ^{14}C]-tryptophan at days 0, 7, 14 and 21 gave high incorporation of the ^{14}C label into the canthin-6-one alkaloids (Fig. 5) while feeding on day 28 resulted in little incorporation. These latter results help to confirm that alkaloid production in *A. altissima* occurs predominantly in the late linear growth phase as seen in Fig. 2 [14]. Experiments (Figs. 4,5) in which L-tryptophan and L-[methylene ^{14}C]-tryptophan were fed at different days during the growth cycle also suggest that native pools of tryptophan are high in the lag phase, thus reducing the incorporation of L-[methylene ^{14}C]-tryptophan at that time, whereas in the linear phase of growth, reduced pools of tryptophan and/or increased enzyme activity result in a rapid increase in incorporation of labelled amino acid. However, feeding L-tryptophan in the lag phase gives better long-term increases in alkaloid production, presumably because tryptophan may also have a role in activating enzymes in the early part of the biosynthetic pathway as has been observed for microorganisms [20].

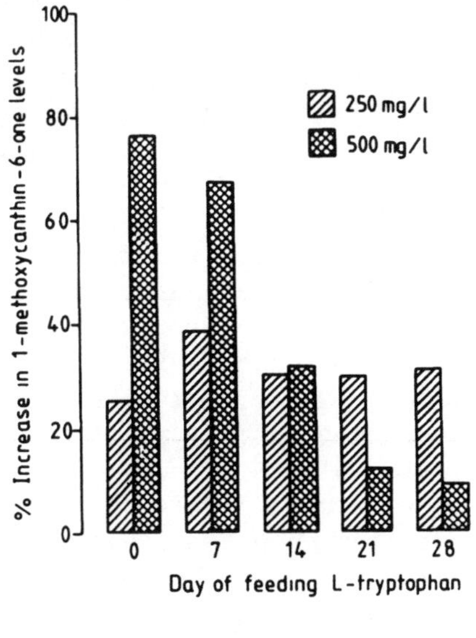

Fig. 4. The effect on 1-methoxycanthin-6-one production of feeding L-tryptophan at various stages of the growth cycle of *Ailanthus altissima*. Cells in media at day 0: 3 g 40 ml^{-1}; at day 30 (harvest day): 15 g 40 ml^{-1}. L-tryptophan, 250 mg or 500 mg l^{-1} fed on the days indicated

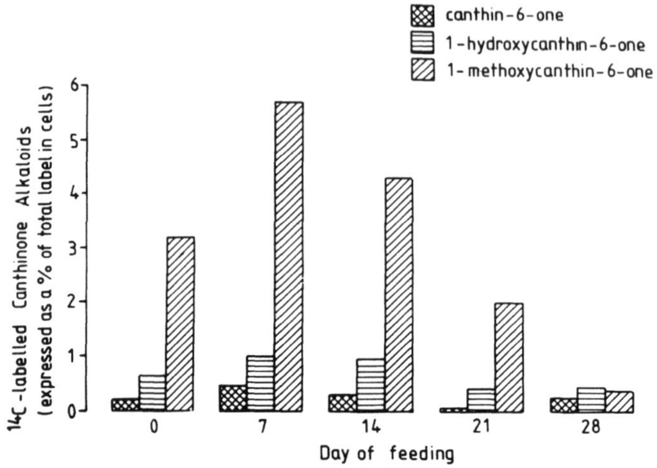

Fig. 5. Incorporation of the radiolabel from L-[methylene ^{14}C]-tryptophan into canthin-6-one alkaloids fed at different stages in the growth cycle of *Ailanthus altissima*. Cells in media at day 0: 3 g 40 ml^{-1}; at day 30 (harvest day): 15 g 40 ml^{-1}. L-[methylene ^{14}C]-tryptophan fed 1 μCi 40 ml^{-1} (specific activity 50 μCi $mmol^{-1}$) to cell cultures on the days indicated. Alkaloids were extracted and analyzed using the methods given in (12, 14)

4.2 Tryptamine

Although tryptamine was not utilized by cell suspension cultures of *A. altissima* developed in the laboratory of Crespi-Perellino et al. [18] our preliminary experiments demonstrated that L-tryptamine (100 mg l^{-1}) fed during the growth cycle increased production of 1-methoxycanthin-6-one by 17–24% (Fig. 6).

Further experiments, which paralleled those with L-tryptophan, demonstrated that [side chain-2-^{14}C]-tryptamine is rapidly taken up by *A. altissima* cells (15 min) in suspension with levels of incorporation into the canthin-6-one alkaloids as shown in Fig. 7. A subsequent experiment in which [side chain-2-^{14}C]-tryptamine was fed at day 15, with cells harvested at day 30, yielded improved incorporation into 1-methoxycanthin-6-one (6.2%) commensurate with incorporation observed with L-[methylene-^{14}C]-tryptophan [21]. Unlike the feeding with L-tryptophan the best results with [side chain-2-^{14}C]-tryptamine were obtained if feeding took place towards the end of the growth cycle although feeding unlabelled tryptamine (100 mg l^{-1}), as with tryptophan, boosted alkaloid production to a greater extent if fed during the early part of the growth cycle (Fig. 6). The results with both unlabelled tryptamine and [side chain-2-^{14}C]-tryptamine suggest that the levels of the enzyme required for the initial step in the conversion of tryptamine to canthin-6-one alkaloid is most active in this latter phase of growth. The difference in the patterns of incorporation of label when tryptophan, as opposed to tryptamine, was fed is attributed to a transient, developmentally induced increase in the presence of L-tryptophan decarboxylase between days 14 and 21 of the growth cycle (Figs. 5,7). The activity of the decarboxylating enzyme as well as that of the tryptamine conversion enzyme apparently provides a significant a degree of regulation of alkaloid production as does the availability of these substrates.

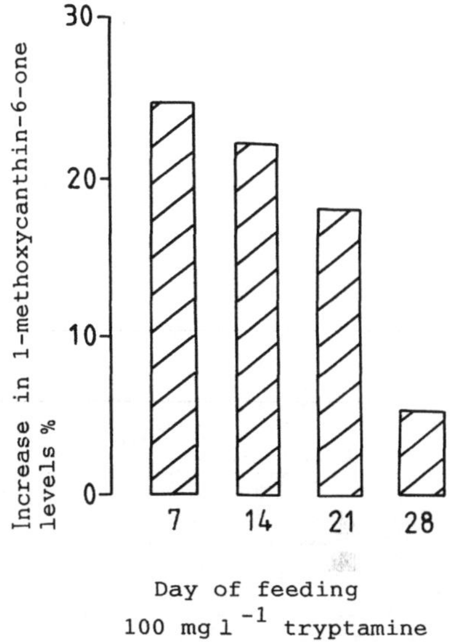

Fig. 6. The effect on 1-methoxycanthin-6-one production of feeding tryptamine to *Ailanthus altissima* cells at various stages of the growth cycle. Experimental conditions as for Fig. 4

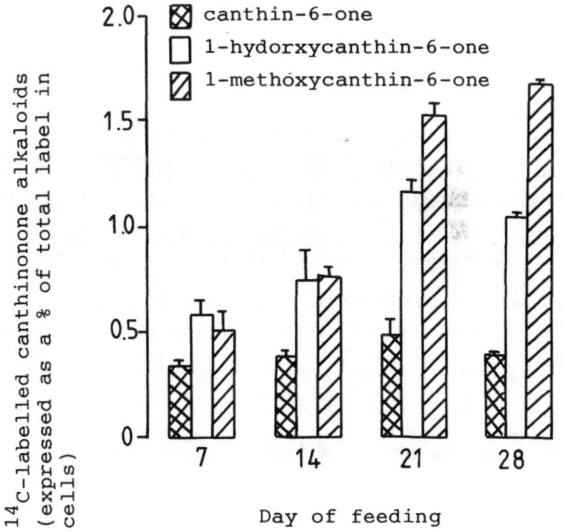

Fig. 7. Incorporation of the radiolabel from [side chain-2-^{14}C]-tryptamine into the canthin-6-one alkaloids fed at different stages of the growth cycle of *Ailanthus altissima*. Experimental conditions as for Fig. 5. [Side chain-2-^{14}C]-tryptamine, 1 μCi 40 ml^{-1} (specific activity 50 μCi mmol^{-1}) was fed to cell cultures

5 The Role of Acetate in Canthin-6-one Synthesis

Since the biosynthetic route via L-tryptamine to the canthin-6-one alkaloids would appear to involve β-carboline-1-propionic acid as an intermediate [16], the possible utilization of acetate was investigated using [1-^{14}C]-acetate. The incorporation of [1-^{14}C]-acetate into the canthin-6-one alkaloids was low (0.14% of total label in the cells) (Fig. 8) as compared with the incorporation levels for L-[methylene ^{14}C]-tryptophan (Fig. 5) and [side chain-2-^{14}C]-tryptamine (Fig.7) and this therefore

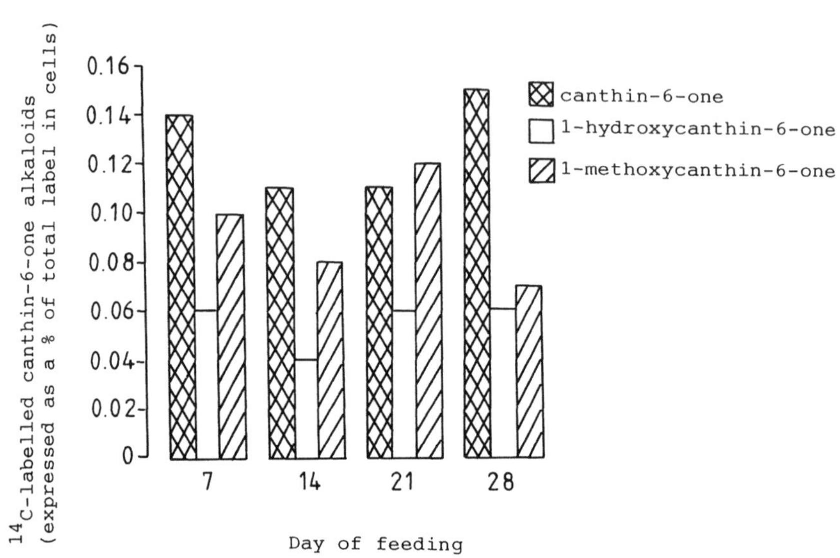

Fig. 8. Incorporation of radiolabel from [1-^{14}C]-acetic acid into canthin-6-one alkaloids fed at different stages in the growth cycle of *Ailanthus altissima*. Experimental details as for Fig. 5. [1-^{14}C]-acetic acid, 1 μCi 40 ml^{-1} (specific activity 50 μCi mmol^{-1}) fed to cell cultures on the days indicated

suggests that either [1-^{14}C]-acetate is not necessarily directly incorporated into the canthin-6-one or is readily utilized in other metabolic pathways. The low incorporation into the canthin-6-one alkaloids of this simple acid precluded the use of ^{13}C-acetate to determine the position of label in the canthin-6-one alkaloids when either ^{14}C- or ^{13}C-labelled acetate was used [21]. Subsequently, Aragozzini et al. [22] have shown a labelling pattern of canthin-6-one after feeding [1-^{13}C]-, [2-^{13}C]- and [1,2-^{13}C]-acetate that is consistent with the involvement of α-ketoglutarate as an intact precursor in the biosynthetic pathway.

6 The Origins of the Methoxy Group of 1-Methoxycanthin-6-one

The fact that 1-methoxycanthin-6-one was the major constituent in our *A. altissima* cultures [12,13], with canthin-6-one and 1-hydroxycanthin-6-one as minor alkaloids, prompted an investigation of the origins of the methyl group. The difference in alkaloid profile from that observed by Crespi-Perellino et al. [19] must reflect differences in the activities of the enzymes responsible for the hydroxylation and subsequent methylation of the parent canthin-6-one to yield the 1-hydroxy and 1-methoxy derivatives respectively.

L-[methyl ^{14}C]-methionine fed to cell suspensions of *A. altissima* was rapidly taken up by the cells (20 min: 95% uptake) if fed from day 7 onwards. Incorporation into 1-methoxycanthin-6-one remained low (ca. 0.15% incorporation) except when fed to cultures in the latter part of the growth cycle. Feeding at days 27 to 29 of the cycle with cells harvested on day 30 produced incorporation of 2–3% into

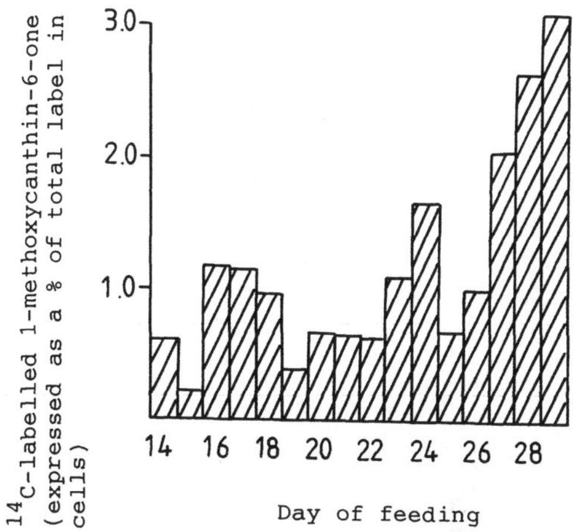

Fig. 9. Incorporation of radiolabel from L-[methyl ^{14}C]-methionine into 1-methoxycanthin-6-one by *Ailanthus altissima* cell cultures fed at various stages of the growth cycle. Experimental details as for Fig. 5. L-[methyl ^{14}C]-methionine 1 μCi 40 ml^{-1} (specific activity 500 μCi $mmol^{-1}$) fed to cultures on the days indicated

1-methoxycanthin-6-one (Fig. 9). Parallel experiments with L-[methyl ^{13}C]-methionine yielded 1-methoxycanthin-6-one with enhancement of the signal at 56.74 ppm (the methoxyl carbon) confirming the involvement of L-methionine in the methylation of 1-hydroxycanthin-6-one.

The results also suggest that L-[methyl ^{14}C]-methionine fed in the early stages of the growth cycle is in competition with higher native levels of L-methionine, whereas in the later stages of the growth cycle, this amino acid is in shorter supply. Measurements of the levels of methyltransferase activity during the growth cycle of the cultures, discussed later, seem to support this view (Fig. 10). It was also noted that other methyltransferases were active within the cell suspensions, since cultures fed L-[methyl ^{14}C]-methionine also produced label in the methoxylated coumarin scopoletin (0.8%). Appraisal of these results and those from other cell suspension cultures examined suggested that within the cells the esculetin content was normally low, and hence scopoletin production is limited [15].

7 The Isolation of 1-Hydroxycanthin-6-one: S-Adenosyl-Methionine (SAM) Methyltransferase

Rapid extractions of cells and absorption of low molecular weight compounds onto Sephadex G-25 produces cell-free extracts for monitoring methyltransferase throughout the growth cycle. Utilizing 1-hydroxycanthin-6-one and [$^{14}CH_3$]-SAM as substrates, methyltransferase activity was shown to vary during the growth cycle of the culture with peaks of activity around days 6–14 and 20 (Fig. 10). Further purification with ammonium sulphate precipitation and chromatography with batch elution on DEAE cellulose (Table 1) yielded a cell-free preparation which showed maximal activity at pH 7.6 and 30°C. The cell-free preparation was found to have the ability to methylate a number of substrates in addition to the O-

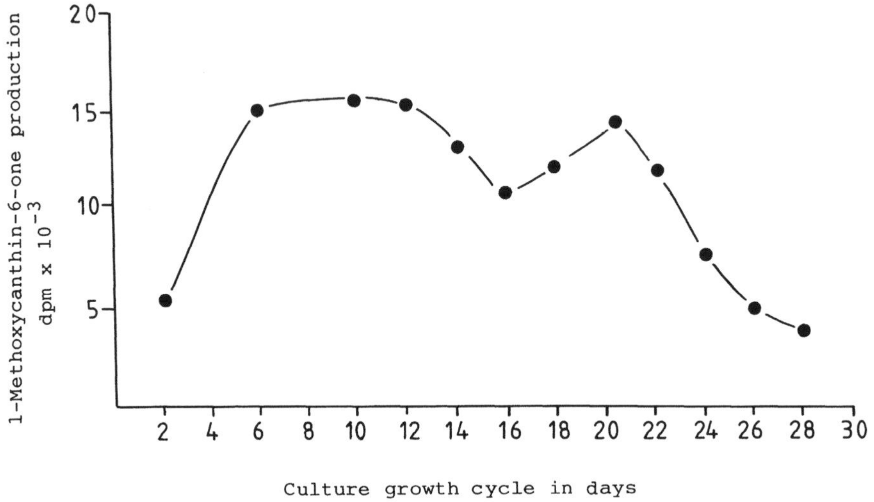

Fig. 10. Activity of 1-hydroxycanthin-6-one:S-adenosyl methionine (SAM) methyltransferase during the growth cycle of a typical cell suspension culture of *Ailanthus altissima*

Table 1. Separation of coumarin and alkaloid methyltransferases

DEAE-cellulose column elution	Methyltransferase activity (pKat g^{-1})			
	Esculetin	Canthin-6-ones		
		1 – OH	8 – OH	10 – OH
1. 0.1 M Phosphate buffer, pH 7.6	36.6	–	–	–
2. 0.1 M Phosphate buffer +0.05 M KCl	–	–	–	–
3. 0.1 M Phosphate buffer +0.15 M KCl	–	–	–	–
4. 0.1 M Phosphate buffer +0.2 M KCl	–	42.1	84.2	21.0

methylation of 1-hydroxycanthin-6-one. These cell-free extracts were also found to methylate 8- and 10-hydroxycanthin-6-one, an activity not previously observed in cell suspension cultures. Further, esculetin was readily methylated to give scopoletin and isofraxin with activity commensurate with that for the methylation of 1-hydroxycanthin-6-one. It had been observed previously that the cells in suspension culture, high yielding in 1-methoxycanthin-6-one and fed L-[methyl ^{14}C]-methionine, also produced low levels of labelled scopoletin compared with labelled 1-methoxycanthin-6-one. However, the preparation of a number of cell-free extracts from *A. altissima* cell suspension cultures suggested that the two pathways were in competition for the available L-methionine and that the rate of incorporation into 1-methoxycanthin-6-one and scopoletin respectively was determined by the availability of the appropriate substrate and the particular methyltransferase and the enzymes' activity. When production by cell cultures of 1-methoxycanthin-6-one was high, that of scopoletin was low, and vice versa. The methyltransferases responsible for the methylation of the hydroxycanthin-6-ones bound to the DEAE cellulose column, whilst that responsible for the conversion of esculetin to scopoletin was not bound to this column.

Further purification of the cell-free extracts with hydroxycanthin-6-one methyltransferase activity using FPLC on a Pharmacia Mono Q HR 5/5 column (Table 2 and Fig. 11) showed that the enzyme responsible for the methylation of 8-hydroxycanthin-6-ones may be separated from those methylating 1- and 10-hydroxycanthin-6-one. The extent to which methylated products occur therefore is mainly determined by the availability of the appropriate hydroxycanthin-6-one.

The ability of cell-free extracts to produce 8- and 10-methoxycanthin-6-ones, not previously reported as occurring in cell cultures, is of particular interest since 10-hydroxycanthin-6-one and its methylated form have not been reported for *Ailanthus* sp. and 8-hydroxycanthin-6-one has only been recorded as occurring in *Ailanthus excelsa* [8]. It will therefore be interesting to investigate the ability of these cultures to methylate other hydroxylated canthin-6-ones reported to occur both in the whole plant and in cell cultures.

Table 2. Purification procedure for 1-hydroxycanthin-6-one: SAM methyltransferases

Purification step	Total activity (pKat)	Total protein (mg)	Specific activity (pKat g^{-1})	Recovery (%)	Purification
Crude dialysate	39.46	105.0	3.76	100	
NH_4SO_4 Precipitation					
40–70%	29.25	28.0	20.89	74.1	5.7-fold
DEAE-cellulose	25.25	3.32	76.06	64.0	20.2-fold
FPLC MonQ	16.22	0.20	227.27	45.0	75.0-fold

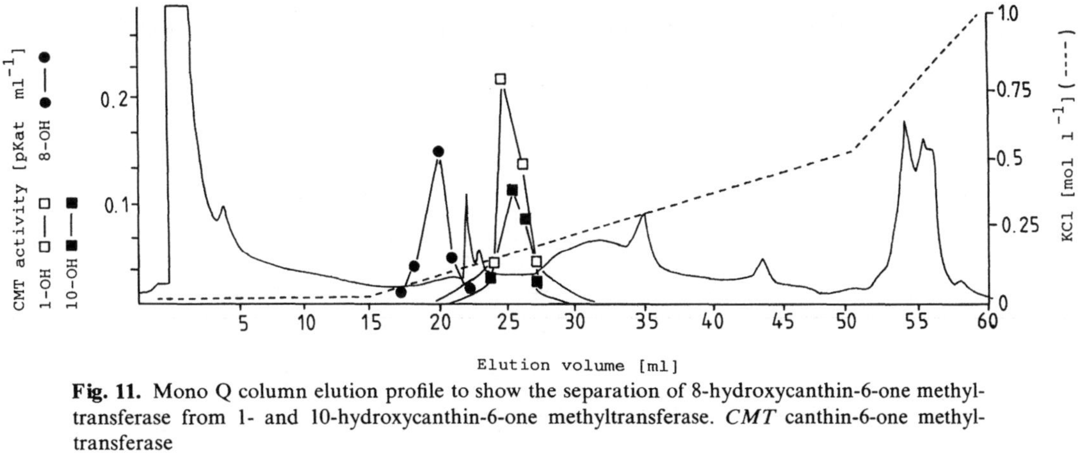

Fig. 11. Mono Q column elution profile to show the separation of 8-hydroxycanthin-6-one methyltransferase from 1- and 10-hydroxycanthin-6-one methyltransferase. *CMT* canthin-6-one methyltransferase

8 Conclusions

A number of factors appear to regulate the production of 1-methoxy-canthin-6-one:

1. The availability of L-tryptophan in the early stages of growth.
2. The activity of the enzymes of the early part of the biosynthetic pathway at the onset of the linear phase of growth.
3. The activity of the hydroxylase system which controls the diversity of alkaloids produced.
4. The activity of enzymes responsible for the further metabolism of the canthin-6-one alkaloids.

References

1. Ohmoto T, Tanaka R, Nikaido T (1976) Studies on the constituents of *Ailanthus altissima* Swingle: On the alkaloid constituents. Chem Pharm Bull 24:1532–1536
2. Varga E, Szendrei K, Reisch J, Maroti G (1980) Indole alkaloids of *Ailanthus altissima*. Planta Med 40:337–339
3. Varga E, Szendrei K, Reisch J, Maroti G (1980) Indole alkaloids of *Ailanthus altissima*. II. Fitoterapia 4:183–186
4. Swingle WT (1916) The early European history and the botanical name of the Tree of Heaven, *Ailanthus altissima*. J Washington Acad Sci 6(14):490–498
5. Suffness M, Douros JD (1980) Miscellaneous natural products with antitumor activity. In: Cassady JM, Douros JD (eds) Anticancer agents based on natural product models. Academic Press, New York London pp 465–476
6. Bray DH, Boardman P, O'Neill MJ, Chan KL, Phillipson JD, Warhurst DC, Suffness M (1987) Plants as sources of antimalarial drugs 5. Activities of *Ailanthus altissima* stem constituents and of some related quassinoids. Phytother Res 1:22–24
7. Fukamiya N, Okano M, Aratani T, Negoro K, McPhail AT, Ju-Ichi M, Lee KH (1986) Antitumor agents, 79. Cytotoxic antileukemic alkaloids from *Brucea antidysenterica*. J Nat Prod 49:428–434
8. Cordell GA, Ogura M, Farnsworth NR (1978) Alkaloid constituents of *Ailanthus excelsa* (Simaroubaceae). Lloydia 41:166–168
9. Mitscher LA, Showalter HDH, Shipchandler MT, Leu RP, Beal JL (1972) Antimicrobial agents from higher plants IV. *Zanthoxylum elephantiasis* isolation and identification of canthin-6-one. Lloydia 35:177–180
10. Sung Y-I, Koike K, Nikaido T, Ohmoto T, Sankawa U (1984) Inhibitors of cyclic AMP phosphodiesterase in *Picrasma quassioides* Bennet, and inhibitory activities of related β-carboline alkaloids. Chem Pharm Bull 32:1872–1877
11. Jaziri M, Homes J, Vanhaeien M (1987) Production of quassinoids by tissue cultures of *Ailanthus altissima*. Phytochemistry 26:999–1000
12. Anderson LA, Harris A, Phillipson JD (1983) Production of cytotoxic canthin-6-one alkaloids by *Ailanthus altissima* cell suspension cultures. J Nat Prod 46:374–378
13. Anderson LA, Roberts MF, Phillipson JD (1987b) Studies on *Ailanthus altissima* cell suspension cultures. The effect of basal media on growth and alkaloid production. Plant Cell Rep 6:239–241
14. Anderson LA, Hay CA, Roberts MF, Phillipson JD (1986) Studies on *Ailanthus altissima* cell suspension cultures. Precursor feeding of L-[methylene-^{14}C]tryptophan and L-tryptophan. Plant Cell Rep 5:387–390
15. Anderson LA, Hay CA, Phillipson JD, Roberts MF (1987a) Studies on *Ailanthus altissima* cell suspension cultures. Uptake of L-[methyl-^{14}C]methionine and incorporation of label into 1-methoxycanthin-6-one. Plant Cell Rep 6:242–243
16. Lang H, Kohlenbach HW (1982) Differentiation of alkaloid cells in cultures of *Macleaya* mesophyll protoplasts. Planta Med 46:78

17. Hegnauer R (1973) Chemotaxonomie der Pflanzen, vol 6. Birkhäuser, Basel, p 387
18. Crespi-Perellino N, Guicciardi A, Malyszko G, Minghetti A (1986b) Biosynthetic relationship between indole alkaloids produced by cell cultures of *Ailanthus altissima*. J Nat Prod 49:814–822
19. Crespi-Perellino N, Guicciardi A, Malyszko G, Arlandini E, Ballabio M, Minghetti A (1986a) Occurrence of indole alkaloids in *Ailanthus altissima* cell cultures. J Nat Prod 49:1010–1014
20. Floss HG, Robbers JG, Heinstein PF (1974) Regulatory control mechanisms in alkaloid biosynthesis. Rec Adv Phytochem 8:141–178
21. Hay CA (1987) Plant tissue culture of *Cinchona ledgeriana* and *Ailanthus altissima*. PhD Thesis, Univ London
22. Aragozzini F, Maconi E, Gualandris R (1988) Evidence for involvement of ketoglutarate in the biosynthesis of canthin-6-one from cell cultures of *Ailanthus altissima* Plant Cell Rep 7:213–215

Part III
Elicitation of Metabolite Biosynthesis

Studies on Alkaloid Formation in Plant Cell Cultures After Treatment with a Yeast Elicitor

P. Brodelius, M.A. Collinge, C. Funk, K. Gügler, and I. Marques[1]

1 Introduction

Interactions between plants and microbial pathogens may result in various defense reactions within the infected plant. Elicitors are important in triggering these reactions. Similar responses may take place in plant cell suspension cultures after addition of pathogen-derived elicitor preparations to the culture medium. An increasing number of studies on the effects of elicitors on secondary metabolism of plant cell cultures may be found in the literature.

The administration of elicitors to cell suspension cultures offers an experimental tool for studies on:

1. Plant defense mechanisms;
2. Regulation of secondary metabolism;
3. Enzymology of secondary metabolism.

Furthermore, elicitation may be employed to increase the yield of a target substance. Our studies have been carried out with an elicitor isolated from yeast extract.

2 Isolation of the Yeast Elicitor

A glucan has been isolated from yeast extract according to a procedure described by Hahn and Albersheim (1978). The isolation of the elicitor is based on ethanolic precipitation of carbohydrates. The procedure also involves a dialysis step resulting in a glucan preparation with a molecular weight larger than 5 kD. The final carbohydrate concentration (determined by the orcinol-sulphuric acid method) is around 6 mg ml^{-1}. From 40 g of yeast extract approx. 600 mg of elicitor is prepared. The elicitor solution is sterilized by filtration before addition to the cell suspension cultures. The optimal amount of elicitor has to be established for each culture and it is normally express as μg carbohydrate per g fresh weight of cells, i.e. related to the biomass within the culture and not the culture volume.

[1]Institute of Biotechnology, Swiss Federal Institute of Technology, Hönggerberg, CH-8093 Zürich, Switzerland

Primary and Secondary Metabolism of Plant Cell Cultures, ed. by W.G.W. Kurz

3 Effects of the Yeast Elicitor on Primary Metabolism

3.1 Protein Synthesis

Addition of the yeast elicitor to a cell suspension culture results in the induction of a large number of proteins, as illustrated for *Thalictrum rugosum* cells in Fig. 1 (Gügler et al. 1988). The induced proteins are most likely enzymes participating in various defense reactions including secondary metabolism.

3.2 Cell Respiration

Treatment of cell cultures with the yeast elicitor results in an increased respiration of the cells (Funk et al. 1987; Collinge and Brodelius 1989). Relatively low concentrations of elicitor (i.e. 5–10 μg carbohydrate/g fresh weight of cells) are required for an increased respiration of *Eschscholtzia californica* cells.

3.3 ATP Levels

Upon treatment with elicitor a rapid increase in the intracellular ATP levels is observed (Häner and Brodelius, unpublished). A two- to three-fold increase in ATP can be measured 1–2 h after addition of elicitor at moderate concentrations. The

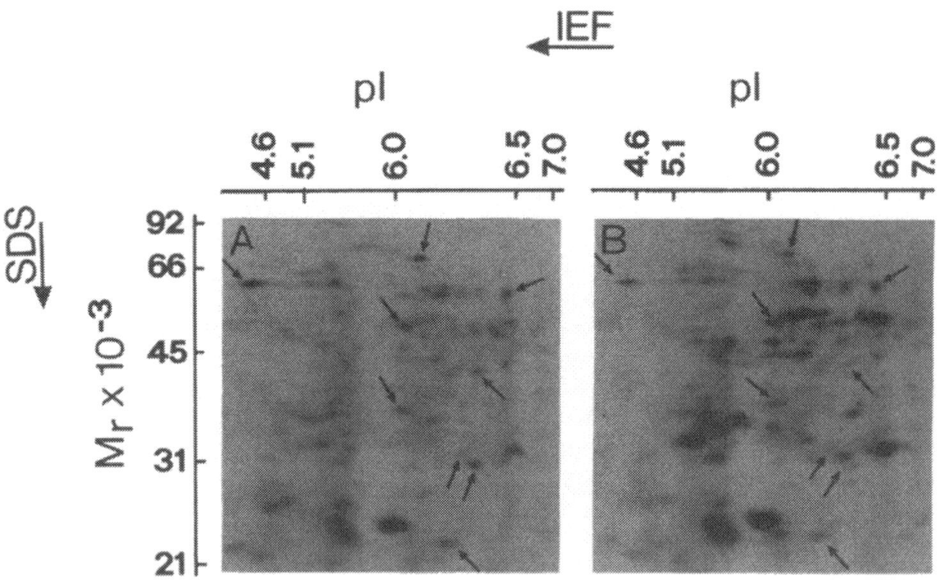

Fig. 1. Fluorography of cell-free extracts of *T. rugosum* cells after two-dimensional electrophoresis. Non-treated and elicitor-treated cells (0.7 g wet weight) were suspended in medium (0.7 ml) and after 6 h [35S]methionine (100 μCi) was added. The cells were harvested 2 h later and extracted. **A** Non-treated cells; **B** elicitor-treated cells. The *arrows* indicate some proteins which were not induced (Gügler et al. 1988)

basic level of ATP is regained after an additional few hours. When the culture is treated with an uncoupler (e.g. CCCP) for 30 min prior to elicitation no such increase in ATP is observed, indicating that oxidative phosphorylation is increased by elicitor treatment (Häner and Brodelius, unpublished).

3.4 Conductivity and pH of Culture Medium

It has been shown that the plasma membrane potential is affected by biotic elicitors (Low and Heinstein 1986) which could influence the transport of solutes over the membrane. To this end we have measured the conductivity and pH of the culture medium after addition of elicitor to cultures of *E. californica*. The conductivity is decreased, indicating that no cell lysis occurs (Collinge and Brodelius 1989). On the contrary, solutes (ions) appear to be taken up by the elicited cells. The increased pH (0.5–1 pH units), which has been observed in a number of cultures after elicitation, may be due to the fact that the ions are taken up in the protonated form. The changes in conductivity and pH are relatively rapid and can be observed within 15–30 min after addition of the yeast elicitor.

3.5 Cell Growth

The growth (biomass increase) of a cell culture is influenced by elicitor treatment. When the cells are left in old medium after treatment, a slight decrease in dry weight is observed as compared to non-treated cultures (Funk et al. 1987). However, if the treated cells are transferred to fresh medium within a short time (e.g. 24 h) after elicitor treatment, a slight increase in the biomass is seen at moderate elicitor concentrations (< 100 μg carbohydrate/g fresh weight of cells) in cultures of *E. californica* (Collinge and Brodelius 1989). At higher elicitor concentrations the cell culture tends to aggregate and a somewhat decreased dry weight is observed. The decreased growth observed for elicitor-treated cells is not reflected in the uptake rate of sucrose from the medium which remains unaffected (Funk et al. 1987).

4 Effects of the Yeast Elicitor on Alkaloid Production

The effects of the yeast elicitor on isoquinoline alkaloid formation have been studied with two cell suspension cultures, i.e. *T. rugosum* and *E. californica*. The former culture produces berberine, while the latter culture produces a number of benzophenanthridine alkaloids. We have studied the formation of sanguinarine, chelerythrine and macarpine but from HPLC analysis of cell-free extracts it may be assumed that other benzophenanthridine alkaloids are present in the culture. It has been reported that a cell culture of *E. californica* produced chelirubine and chelilutine after treatment with a similar elicitor isolated from yeast extract (Schumacher et al. 1987). A number of reports on the elicitor-induced formation of isoquinoline alkaloids in plant cell cultures have been published (Eilert et al. 1985; Heinstein 1985; Cline and Coscia 1988).

4.1 Elicitor Concentration

The optimal concentration of the yeast elicitor for alkaloid formation differs considerably for the two cell cultures as illustrated in Fig. 2A. The optimal concentrations for the biosynthesis of berberine and chelerythrine are 200 and 30 μg elicitor (carbohydrate) per g fresh weight of cells respectively.

4.2 Incubation Time

Also, the incubation time required for maximum alkaloid formation varies for different products and different cell cultures as shown in Fig. 3. During the incubation, the amounts of alkaloids in non-treated cells were not significantly changed. The cells of *E. californica* respond rapidly to the elicitor treatment and

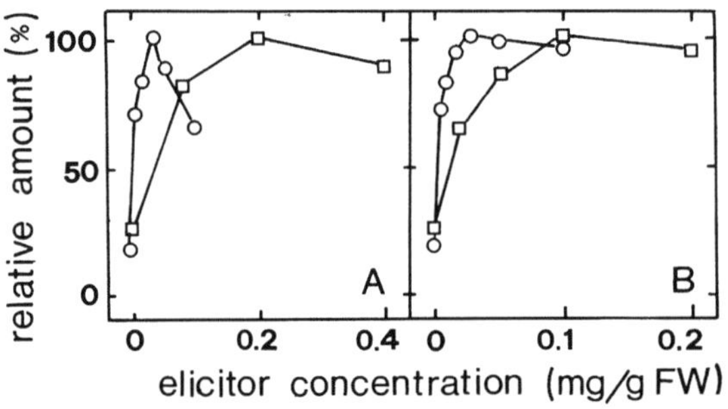

Fig. 2. Relative amount of alkaloids and TDC induced in cell cultures as a function of elicitor concentration. **A** (□) berberine in *T. rugosum* cells and (○) chelerythrine in *E. californica* cells extracted 96 and 48 h after addition of elicitor respectively; **B** (□) TDC in *T. rugosum* cells and (○) TDC in *E. californica* cells extracted 16 and 5 h after addition of elicitor respectively (Data from Funk et al. 1987; Gügler et al. 1988; Collinge and Brodelius 1989; Marques and Brodelius 1988a)

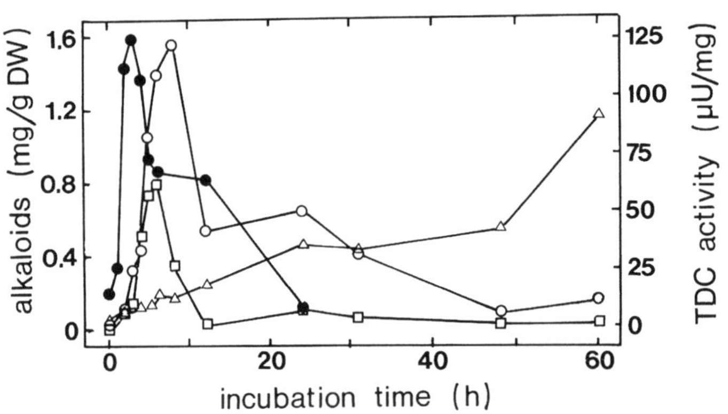

Fig. 3. Time course of TDC and alkaloid formation in a cell suspension culture of *E. californica* after addition of the yeast elicitor. (●) TDC activity; (□) sanguinarine; (○) chelerythrine; (△) macarpine (Data from Collinge and Brodelius 1989)

within 6–8 h maximum levels of sanguinarine and chelerythrine are observed. These alkaloids disappear within a few hours and the related alkaloid macarpine is found in increasing amounts in the culture. The fact that sanguinarine and chelerythrine are only present intermittently may be explained by the fact that these alkaloids are precursors of other alkaloids (Takao et al. 1983). Sanguinarine may be methoxylated in positions 5 and 12 to yield macarpine, and chelerythrine may be methoxylated in position 12 to yield chelilutine.

While berberine was found within the cells of *T. rugosum* throughout an incubation, a major portion of the benzophenanthridine alkaloids within cultures of *E. californica* was extracellular. Figure 4 demonstrates this for chelerythrine. At high chelerythrine concentrations (5–12 h after addition of elicitor) around 75% of the alkaloid could be isolated from the culture medium. Similar results were obtained for sanguinarine and at the end of the incubation (60 h) over 50% of the macarpine formed was isolated from the medium. Apparently the cells are actively excreting the alkaloids into the medium since no cell lysis occurs (cf. Sect. 3.4). Furthermore, it is interesting to note that if cells and medium are separated, the levels of alkaloids in the medium remain unchanged, indicating that no catabolyzing enzymes (e.g. peroxidases) are present in the medium. Feeding experiments will show if the cells take up the intermittently formed alkaloids from the medium for further conversions or if cell-bound enzymes are responsible for the disappearance of sanguinarine and chelerythrine. If it is found that the cells reutilize the excreted alkaloids, this is an interesting example of how alkaloids are intermittently "stored" in the culture medium and not within the vacuoles of the cells.

4.3 Culture Age

Finally, the age of the culture has a great influence on the response of the cells to elicitor treatment. Both *T. rugosum* and *E. californica* cultures show the highest alkaloid production when elicited in late exponential or early stationary growth stage as illustrated in Fig. 5A for berberine formation in *T. rugosum* cells (Funk et al. 1987). Elicitor treatment in early growth stages does not result in a significantly

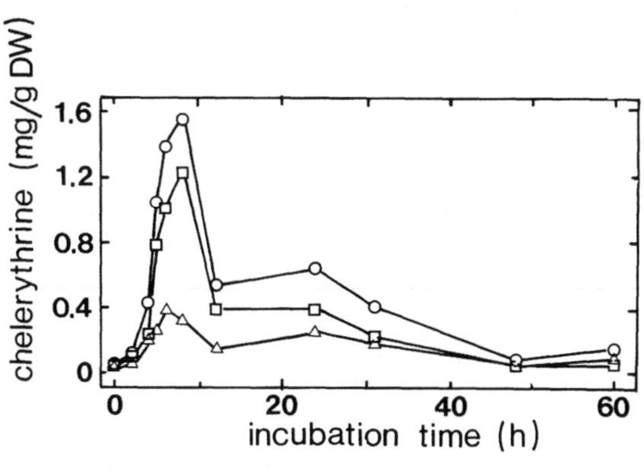

Fig. 4. Time course of chelerythrine formation in elicitor-treated cells of *E. californica*. (○) Total extractable chelerythrine; (□) chelerythrine extracted from the culture medium: (△) chelerythrine extracted from the cells (Data from Collinge and Brodelius 1989)

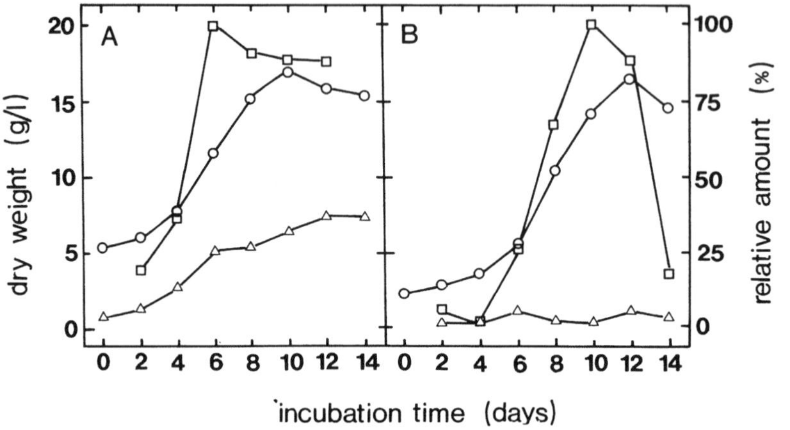

Fig. 5. Induction of berberine and TDC at various growth stages of *T. rugosum* cells. **A** (○) Dry weight; (△) berberine in non-treated cells; (□) berberine in elicitor-treated cells. **B** (○) Dry weight; (△) TDC in non-treated cells; (□) TDC in elicitor-treated cells. The points representing elicited samples have been plotted against the incubation time at which they were elicited. Extractions were made 120 and 16 h after elicitation for berberine and TDC determinations respectively (Data from Funk et al. 1987; Gügler et al. 1988)

increased production of berberine. Similar results have been obtained for the production of benzophenanthridine alkaloids in cultures of *E. californica* (Collinge and Brodelius 1989).

4.4 Optimal Conditions

Table 1 summarizes the optimal conditions for the formation of berberine and benzophenanthridine alkaloids in cell suspension cultures of *T. rugosum* and *E. californica* respectively. From Table 1 it is clear that the amount of alkaloid produced may be improved considerably by transferring the elicited cells to fresh medium. Obviously, when energy is available the elicited cells are capable of an increased secondary metabolism.

Table 1. Alkaloid production in plant cell cultures after addition of the yeast elicitor (Data from Funk et al. 1987; Collinge and Brodelius 1989)

Cell culture	Product	Elicitor (μg g^{-1} FW)	Alkaloid[a] (mg g^{-1} DW)		Incubation (h)
			–	+	
T. rugosum	Berberine	200	50	200	120
E. californica	Sanguinarine	50	0	0.8	6
	Chelerythrine	50	0.1	1.6	8
	Macarpine	50	0.1	1.1	60
	Macarpine[b]	100	0.1	6.0	168

[a](–) = Without addition of elicitor; (+) = with addition of elicitor.
[b]The cells were transferred to fresh medium 24 h after addition of elicitor.

5 Induction of Tyrosine Decarboxylase

As shown in Fig. 1A large number of proteins are induced by the yeast glucan. Some of these proteins are most likely enzymes involved in alkaloid biosynthesis. Isoquinoline alkaloids are synthesized from two molecules of tyrosine and therefore we have been involved in studies to establish whether an enzyme with tyrosine as substrate is induced by the elicitor.

To this end radiolabelled tyrosine was fed to non-elicited and elicited cultures of *T. rugosum* (Gügler et al. 1988). The tyrosine uptake rate was not influenced by the elicitor treatment. From Table 2 it is evident that tyrosine is rapidly decarboxylated to tyramine in elicited cells. The tyramine is subsequently hydroxylated to dopamine which is the precursor of the "upper" part of isoquinoline alkaloids (Zenk et al. 1985). The induction of tyrosine decarboxylase (TDC) in cell cultures of *T. rugosum* and *E. californica* by the yeast elicitor has been further studied.

Measurements of TDC activity in cell-free extracts of *T. rugosum* and *E. californica* have shown that the induction is dependent on the concentration of elicitor (Fig. 2B) and age of culture (Fig. 5B) in a very similar way as described above for alkaloid formation. The time course of the induction (Fig. 3) preceed that of alkaloid formation. All these results indicate that TDC may be a key enzyme in the early steps of isoquinoline alkaloid biosynthesis.

The optimal conditions for induction of TDC in the two cell cultures are summarized in Table 3. The TDC activity in elicited cells is normally 10- to 20-fold higher than in non-elicited cells. Addition of transcriptional (e.g. actinomycin D) or translational (e.g. cycloheximide) inhibitors to the culture prior to elicitation prevents the induction of TDC, indicating that the regulation of this enzyme is on a transcriptional level.

Table 2. Relative composition of radiolabelled metabolites after addition of tyrosine to cell cultures of *T. rugosum*. Tyrosine was added 12 h after the addition of elicitor and the cells were extracted 7 h later (Data from Gügler et al. 1988)

Metabolite	Relative amount (%)	
	Non-elicited cells	Elicited cells
Tyrosine	68	15
Tyramine	24	66
DOPA	1	2
Dopamine	7	17

Table 3. Optimal conditions for induction of TDC in cell suspension cultures by treatment with a yeast elicitor (Data from Gügler et al. 1988; Marques and Brodelius 1988a)

	T. rugosum	*E. californica*
Age of culture (days)	10–11	7–8
Incubation after elicitation (h)	16–20	3–5
Elicitor concentration ($\mu g\ g^{-1}$ FW)	100–200	30–40
TDC activity in crude extract ($mU^a\ g^{-1}$ cells)	0.66	2.2
Specific TDC activity in crude extract ($mU^a\ mg^{-1}$ protein)	0.18	1.7

[a] One unit (U) = μmol CO_2 produced per min at 30°C.

6 Purification and Partial Characterization of Tyrosine Decarboxylase

In order to determine the possible regulatory function of TDC in isoquinoline alkaloid biosynthesis, more extensive studies on this enzyme have been initiated.

TDC has been purified from elicitor-treated cell cultures of *T. rugosum* and *E. californica* (Marques and Brodelius 1988a). The purification procedure involved ammonium sulphate fractionation, anion-exchange fast-protein liquid chromatography (MonoQ), ultrafiltration and hydrophobic interaction chromatography (Phenyl-Superose).

The purified enzymes have been partly characterized as summarized in Table 4 (Marques and Brodelius 1988b). TDC requires pyridoxal phosphate (PLP) as coenzyme and it is also stabilized by PLP. Tyrosine is the best substrate but also DOPA is decarboxylated efficiently while phenylalanine and tryptophan are essentially unaffected by TDC.

7 Concluding Remarks

The elicitor isolated from the non-pathogenic organism *Saccharomyces cerevisiae* is efficient in inducing metabolic changes in various plant cell suspension cultures. This elicitor is readily isolated in large quantities from a commercially available yeast extract.

The effects of the elicitor are highly dependent on elicitor concentration and age of the culture. Optimal conditions must be empirically determined for each culture studied. The time course of the elicitation is another important factor. As shown for cells of *E. californica* certain alkaloids (sanguinarine and chelerythrine) are only intermittently present in the culture and therefore a "one-point experiment" may lead to a false conclusion regarding which secondary product(s) can be induced by elicitation.

Elicitation is a powerful tool to study the enzymology of an induced, secondary biosynthetic pathway. Our studies on isoquinoline alkaloid producing cell cultures have led to the suggestion that tyrosine decarboxylase and not tyrosine hydroxylase [as previously assumed (Zenk et al. 1985)] is the enzyme linking primary and secondary metabolism. Further, studies will show how important this enzyme is in the regulation of isoquinoline alkaloid biosynthesis.

Table 4. Some characteristics of purified TDC (Data from Marques and Brodelius 1988a,b)

	T. rugosum	*E. californica*
Specific activity (U mg^{-1} protein)	0.18	3.0
K_m (Tyrosine) (mM)	0.27	0.25
pH Optimum	8.4	8.4
Isoelectric point	5.4	5.2
Number of subunits	2	2
Molecular weight (kD)	112	112

References

Cline SD, Coscia CJ (1988) Stimulation of sanguinarine production by combined fungal elicitation and hormonal deprivation in cell suspension cultures of *Papaver bracteatum*. Plant Physiol 86:161

Collinge MA, Brodelius PE (1989) Dynamics of benzophenanthridine alkaloid production in suspension cultures of *Eschscholtzia californica* after treatment with a yeast elicitor. Phytochemistry 28:1101

Eilert U, Kurz WGW, Constabel F (1985) Stimulation of sanguinarine accumulation in *Papaver somniferum* cell cultures by fungal elicitors. J Plant Physiol 119:65

Funk C, Gügler K, Brodelius P (1987) Increased secondary product formation in plant cell suspension cultures after treatment with a yeast carbohydrate preparation (elicitor). Phytochemistry 26:401

Gügler K, Funk C, Brodelius P (1988) Elicitor-induced tyrosine decarboxylase in berberine-synthesizing suspension cultures of *Thalictrum rugosum*. Eur J Biochem 170:661

Hahn MG, Albersheim P (1978) Host-pathogen interactions XIV. Isolation and partial characterization of an elicitor from yeast extract. Plant Physiol 62:107

Heinstein P (1985) Future approaches to the formation of secondary natural products in plant cell suspension cultures. J Nat Prod 48:1

Low PS, Heinstein P (1986) Elicitor stimulation of the defense responses in cultured plant cells monitored by fluorescent dyes. Arch Biochem Biophys 249:472

Marques I, Brodelius P (1988a) Elicitor-induced L-tyrosine decarboxylase from plant cell suspension cultures. I. Induction and purification. Plant Physiol 88:46

Marques I, Brodelius P (1988b) Elicitor-induced L-tyrosine decarboxylase from plant cell suspension cultures. II. Partial characterization. Plant Physiol 88:52

Schumacher H-M, Gundlach H, Fiedler F, Zenk MH (1987) Elicitation of benzophenanthridine alkaloid synthesis in *Eschscholtzia* cell cultures. Plant Cell Rep 6:410

Takao N, Kamigauchi M, Okada M (1983) Biosynthesis of benzo[c]phenanthridine alkaloids sanguinarine, chelirubine and macarpine. Helv Chim Acta 66:473

Zenk MH, Rueffer M, Amann M, Deus-Neumann B (1985) Benzylisoquinoline biosynthesis by cultivated plant cells and isolated enzymes. J Nat Prod 48:725

Studies on Benzophenanthridine Alkaloid Production in Elicited Cell Cultures of *Papaver somniferum* L.*

R.T. Tyler[1], U. Eilert[2], C.O.M. Rijnders[3], I.A. Roewer[1], C.K. McNabb[1], and W.G.W. Kurz[1]

1 Introduction

Benzophenanthridine alkaloids occur widely in members of the Papaveraceae (poppy family) and have been detected in cell cultures of several species (Furuya et al. 1972, Ikuta et al. 1979, Berlin et al. 1983, Krane et al. 1984). Aspects of production in elicited cultures of *Eschscholtzia californica* (Schumacher et al. 1987) and *Papaver bracteatum* (Cline and Coscia 1988) have recently been described. In this regard, our laboratory has demonstrated that a cell line (PBI# 2009) of *Papaver somniferum* L. initiated approximately 10 years ago (Tam et al. 1980) synthesizes and accumulates sanguinarine, a quaternary benzophenanthridine alkaloid, when treated with any of a number of elicitor preparations (Eilert et al. 1985). A portion of the sanguinarine is released into the culture medium. This cell line normally accumulates trace quantities, at best, of benzophenanthridines in unelicited suspension culture.

The significant market potential for sanguinarine in oral hygiene products (Southard et al. 1984, Anonymous 1985) justified further study, from a biotechnological perspective, of the production of benzophenanthridine alkaloids in elicited *P. somniferum* cultures. The results of our studies, both completed and in progress, form the basis of this chapter.

2 The *Papaver somniferum* – Botrytis System

Eilert et al. (1985) treated poppy cell line PBI# 2009 with a variety of defined and undefined (primarily fungal) elicitor preparations. In all cases, sanguinarine was the predominant and, essentially, only alkaloid produced. Alkaloid accumulation was accompanied by culture browning, presumably due to oxidation and polymerization of phenolic compounds also produced in response to elicitation. *Botrytis* culture homogenate was judged the most effective of the elicitor preparations tested and was employed in more detailed experiments. The optimum

[1]Plant Biotechnology Institute, National Research Council of Canada, Saskatoon, Saskatchewan S7N OW9, Canada
[2]Institut für Pharmazeutische Biologie, Technische Universität Braunschweig, D-3300 Braunschweig, FRG
[3]Centre for Biopharmaceutical Sciences, Einsteinweg 5, 2300 RA Leiden, The Netherlands
*Publication No. NRCC 29147 from the National Research Council of Canada

Primary and Secondary Metabolism of Plant Cell Cultures, ed. by W.G.W. Kurz

concentration of *Botrytis* homogenate was determined to be 1 ml per 100 ml culture; higher concentrations resulted in a stronger browning reaction, reduced sanguinarine accumulation, and relatively high culture pH values, indicative of cell damage. Cultures treated with 1 ml homogenate showed signs of browning 6 h after treatment, significant sanguinarine accumulation after 12 h, and after 79 h had accumulated sanguinarine equivalent to 3% of cell dry weight, with approximately 15% of the sanguinarine released into the culture medium. The relatively modest release of sanguinarine into the medium and the slight increase in culture pH observed suggested that cells generally remained intact after 79 h of exposure to the elicitor. In a parallel study, Eilert and Constabel (1985) provided visual evidence that treatment of poppy cells with *Botrytis* homogenate did not necessarily result in cell lysis despite culture browning and significant sanguinarine production and release. It appeared that sanguinarine release involved more than just simple diffusion from lysed cells. In a subsequent study, Eilert and Constabel (1986) confirmed that accumulation of sanguinarine in response to treatment with *Botrytis* homogenate was indeed an induction process by demonstrating incorporation of L-tyrosine [C-14(U)] into sanguinarine by elicited cultures and inhibition of both sanguinarine accumulation and incorporation of label by actinomycin D and cycloheximide.

The observation that elicited poppy cells could be maintained in a viable state, particularly if transferred to fresh medium shortly (within approximately 72 h) after elicitation (Kurz et al. 1987), coupled with the release of a significant proportion of the sanguinarine from elicited cells into the culture medium, suggested that a plant cell culture process utilizing these phenomena might be worthy of consideration. Consequently, the merits of semi-continuous culture, whereby sanguinarine would be recovered from spent medium only and cells supplied with fresh medium and re-elicited at appropriate intervals, were examined and a process, for which patent protection has been sought, was developed (Constabel et al. 1986).

3 Semi-Continuous Production of Benzophenanthridine Alkaloids by Elicited *Papaver somniferum* Cultures

Laboratory-scale, semi-continuous culture is typified in Fig. 1. Inoculum size governs the physiological state (i.e. stage of growth) of the culture at the time of elicitation. Medium exchange serves several purposes in addition to permitting recovery of alkaloids from spent medium: (1) the concentration of elicitor is reduced, which aids in maintaining cell viability, as does (2) the nutrition afforded by fresh medium, and (3) inhibition (if any) of product synthesis by concentration of product in the medium is minimized. When alkaloid production drops below a practical level, the culture may either be re-elicited in its entirety or, when culture density is at or near a maximum, a portion of the cell mass may be used as inoculum to initiate a new, complete elicitation cycle.

The semi-continuous process has been evaluated over several cycles of elicitation/re-elicitation with *Botrytis* culture homogenate (Tyler et al. 1988).

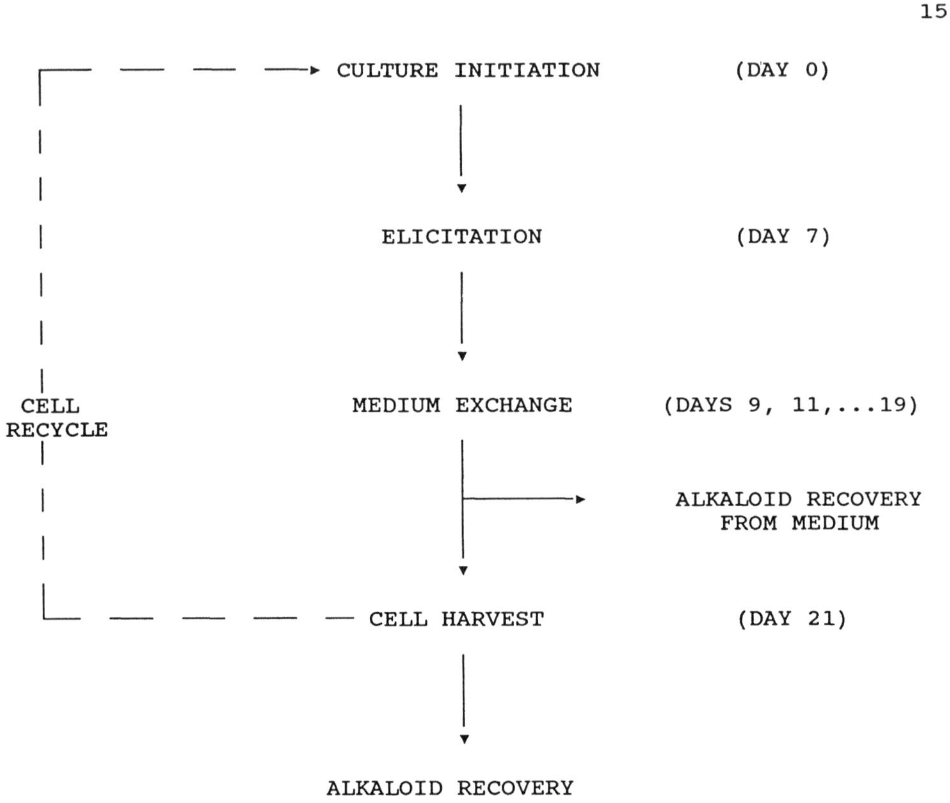

Fig. 1. Typical laboratory-scale, semi-continuous culture protocol applied to the production of benzophenanthridine alkaloids in elicited *Papaver somniferum* cultures (cf. Tyler et al. 1988)

Cultures (100 ml) were derived from 6 g fresh weight of 4-day-old poppy cells in 1B5C medium (Gamborg et al. 1968) and elicited 4 days after inoculation with 0.5 ml *Botrytis* homogenate. The medium was exchanged at 2-day intervals over a post-elicitation culture period of 14 days. Subsequent elicitation cycles were similar to the first, employing cells from the previous cycle as culture inoculum. Elicited cultures were slightly brown within 12 h and synthesized both sanguinarine and dihydrosanguinarine. Non-elicited controls never accumulated more than trace levels of either alkaloid. Replicate cultures were quite variable with respect to the quantities of sanguinarine and dihydrosanguinarine produced. Cell growth was not affected by the first elicitor treatment, although subsequent elicitations resulted in progressively greater impedance of cell growth. Cultures elicited a third time exhibited a significantly slower growth rate and, when subcultured, demonstrated growth but did not survive a fourth elicitation. Alkaloid production increased markedly with each elicitation, with dihydrosanguinarine quantitatively the major alkaloid in each case (Table 1). Combined yields of sanguinarine and dihydrosanguinarine in cells and medium for the first, second, and third elicitation cycles were approximately 0.8, 2.1, and 3.2% of cell dry weight respectively, equivalent to

Table 1. Alkaloid yields (mg/100 ml culture) from semi-continuous culture of *Papaver somniferum* cells elicited with *Botrytis* culture homogenate (Tyler et al. 1988)

Elicitation Cycle	DSGE[a]			SGE[a]			Total		
	Cells	Medium	Total	Cells	Medium	Total	Cells	Medium	Total
1	6.2(63)[b]	3.7(37)	9.9	2.4(41)	3.5(59)	5.9	8.6(54)	7.2(46)	15.8
2	27.2(78)	7.9(22)	35.1	3.5(54)	3.0(46)	6.5	30.7(74)	10.9(26)	41.6
3	37.2(84)	7.1(16)	44.3	15.8(75)	5.3(25)	21.1	53.0(81)	12.4(19)	65.4

[a] DSGE = Dihydrosanguinarine; SGE = sanguinarine.
[b] Figures in parentheses = percent of total.

50, 125, and 200 mg l^{-1} of culture medium. Variable proportions of the sanguinarine and dihydrosanguinarine were released into the culture medium. For all elicitation cycles, a greater proportion of the sanguinarine than of the dihydrosanguinarine was released, and for both alkaloids the proportion declined with each re-elicitation. We attribute the production of dihydrosanguinarine in addition to sanguinarine, which is in contrast to results from previous studies, to the relatively mild elicitor treatment employed (a requirement for sustained cell viability), based on the assumption that milder elicitation produces a less "oxidizing" culture environment which results in the appearance of the reduced product, dihydrosanguinarine. Whether or not a dihydrobenzophenanthridine oxidase enzyme such as that described by Schumacher and Zenk (1988) is responsible for the conversion of dihydrosanguinarine to sanguinarine in our cell cultures is unknown. Since most cells apparently survived elicitation, it seems that a release mechanism other than, or in addition to, diffusion from dead or damaged cells must account for the appearance of sanguinarine and dihydrosanguinarine in the culture medium. Possibilities would include active transport, as described by Yamamoto et al. (1987) for *Thalictrum minus*, and temporary loss of membrane integrity.

The semi-continuous process as employed was deficient in that alkaloid release was incomplete, elicitor treatment sufficiently mild to enable continued cell viability was not effective in restricting cell growth such that subculture was required prior to re-elicitation, and maximum alkaloid yield was realized only in repeatedly elicited, "sensitized" cells. Subsequent work has revealed that although premature cell death (within 3–4 days), such as when medium exchange is omitted or high elicitor concentrations are employed, results in markedly reduced alkaloid production, it is probably not necessary or practical to maintain cell viability beyond a single elicitation cycle. For example, cultures subjected to a more severe, first elicitor treatment (1 vs 0.5 ml of *Botrytis* homogenate) were not viable by the end of the elicitation cycle but produced more alkaloids (over 70 mg per 100 ml culture) than previously recovered from even the third elicitation cycle (Table 1).

4 Elicitation Parameters

It is generally appreciated that the magnitude, and sometimes the nature, of the elicitation response depends upon a number of factors including the cell culture line, the nature and concentration of the elicitor, the timing of elicitation, and the culture environment (Eilert 1987). The effects of a number of such variables on the elicitation response of *Papaver somniferum* cultures have been investigated (Rijnders 1988).

4.1 Effect of Cell Line

Our elicitation studies have been performed primarily with a relatively old, "de-differentiated" poppy cell line, PBI # 2009, a fine-textured, slightly yellowish, fast-growing culture. Newly established lines of *P. somniferum*, which tended to be granular, pigmented, and much slower-growing, accumulated sanguinarine and dihydrosanguinarine when elicited, but showed a rather weak response. In addition, the alkaloid profile of younger cultures was more complex, and significant constitutive production (i.e. in the absence of elicitation) occurred. Since elicitation was cell line-specific in *P. somniferum*, the response of cell line # 2009 only will be considered in the discussion to follow.

4.2 Nature and Concentration of the Elicitor

Eilert et al. (1985) have clearly shown that, quantitatively, the elicitor response was dependent on the nature of the elicitor employed. Complicating any determination of the "usefulness" of a particular elicitor, however, is the existence of an optimum elicitor concentration, as shown in Fig. 2 for *Botrytis* culture homogenate, which we attribute mainly to the culture browning reaction which accompanies benzophenanthridine alkaloid production in *P. somniferum*, i.e. beyond the optimal concentration of elicitor, the severity of the browning reaction affects the viability of elicited cultures and, therefore, inhibits alkaloid production. In practical terms, screening of potential elicitors requires the use of a concentration series for each preparation.

Although Eilert et al. (1985) selected *Botrytis* culture homogenate as the most "effective" elicitor for inducing benzophenanthridine alkaloid synthesis in poppy cultures, homogenates suffer a number of shortcomings including variation in composition among batches of fungus, poor reproducibility of sampling from suspensions, and the strong browning reaction which accompanies alkaloid production. We have initiated studies (unpublished) aimed at isolating an eliciting fraction or fractions from *Botrytis* which would circumvent at least some of these shortcomings. We have determined that although elicitor activity is released into fungal culture filtrate by autoclaving the entire culture, elicitor activity remains primarily associated with the mycelial fraction and is maintained in cell wall preparations. Furthermore, acid hydrolysates of purified cell wall show considerable promise in that a soluble, effective elicitor which causes little or no culture browning can be produced by partial hydrolysis. At this point, however,

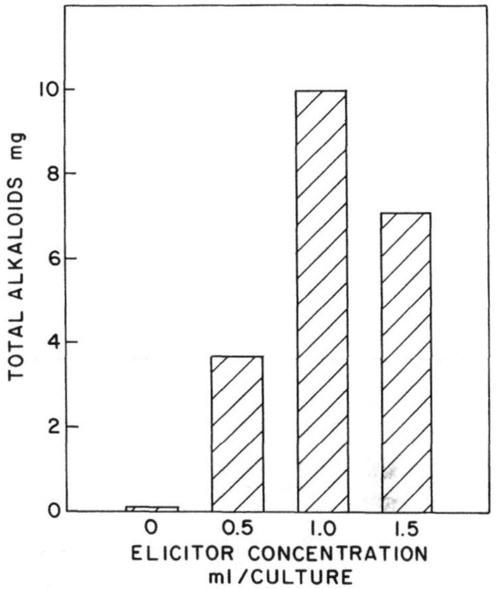

Fig. 2. Effect of elicitor (*Botrytis* culture homogenate) concentration on total alkaloid (sanguinarine and dihydrosanguinarine) yields from 100 ml *Papaver somniferum* cell suspension cultures

production of this elicitor remains unpredictable. Interestingly, use of hydrolysate elicitors has shown that when little or no culture browning occurs, dihydrosanguinarine is essentially the sole product. Fortunately, since oxidation of dihydrosanguinarine to sanguinarine is facile (Stipanovic et al. 1972) accumulation of dihydrosanguinarine when sanguinarine is the desired product is not of practical significance.

4.3 Timing of Elicitation

In most systems studied, the elicitation response has been dependent on the growth stage of the plant cell culture (Eilert 1987). The response may be affected either quantitatively or qualitatively in some systems. Generally, cultures respond to elicitation during the growth phase only, and optimally near the end of the growth phase. In the case of *P. somniferum* cell cultures, we also see the best response late in the growth phase. The elicitation "window" is quite narrow for the 2009 cell line in that we see a marked decline in alkaloid yield if cultures are elicited either shortly prior to or shortly after the optimum elicitation point.

4.4 The Culture Environment

Some significant observations on the effects of culture parameters have been made. Culture of cells in even moderate light (40 μE m^{-2} s^{-1}) promotes culture browning at the expense of alkaloid production as compared to culturing in darkness or dim light. The growth rate of #2009 cells also affects their elicitability. Cells responded to frequent transfer using heavy inocula (such as might be employed to speed up production of biomass) by increasing their rate of growth; however, such cultures exhibited a greater susceptibility to browning and a marked reduction in alkaloid yield, ultimately becoming essentially non-productive. Cultures maintained as suspensions also exhibited a gradual (noticeable over a period of several months) decline in alkaloid yield and an increased browning response, which could be overcome by initiating new suspension cultures from intact callus tissue. This phenomenon is again probably related to a gradual selection for growth at the expense of alkaloid production during repeated transfer of suspension cultures.

5 Conclusions

The sensitivity of *Papaver somniferum* cell line PBI # 2009 to elicitation results in the accumulation of the benzophenanthridine alkaloids sanguinarine and dihydrosanguinarine at levels of 3% or more of cell dry weight, i.e. levels of interest from a commercial point of view. However, a great many obstacles must be cleared in developing a quantitative, controllable process from laboratory-scale, largely qualitative observations. Fundamental aspects such as cell line stability, managing (and minimizing) variability in the response to elicitation, and establishing culture conditions and protocols for maximizing product yield still require considerable research before questions such as "what type of process lends itself best to large-scale production?" and "can the partial release of product into the medium be exploited?" can be addressed logically and the economics of production be given any consideration. Hopefully, the answers to such questions derived from future elicitation studies on *P. somniferum* can be related to other plant cell – elicitor systems for the benefit of plant cell culture technology as a whole.

Acknowledgements. The technical assistance of K.B. Chatson, D.K. Dymond, and B.D. Panchuk and the expert advice of F. Constabel were greatly appreciated.

References

Anonymous (1985) Annual report (1984–85) of Vipont Laboratories Inc, Fort Collins, CO

Berlin J, Forche E, Wray V, Hammer J, Hösel W (1983) Formation of benzophenanthridine alkaloids by suspension cultures of *Eschscholtzia californica*. Z Naturforsch 38c:346–352

Cline SD, Coscia CJ (1988) Stimulation of sanguinarine production by combined fungal elicitation and hormonal deprivation in cell suspension cultures of *Papaver bracteatum*. Plant Physiol 86:161–165

Constabel F, Kurz WGW, Eilert U (1986) Pat Appl 496984 (Can), 06/889247 (USA), 86309182.3 (Eur), 5762/86 (Den), 289735/86 (Jpn)

Eilert U (1987) Elicitation: methodology and aspects of application. In: Constabel F, Vasil IK (eds) Cell culture and somatic cell genetics of plants, vol 4. Academic Press, New York London, pp 153–196

Eilert U, Constabel F (1985) Ultrastructure of *Papaver somniferum* cells cultured in vitro and treated with fungal homogenate eliciting alkaloid production. Protoplasma 128:38–42

Eilert U, Constabel F (1986) Elicitation of sanguinarine accumulation in *Papaver somniferum* cells by fungal homogenates – an induction process. J Plant Physiol 125:167–172

Eilert U, Kurz WGW, Constabel F (1985) Stimulation of sanguinarine accumulation in *Papaver somniferum* cell cultures by fungal elicitors. J Plant Physiol 119:65–76
Furuya T, Ikuta A, Syono K (1972) Alkaloids from callus tissue of *Papaver somniferum*. Phytochemistry 11:3041–3044
Gamborg OL, Miller RA, Ojima K (1968) Nutrient requirements of suspension cultures of soybean root cells. Exp Cell Res 50:151–158
Ikuta A, Syono K, Furuya T (1979) Alkaloids of callus tissues and redifferentiated plantlets in the Papaveraceae. Phytochemistry 13:2175–2179
Krane BD, Fagbule MO, Shamma M (1984) The benzophenanthridine alkaloids. J Nat Prod 47:1–43
Kurz WGW, Constabel F, Eilert U, Tyler RT (1987) Elicitor treatment: a method for metabolite production by plant cell cultures in vitro. In: Breimer DD, Speiser P (eds) Topics in pharmaceutical sciences 1987. Elsevier, Amsterdam New York, pp 283–290
Rijnders COM (1988) *Catharanthus roseus* and *Papaver somniferum* cell cultures. Licentiate Thesis, Leiden Univ
Schumacher H-M, Zenk MH (1988) Partial purification and characterization of dihydrobenzophenothridine oxidase from *Eschscholtzia california* cell suspension cultures. Plant Cell Rep 7:43–46
Schumacher H-M, Gundlach H, Fiedler F, Zenk MH (1987) Elicitation of benzophenanthridine alkaloid synthesis in *Eschscholtzia* cell cultures. Plant Cell Rep 6:410–413
Southard GL, Boulware RT, Walborn DR, Groznik WJ, Thorne EE, Yankell SL (1984) Sanguinarine, a new antiplaque agent: retention and plaque specificity. J Am Dent Assoc 108:338–341
Stipanovic RD, Howell CR, Bell AA (1972) Sanguinarine: a simplified method of isolation. J Hetero Chem 9:1453–1454
Tam WHJ, Constabel F, Kurz WGW (1980) Codeine from cell suspension cultures of *Papaver somniferum*. Phytochemistry 19:486–487
Tyler RT, Eilert U, Rijnders COM, Roewer IA, Kurz WGW (1988) Semi-continuous production of sanguinarine and dihydrosanguinarine by *Papaver somniferum* L. cell suspension cultures treated with fungal homogenate. Plant Cell Rep 7:410–413
Yamamoto H, Suzuki M, Suga Y, Fukui H, Tabata M (1987) Participation of an active transport system in berberine-secreting cultured cells of *Thalictrum minus*. Plant Cell Rep 6:356–359

Elicitation and Suppression of Isoflavones and Pterocarpan Phytoalexins in Chickpea (*Cicer arietinum* L.) Cell Cultures

W. Barz, W. Bless, S. Daniel, W. Gunia, W. Hinderer, U. Jaques, H. Kessmann, D. Meier, K. Tiemann, and U. Wittkampf[1]

1 Introduction

Plants possess sophisticated mechanisms to defend themselves against pathogenic microorganisms. In addition to various constitutively formed, preinfectional mechanisms, the wide range of active defence reactions is of greatest importance (Keen 1986; Collinge and Slusarenko 1987; Fritig et al. 1987). Such reactions are induced after perception of a chemical signal provided by the invading pathogen. The de novo synthesis and differential accumulation in the plant of antimicrobial phytoalexins (Bailey and Mansfield 1982) in incompatible and compatible plant – pathogen interactions play crucial roles in the specificity of host resistance (Dixon 1986; Wood 1986).

Plant cell suspension cultures have been extensively used for the study of the biochemical reactions involved in phytoalexin formation because such systems of reduced complexity greatly facilitate analyses of inducible pathways (Dixon 1986; de Wit 1986; Barz et al. 1988a, b).

Our investigations on the biochemical and molecular biological aspects of the interaction between the chickpea (*Cicer arietinum* L.) plant and its main fungal pathogen *Ascochyta rabiei* (Deuteromycete) aim at elucidating the molecular basis of both plant resistance and fungal virulence (Barz et al. 1988 a, b). In this chapter we shall present recent findings on the elicitation and suppression of phytoalexin formation in chickpea cells as caused by polymeric substances of fungal origin.

2 Elicitation of Pterocarpan Phytoalexins

Accumulation of the phytoalexins medicarpin and maackiain (structures, Fig. 1) in chickpea tissue as an active response against the fungal parasite seems to play a pivotal role (Weigand et al. 1986). The *A. rabiei*-resistant cultivar ILC (International Line of Chickpea) 3279 accumulates large amounts of both phytoalexins rapidly and for prolonged periods of time, whereas the highly susceptible cultivar ILC 1929 produced only very small amounts of medicarpin. This decisive cultivar-specific difference in phytoalexin accumulation has similarly been measured in sliced cotyledons (Kessmann and Barz 1986) and in cell suspension cultures of both cultivars when treated with various exogenous elicitors (Kessmann

[1]Lehrstuhl für Biochemie der Pflanzen, Westfälische Wilhelms – Universität, Hindenburgplatz 55, 4400 Münster, FRG

Primary and Secondary Metabolism of Plant Cell Cultures, ed. by W.G.W. Kurz

and Barz 1987; Barz et al. 1988a, b; Kessmann 1988; Kessmann et al. 1988b). The cell suspension cultures of the two chickpea cultivars have extensively been used with great benefit for biochemical and molecular biological investigations on the expression of elicitor-induced defence reactions. Though different in their ability to accumulate the pterocarpan phytoalexins, the two cell culture lines show a qualitatively and quantitatively similar ability to accumulate the isoflavones, formononetin and biochanin A, together with their 7-0-glucoside and their 7-0-glucoside-6″-malonate conjugates (FGM and BGM). This identical pattern of isoflavones and isoflavone conjugates of the chickpea cultivars had previously been determined at the level of the intact plants (Weigand et al. 1986).

Using the cell suspension culture ILC 3279, the metabolic linkage between the elicitor-induced accumulation of pterocarpan phytoalexins and the concomitant expression of enzyme activities of the energy yielding primary metabolism and of the general phenylpropane pathway from L-phenylalanine to the isoflavone formononetin and further to the malonyl conjugate has been investigated. These studies were based on our detailed knowledge of the enzymes involved in the formation and the turnover of FGM in chickpea (review Barz et al. 1988a). Altogether 14 enzymes of primary and secondary metabolism were assayed for such elicitor-caused changes in enzyme activity and only a few were found to respond to the elicitor either concomitant with or prior to phytoalexin formation (Daniel et al. 1988). The data are summarized in Fig. 1. Among the enzymes of primary metabolism only glucose-6-phosphate dehydrogenase showed an increase in enzyme activity, whereas numerous other enzymes of energy-producing pathways remained unchanged. In the general phenylpropane pathway, which is constitutively expressed in these cell cultures, phenylalanine ammonia lyase (PAL), cinnamic acid-4-hydroxylase, chalcone synthase and the isoflavone-specific 7-0-glucosyltransferase showed a market increase in their activities (Daniel et al. 1988; Gunia, Hinderer, Flentje and Barz unpublished). The data support the assumption that enzymes at the beginning and at branching points of pathways are especially prone to elicitor induction. The increased activity of the isoflavone 7-0-glucosyltransferase is readily explained by the observation that formononetin produced in excess is converted into the conjugate pool. An elicitor-caused increase in the level of FGM has also been measured in intact tissue of chickpea plants (Kessmann et al. 1988b).

The maxima of the transiently occurring increase in enzyme activity and in phytoalexin accumulation were measured some 4–8 h after elicitor application. In general, the increase in enzyme activities was observed a few hours before maximum phytoalexin accumulation occurred. Furthermore, in chickpea cells the phenylpropane pathway and the branch leading to the isoflavone conjugates are a metabolic system composed of both constitutively formed and elicitor – induced enzyme activities; future investigations will have to show whether new enzymes are formed by the activation of new genes or whether the increased enzyme activities have to be explained by acceleration of enzyme production regulated by genes which are also active in the absence of an elicitor.

Chickpea cells excrete phytoalexins into the culture medium where they are oxidatively metabolized by extracellular peroxidases. This excretion of phytoalexins cannot be explained by cell lysis because the concomitantly accumulating

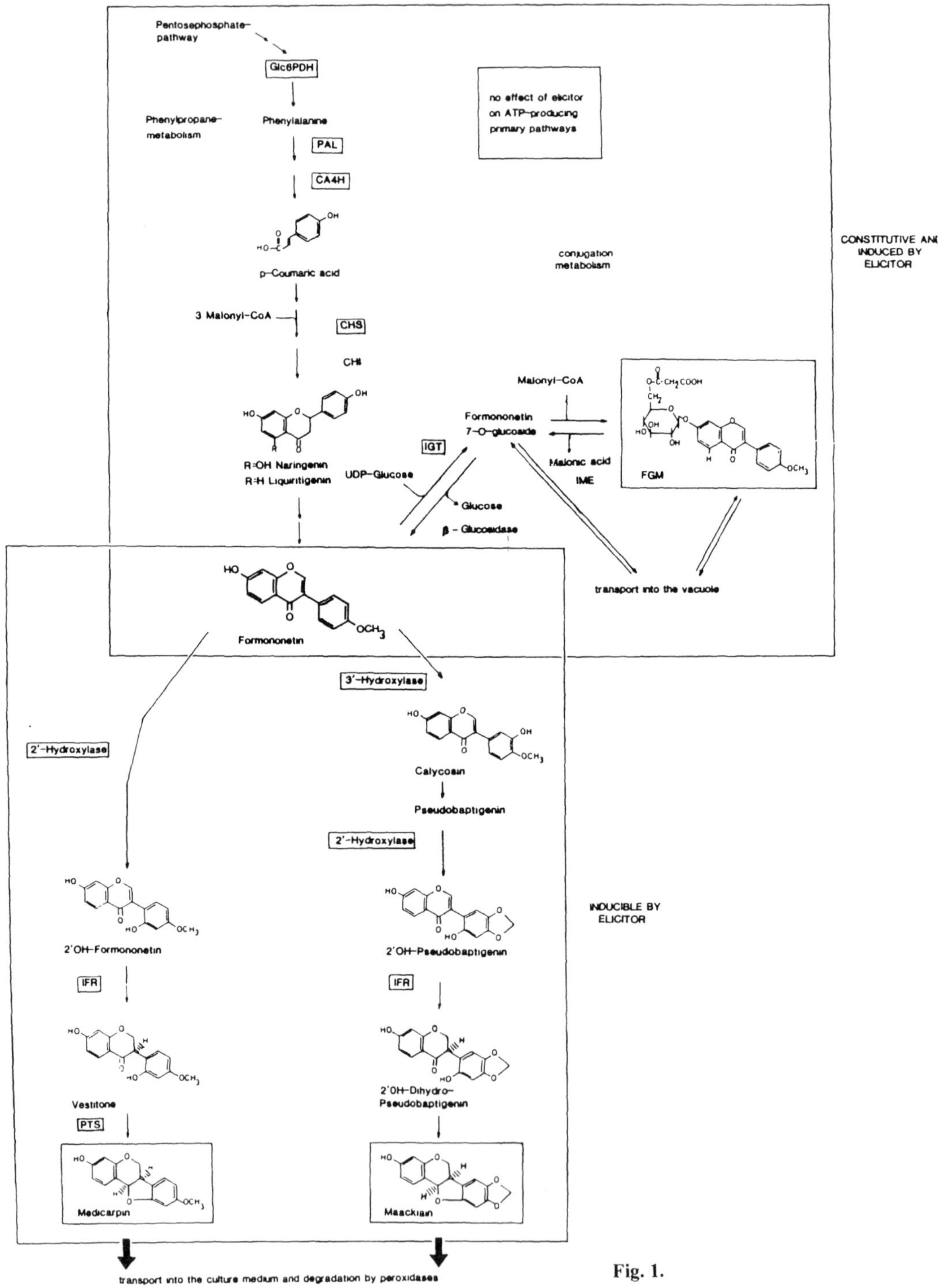
Pentosephosphate-pathway
Glc6PDH
Phenylpropane-metabolism
Phenylalanine
PAL
CA4H
p-Coumaric acid
no effect of elicitor on ATP-producing primary pathways
CONSTITUTIVE AND INDUCED BY ELICITOR
conjugation metabolism
3 Malonyl-CoA
CHS
CHI
R=OH Naringenin
R=H Liquiritigenin
Malonyl-CoA
Formononetin 7-O-glucoside
Malonic acid
IME
FGM
IGT
UDP-Glucose
Glucose
β - Glucosidase
transport into the vacuole
Formononetin
3'-Hydroxylase
Calycosin
Pseudobaptigenin
2'-Hydroxylase
2'-Hydroxylase
2'OH-Formononetin
2'OH-Pseudobaptigenin
INDUCIBLE BY ELICITOR
IFR
IFR
Vestitone
2'OH-Dihydro-Pseudobaptigenin
PTS
Medicarpin
Maackiain
transport into the culture medium and degradation by peroxidases

Fig. 1.

formononetin and especially FGM are not found in the culture fluid but in the cell vacuoles (Barz et al. 1988a). Phytoalexin accumulation in the growth media of cell cultures and not in the producing cells seems to be a widespread phenomenon (Barz et al. 1988b).

3 The Pterocarpan-Specific Biosynthetic Pathway

The isoflavone formononetin is the central intermediate in the biosynthesis of the pterocarpans medicarpin and maackiain (Smith and Banks 1986). This isoflavone may, therefore, be regarded as the metabolic link between the constitutively accumulating formononetin glucosylconjugates and the inducibly formed phytoalexins in chickpea (Fig. 1). Using chickpea cell suspension cultures ILC 3279, the biosynthetic sequence and the enzymology of pterocarpan formation have recently been elucidated in our laboratory. Pterocarpan accumulation can be elicited in such cultures by various elicitors such as yeast extract, yeast elicitor, *Ascochyta rabiei* elicitor or heavy metal ions (Barz et al. 1988a, b; Kessman et al. 1988b; Jaques and Barz unpublished). Elicitation of medicarpin and maackiain formation requires de novo protein synthesis as demonstrated by the pronounced inhibition of phytoalexin accumulation by molecular inhibitors of transcription and translation (Barz et al. 1988a).

The first step in the pterocarpan-specific pathway is catalyzed by a set of microsomal cytochrome P 450 monooxygenases which hydroxylate formononetin in position 2′ and 3′ (Hinderer et al. 1987). Formononetin 2′- and 3′-hydroxylase are clearly elicitor-induced and are different with regard to various biochemical and kinetic properties. The enzymes require molecular oxygen and NADPH as cosubstrates and they are further characterized by carbon monooxide inhibition of enzyme activity, reversion of this inhibition with blue light and inhibition of enzyme activity with various established cytochrome P 450 monooxygenase molecular inhibitors (Flentje and Barz unpublished). All available data lead to the assumption that formononetin 2′-hydroxylase is the rate-limiting enzyme step in pterocarpan phytoalexin formation in chickpea cells. Furthermore, the highly differential accumulation of phytoalexins in chickpea cell cultures ILC 3279 and ILC 1929 (Kessmann et al. 1988b) is at least in part due to the dramatically different

Fig. 1. Scheme of three sets of metabolically linked pathways (phenylpropane metabolism, isoflavone conjugation reactions, pterocarpan-specific biosynthetic branch) in elicitor-treated chickpea cell suspension cultures with formononetin as the central intermediate. Enzymes and compounds shown in *frames* are subject to elicitor-caused increases in specific activity or de novo synthesis and higher accumulation respectively. The *unframed* enzymes and compounds indicate that no elicitor effect could be measured. Abbreviated enzymes are: *Glc6PDH*, glucose-6-phosphate dehydrogenase; *PAL*, L-phenylalanine ammonia lyase; *CA4H*, cinnamic acid 4-hydroxylase; *CHS*, (naringenin) chalcone synthase; *CHI*, chalcone isomerase; *IGT*, isoflavone-specific UDP-glucose: 7-0-glucoslytransferase; *IME*, isoflavone malonylglucoside-specific:malonylesterase; 2′-hydroxylase, NADPH:isoflavone 2′-hydroxylase; 3′-hydroxylase, NADPH:isoflavone 3′-hydroxylase; *IFR*, NADPH:2′-hydroxyisoflavone oxidoreductase; *PTS*, pterocarpan synthase (After Barz et al. 1988b, modified)

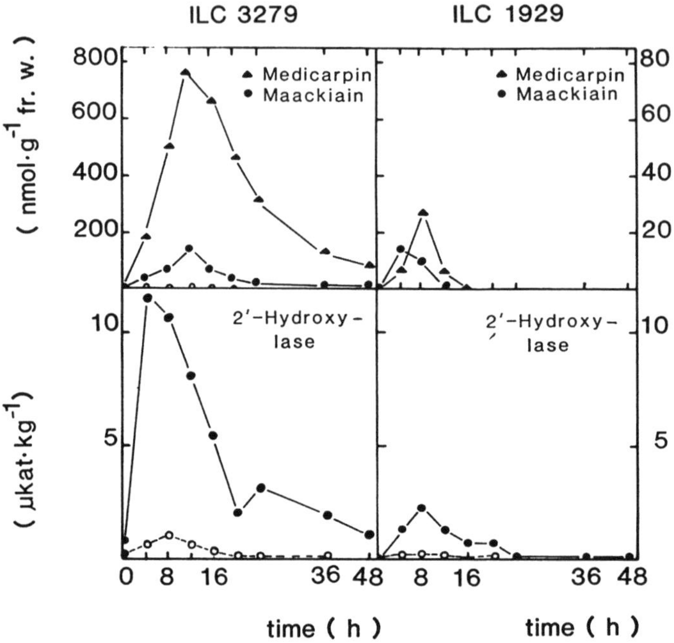

Fig. 2. Time course of the differential induction of the phytoalexins medicarpin and maackiain and of the isoflavone (formononetin) 2′-hydroxylase in chickpea cell suspension cultures ILC 3279 and ILC 1929 with a polysaccharide elicitor from *Saccharomyces cerevisiae*

expression of this enzyme activity in the two cell culture lines (Fig. 2). Other cytochrome P 450 monooxygenases such as cinnamic acid- 4-hydroxylase and formononetin 3′-hydroxylase, which also respond in chickpea cell cultures to elicitor application, do not appear to be involved in the pronounced cultivar-specific difference in phytoalexin formation (Gunia and Barz unpublished). In general, Fig. 1 can be used to document that cell suspension cultures ILC 3279 and ILC 1929 do not differ in their ability to express the enzymes of the general phenylpropane biosynthetic pathway (Daniel, Tiemann and Barz unpublished) or in their ability to accumulate the isoflavones formononetin and biochanin A together with their glucosylconjugates (Kessman and Barz 1987; Kessman 1988). It is hypothesized that one of the essential differences is the introductory hydroxylation step of the pterocarpan-specific pathway.

The second enzyme reaction in the pterocarpan-specific pathway is catalyzed by an NADPH:isoflavone oxidoreductase which is also elicitor-induced. Enzyme activity appears concomitant with or slightly before phytoalexin accumulation can be observed (Tiemann et al. 1987). The enzyme shows a strict specificity for 2′-hydroxyformononetin and 2′-hydroxypseudobaptigenin (cf. Fig. 1). This enzyme provides the chiral centre at the carbon atom 3 in the 2′-hydroxyisoflavone and with regard to the specificity of the hydride transfer from NADPH to carbon atom 2 of the resulting isoflavanone, it belongs to the A-group of the oxidoreductases (Schlieper, Tiemann and Barz unpublished). This enzyme has meanwhile been purified to near homogeneity and it will be used as a molecular target for future molecular biological investigations on the induction and the regulation of the pterocarpan-specific biosynthetic pathway.

The terminal enzyme of pterocarpan biosynthesis has recently been isolated from elicitor-treated chickpea cell suspension cultures ILC 3279 and it was found to be an NADPH requiring pterocarpan synthase (Bless and Barz 1988). The enzyme shows the same elicitor-caused accumulation kinetics as the isoflavone oxidoreductase and it only converts this stereoisomer of vestitone (Fig. 1) which has been formed in the preceding isoflavone oxidoreductase step. This observation is in line with the stereochemistry of the chickpea pterocarpan phytoalexins which possess the 6aR:11aR configuration (Kessmann et al. 1988a).

After the successful isolation of the enzymes of the pterocarpan-specific biosynthesis the complete pathway to medicarpin as shown in Fig. 1 can now be studied at the enzyme level. Except for the 5′-deoxychalcone synthase and the enzyme involved in the formation of the methylenedioxy ring of maackiain, all enzymes from PAL to medicarpin and maackiain are known. This represents the first example of a completely elucidated pterocarpan biosynthetic pathway.

4 Pterocarpan Formation from Constitutive Isoflavone Conjugates

The present concept of elicitor-induced phytoalexin formation holds that these compounds are synthesized de novo from early precursors of primary metabolism by newly induced enzymes (Dixon 1986). In chickpea tissue and cell suspension cultures pterocarpan formation is accompanied by a substantial increase in the level of various phenolic compounds such as intermediates of phytoalexin formation and isoflavone conjugates (Kessmann and Barz 1986; Jaques et al. 1987) among which FGM is especially effected. Similar observations on an elicitor-induced increased accumulation of isoflavone conjugates have been reported for other cell culture systems (Hattori and Ohta 1985). The consumption of such constitutively formed or elicitor-induced glucosyl conjugates for phytoalexin biosynthesis has so far not been reported. The various glucosyl and especially malonylglucosyl conjugates of isoflavones have been considered to be end products stored in vacuoles (Barz et al. 1985).

Recent investigations on the metabolism of isoflavone 7-0-glucoside-6″-0-malonates in chickpea tissue provided evidence for the rapid turnover of formononetin conjugates in contrast to biochanin A derivatives (Jaques et al. 1985). Labelling experiments and studies with a molecular inhibitor of phenalalanine ammonia lyase (AOPP) in chickpea cell suspension cultures have now shown that the formononetin moiety of FGM can be used for pterocarpan phytoalexin biosynthesis in these cells under specific conditions. Cell cultures ILC 3279 in which the FGM pool had been labelled with ^{14}C-formononetin were used. Upon elicitation and concomitant inhibition of PAL with AOPP an almost stoichiometric conversion of formononetin from FGM to the pterocarpan phytoalexins was measured. Determination of the total and the specific radioactivity of FGM and the phytoalexins showed the direct and undiluted conversion of the isoflavone from the conjugate into the pterocarpans. Control experiments with either AOPP and/or elicitor substantiated that the consumption of FGM for phytoalexin biosynthesis depends on the simultaneous effects of elicitation of pterocarpan biosynthesis and the inhibition of PAL as the introductory enzyme of this biosynthetic sequence.

These investigations indicate that such an auxiliary pathway for pterocarpan formation provides for the plant an anapleurotic sequence which can cover the demand of suitable carbon skeletons for phytoalexin formation if pterocarpan biosynthesis can no longer be provided with sufficient amounts of early precursors or if phytoalexins should be in very high demand. These investigations also indicate that elicitation of chickpea cells must lead to changes in vacuolar efflux of phenolic conjugates. Furthermore, the data assign a biological function to the highly substrate-specific hydrolytic enzymes which convert FGM into the aglycone (Hinderer et al. 1986). The enzymes involved in isoflavone conjugation and conjugate hydrolysis have all been found to occur in these chickpea cell suspension cultures (Fig. 1) (Daniel et al. 1988).

5 Suppression of Isoflavone and Pterocarpan Phytoalexin Accumulation

Suppressors are non-phytotoxic, polymeric fungal compounds which inhibit, delay or prevent the elicitor-caused formation of plant defence reactions (Tepper and Anderson 1984; Collinge and Slusarenko 1987). A limited number of reports have demonstrated that fungal glycoproteins, peptides or glucans lead to an inhibition of the elicitor-induced accumulation of phytoalexins or the hypersensitive reaction (Ziegler and Pontzen 1982; Oku et al. 1987; Doke et al. 1987). The exact structure and the mode of action of such repressors remain to be elucidated. In general, however, suppressors are believed to play a key role in some plant-parasite interactions and to be responsible for the phenomenon of "induced susceptibility" (Keen 1986).

A suppressor preparation previously isolated from the culture fluid of *A. rabiei* caused substantial suppression of the accumulation of the isoflavones, the isoflavone glucosylconjugates and the pterocarpan phytoalexins in sliced cotyledons of *Cicer arietinum* as caused by endogenous or exogenous elicitors (Kessmann and Barz 1986).

In a more refined purification procedure suppressor preparations have now been obtained by extraction of germinating *A. rabiei* spores with LiCl and subsequent steps of dialysis, ultrafiltration and high performance liquid chromatography on reversed-phase material. The most active suppressor turned out to be a glycoprotein (molecular mass 10 000–25 000 daltons) with glucose as the quantitatively most important sugar. Polysaccharide fractions with suppressor activity were also obtained. Chemical and enzymatic degradation studies with subsequent assays of the biological activity revealed that the glycoportion of the glycoprotein was responsible for the suppressor activity (Kessmann 1988).

In the cotyledon assay the purified suppressor inhibited in a dose-dependent mode the accumulation of the isoflavone aglycones, the isoflavone glucosylconjugates and the two pterocarpan phytoalexins (Fig. 3). Suppressor-elicitor antagonism could be observed in the case of both the endogenous elicitor liberated by slicing of the cotyledons or the exogenously applied *A. rabiei* elicitor (Fig. 3). Suppressor-caused inhibition of isoflavonoid accumulation required that the fungal inhibitor was applied between 0–12 h after elicitation. Suppressor-elicitor counteraction was demonstrated in experiments where constant amounts of suppressor per assay, together with increasing amounts of elicitor, were used. Alternative experiments with a stepwise increase of the suppressor at constant elicitor concentrations also demonstrated the inhibitory activity of this material on isoflavone and pterocarpan accumulation (Fig. 3). Mathematical evaluation of such competition experiments seems to indicate that elicitor and suppressor do not interact in a simple competitive relationship.

An especially important result of our present investigations on the mode of action of this *A. rabiei*-derived material is depicted in Fig. 4. The aforementioned inhibitory activity could so far only be demonstrated in cotyledon tissue from the highly susceptible cultivar ILC 1929, whereas tissue derived from the *A. rabiei*-resistant cultivar ILC 482 failed to show this susceptibility. Our future studies are based on the hypothesis that the *A. rabiei* suppressor represents a plant cultivar-specific compound which will not exercise its inhibitory potential on *A. rabiei*-

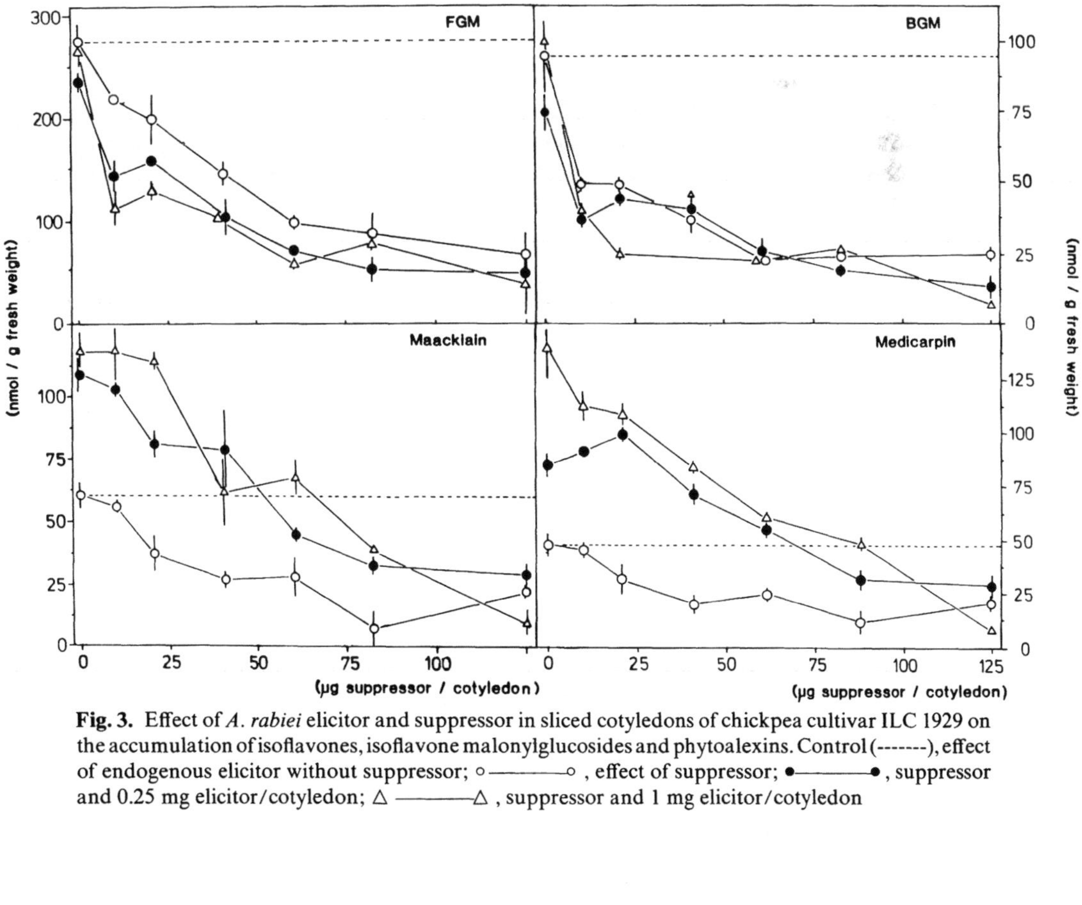

Fig. 3. Effect of *A. rabiei* elicitor and suppressor in sliced cotyledons of chickpea cultivar ILC 1929 on the accumulation of isoflavones, isoflavone malonylglucosides and phytoalexins. Control (-------), effect of endogenous elicitor without suppressor; ○———○, effect of suppressor; ●———●, suppressor and 0.25 mg elicitor/cotyledon; △———△, suppressor and 1 mg elicitor/cotyledon

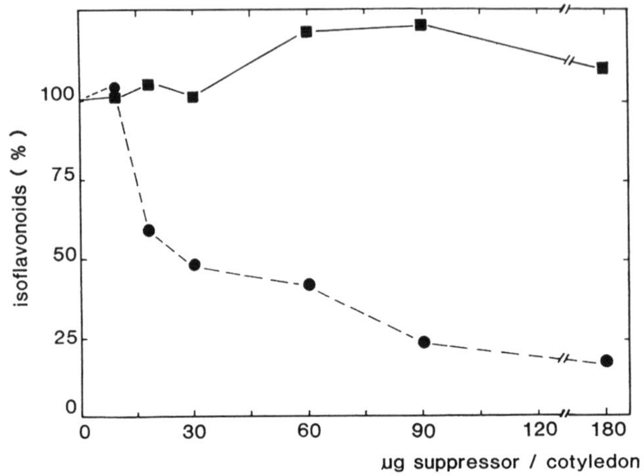

Fig. 4. Dose response of *A. rabiei* suppressor on the accumulation of isoflavonoids (biochanin A, formononetin, BGM, FGM, medicarpin and maackiain) in sliced cotyledons of chickpea cultivars ILC 482 (■——■, *A. rabiei*-resistant) and ILC 1929 (●——●, *A. rabiei*-susceptible)

resistant chickpea genotypes. This assumption appears to be in agreement with the present phytopathological concept of a suppressor as part of a host-parasite interaction (Keen 1986; Collinge and Slusarenko 1987; Doke et al. 1987; Kessmann 1988).

Current investigations with chickpea cell suspension cultures ILC 3279 and ILC 1929 are designed to demonstrate whether the inhibitory activity of the *A. rabiei* suppressor can also be measured in cultured cells. Preliminary data indicate that in the suspension culture ILC 1929 elicitor-induced phytoalexin accumulation can also be partially inhibited (Meier and Barz unpublished). Such a system would provide a more practical experimental model to investigate the mode of action and to demonstrate the actual target of a fungal suppressor in plant cells.

6 Conclusions

Our investigations have shown that chickpea cell suspension cultures allow the study of plant cultivar-specific differences involved in a host-parasite interaction. Phytoalexin formation formally represents a terminal branch of the constitutively expressed phenylpropane isoflavone biosynthetic pathway, although the data seem to indicate that this part of secondary metabolism is also significantly effected by an elicitor. The pterocarpan-specific biosynthetic sequence from the isoflavone formononetin is catalyzed by inducibly formed enzymes which will only occur in elicited cells. Under specific physiological conditions constitutively accumulating isoflavone conjugates such as FGM can be funnelled into phytoalexin formation. The actual level of pterocarpan phytoalexin accumulation may be regulated by an antagonism between an elicitor and a suppressor in which the latter could be of greater importance due to its demonstrated cultivar specificity. As a final result of our recent investigations it should be emphasized that the complete pterocarpan biosynthetic pathway has now been elucidated at the enzymatic level. Molecular biological investigations of this reaction sequence will contribute to a better understanding of the specific role of fungal elicitors and fungal suppressors in regulating plant antimicrobial defence mechanisms.

Acknowledgements. Our investigations were generously supported by the Deutsche Forschungsgemeinschaft and Fonds der Chemischen Industrie. We thank the management of ICARDA, Aleppo, Syria, for chickpea seeds.

References

Bailey JA, Mansfield JA (1982) Phytoalexins. Blackie, Glasgow London

Barz W, Köster J, Weltring KM, Strack W (1985) Recent advances in the metabolism and degradation of phenolic compounds in plants and animals. Annu Proc Phytochem Soc Eur 25:307–347

Barz W, Daniel S, Hinderer W, Jaques U, Kessmann H, Köster J, Tiemann K (1988a) Elicitation and metabolism of phytoalexins in plant cell cultures. In: Pais MSS (ed) Plant cell biotechnology. Springer, Berlin Heidelberg New York Tokyo, pp 211–230

Barz W, Daniel S, Hinderer W, Köster J, Otto C, Tiemann K (1988b) Elicitation and metabolism of phytoalexins in plant cell cultures. In: Yamada Y (ed) Application of plant cell and tissue culture (CIBA Found Symp 137). Wiley & Sons, New York Chichester, pp 178–191

Bless W, Barz W (1988) Isolation of pterocarpan synthase, the terminal enzyme of pterocarpan phytoalexin biosynthesis in cell suspension cultures of *Cicer arietinum.* FEBS Lett 235:47–50

Collinge DB, Slusarenko AJ (1987) Plant gene expression in response to pathogens. Plant Mol Biol 9:389–410

Daniel S, Hinderer W, Barz W (1988) Elicitor-induced changes of enzyme activities related to isoflavone and pterocarpan accumulation in chickpea (*Cicer arietinum*) cell suspension cultures. Z Naturforsch Ser C 43:536–544

De Wit PJGM (1986) Elicitation of active resistance mechanisms. In: Bailey JA (ed) Biology and molecular biology of plant-pathogen interactions. Springer, Berlin Heidelberg New York Tokyo, pp 149–169

Dixon RA (1986) The phytoalexin response: elicitation, signalling and control of host gene expression. Biol Rev 61:239–291

Doke N, Chai HB, Kawaguchi A (1987) Biochemical basis of triggering and suppression of hypersensitive cell response. In: Nishimura S, Vance CP, Doke N (eds) Molecular determinants of plant disease. Springer, Berlin Heidelberg New York London Paris Tokyo, pp 235–252

Fritig B, Kauffmann S, Dumas B, Geoffroy P, Kopp M, Legrand M (1987) Mechanism of the hypersensitive reaction of plants. In: Evered D, Harnett S (eds) Plant resistance to viruses (CIBA Found Symp 133). Wiley & Sons, New York Chichester, pp 92–108

Hattori T, Ohta Y (1985) Induction of phenylalanine ammonia lyase activation and isoflavone glucoside accumulation in suspension cultured cells of red bean, *Vigna angularis*, by phytoalexin elicitors, vanadate and elevation of medium pH. Plant Cell Physiol 26:1101–1110

Hinderer W, Köster J, Barz W (1986) Purification and properties of a specific isoflavone 7-0-glucoside-6″-malonate malonylesterase from roots of chickpea (*Cicer arietinum*). Arch Biochem Biophys 248:570–578

Hinderer W, Flentje U, Barz W (1987) Microsomal isoflavone 2′-and 3′-hydroxylases from chickpea (*Cicer arietinum*) suspension cultures induced for pterocarpan phytoalexin formation. FEBS Lett 214:101–106

Jaques U, Köster J, Barz W (1985) Differential turnover of isoflavone 7-0-glucoside-6″-0-malonates in *Cicer arietinum* roots. Phytochemistry 24:949–951

Jaques U, Kessmann H, Barz W (1987) Accumulation of phenolic compounds and phytoalexins in sliced and elicitor-treated cotyledons of *Cicer arietinum*. Z Naturforsch 42c:1171–1178

Keen NT (1986) Pathogenic strategies of fungi. In: Lugtenberg B (ed) Recognition in microbe-plant symbiotic and pathogenic interactions. Springer, Berlin Heidelberg New York London Paris Tokyo, pp 171–188

Kessmann H (1988) Biochemische Untersuchungen zur Interaktion des phytopathogenen Deuteramyceten *Ascochyta rabiei* (pass.) Lab. und seines Wirtes, *Cicer arietinum* L. unter besonderer Berücksichtigung der beteiligten Signalsubstanzen. Doct Thesis, Münster Univ

Kessmann H, Barz W (1986) Elicitation and suppression of phytoalexin and isoflavone accumulation in cotyledons of *Cicer arietinum* L. as caused by polymeric compounds from the fungus *Ascochyta rabiei*. J Phytopathol 117:321–335

Kessmann H, Barz W (1987) Accumulation of isoflavones and pterocarpan phytoalexins in cell suspension cultures of different cultivars of chickpea (*Cicer arietinum*). Plant Cell Reop 6:55–59

Kessmann H, Tiemann K, Jansen JR, Reuscher H, Bringmann G, Barz W (1988a) In-vivo characterization of NADPH: 2′-hydroxyisoflavone oxidoreductase in elicitor treated chickpea cell cultures and stereochemical aspects of the phytoalexins medicarpin and maackiain. In: Pais MSS (ed) Plant cell biotechnology. Springer, Berlin Heidelberg New York Tokyo 231–238

Kessmann H, Daniel S, Barz W (1988b) Elicitation of pterocarpan phytoalexins in cell suspension cultures from different cultivars of chickpea (*Cicer arietinum*) by an elicitor from the fungus *Ascochyta rabiei*. Z Naturforsch Ser C 43:529–535

Oku H, Shiraishi T, Ouchi S (1987) Role of specific suppressors in pathogenicity of *Mycosphaerella* species. In: Nishimura S, Vance CP, Doke N (eds) Molecular determinants of plant disease. Springer, Berlin Heidelberg New York London Paris Tokyo, pp 145–156

Smith DA, Banks SW (1986) Biosynthesis, elicitation and biological activity of isoflavonoid phytoalexins. Phytochemistry 25:979–995

Tepper CS, Anderson AJ (1984) The genetic basis of plant-pathogen interaction. Phytopathology 74:1143–1145

Tiemann K, Hinderer W, Barz W (1987) Isolation of NADPH: isoflavone oxidoreductase, a new enzyme of pterocarpan phytoalexin biosynthesis in cell suspension cultures of *Cicer arietinum*. FEBS Lett 213:324–328

Weigand F, Köster J, Weltzien HC, Barz W (1986) Accumulation of phytoalexins and isoflavone glucosides in a resistant and a susceptible cultivar of *Cicer arietinum* during infection with *Ascochyta rabiei*. J Phytopathol 115:214–221

Wood RKS (1986) Introductory comments on host-parasite interaction. In: Bailey JA (ed) Biology and molecular biology of plant-pathogen interactions. Springer, Berlin Heidelberg New York Tokyo, pp 1–13

Ziegler E, Pontzen R (1982) Specific inhibition of glucan-elicited glyceollin accumulation in soybeans by an extracellular mannan-protein of *Phytophthora megasperma* f sp. *glycinea*. Physiol Plant Pathol 20:321–331

Elicitor Induction of Secondary Metabolism in Dedifferentiated and Differentiated in Vitro Systems of *Ruta graveolens*

U. Eilert[1]

1 Introduction

In response to fungal elicitors dedifferentiated callus and suspension cultures of *Ruta graveolens* rapidly accumulate rutacridone epoxide (RE) and hydroxy-rutacridone epoxide (OH-RE; Eilert et al. 1984, Wolters and Eilert 1982). With their mode of accumulation (kinetics) and their antibiotic properties (Wolters and Eilert 1981) acridone epoxides fit very well the conditions postulated for a phytoalexin (Kuc 1972). The concept of phytoalexin induction, however, is a phytopathological one and thus inseparably linked to the differentiated plant. Consequently, to integrate the process into its plant physiological context, studies with differentiated *Ruta graveolens* plants are required.

What is known about elicitor-induced secondary metabolism in differentiated versus dedifferentiated systems? Cell culture systems were developed by the plant pathologist in order to facilitate the investigation of particular questions related to elicitor-induced secondary metabolite formation in plants. Generally, the respective elicitor-induced phytoalexin formation was achieved in callus or suspension cultures derived from phytoalexin-forming plant species (for review, Eilert 1987).

In the case of elicitor-induced acridone epoxide formation, elicitor induction of secondary metabolite accumulation was observed for the first time in a cell culture system with the reaction of the plant unknown. Can one conclude from the response of a cell culture to elicitation what the response of the differentiated plant will be? Elicitor induction of sanguinarine formation in cell cultures of *Papaver somniferum* (Eilert et al. 1985) and of monoterpene indolalkaloid formation in cell cultures of *Catharanthus roseus* (Eilert et al. 1986) proved to be processes which do not occur in the differentiated plant. Compounds, which belong to the pool of constitutive secondary metabolites in a plant, may be inducible in the respective cell culture. This transition from constitutive to inducible status of pathways is not all that surprising, as within the plant in some tissues a pathway may also be expressed constitutively while in others it is inducible, e.g. stilbene accumulation in conifers (Hart 1981) or phenolic and steroidal alkaloid accumulation in potatoes (Kuc 1972).

Ruta graveolens forms an unusually rich spectrum of secondary metabolites. Accordingly, a more general investigation of the effect of fungal elicitors on

[1]Institut für Pharmazeutische Biologie, der Technischen Universität Braunschweig, Mendelssohnstraße 1, D-3300 Braunschweig, FRG

Primary and Secondary Metabolism of Plant Cell Cultures, ed. by W.G.W. Kurz

different pools of secondary metabolites was appropriate. Besides acridone epoxide levels, the content of furoquinoline alkaloids and furanocoumarins, both also groups comprising compounds with antibiotic and/or phototoxic properties (Wolters and Eilert 1981), were monitored in the experiments.

Elicitor-stimulated acridone alkaloid accumulation was shown to be an induction process (Eilert and Wolters 1987), which allows one to monitor the effect of elicitation not only on the product level but also on the level of biosynthetic enzyme activity. The activity of S-adenosyl-L-methionine:anthranilic acid N-methyltranferase (NMT), the first step in acridone alkaloid biosynthesis (Baumert et al. 1983), was selected.

2 Experimental Methods

2.1 In Vitro Culture Systems

1. R-MS Culture: A long established suspension culture of *R. graveolens*. This suspension culture was established from a light-grown, green callus in 1981, which was initiated by Reinhard et al. (1968) in 1965. The cells were grown in 250-ml Erlenmeyer flasks containing 75 ml nutrient medium on a gyratory shaker (120 rpm) and subcultured every 4 weeks. The nutrient medium used was a Murashige-Skoog (MS) medium (Murashige and Skoog 1962) containing 0,1 mg l^{-1} 2,4-D (2,4-dichlorophenoxy acetic acid), 0,25 mg l^{-1} kinetin and 3% sucrose. Additional culturing conditions were a room temperature of 25°C and a 16 h day^{-1} photoperiod. Cells of this cell line form large, compact clumps and are light to dark green in appearance.

2. RN Culture: A newly established suspension culture of *R. graveolens*. In 1986 this cell line was established from explants of sterile seedlings. Culturing was performed on B5 medium (Gamborg et al. 1968) supplemented with 1 mg l^{-1} 2,4-D and 2% sucrose. Culturing conditions were otherwise the same as with R-MS cultures. The RN cell line only forms small, coarse, yellow-greenish aggregates and exhibits a strong tendency to redifferentiate into small, leaflike structures.

3. RAT-Shoot Teratoma Culture: A shoot teratoma culture of *R. graveolens* was initiated by transformation with *Agrobacterium tumefaciens* CT 58 following the protocol as given in Eilert et al. (1987). The teratoma culture was grown on hormone-free B5 medium (OB5 medium) with 2% sucrose under the same culture conditions as for R-MS cultures.

4. Hydroponically and Aseptically Grown Plants (ROH): Aseptically grown seedlings were transferred to liquid, hormone-free B5 medium containing 2% sucrose and cultured under the same conditions as the other cultures. Subculturing was performed every 6–8 weeks by cutting off rooted shoots. Compared to RAT cultures growth was much slower and the tissue consisted of properly developed roots and shoots with leaves only. Callus formation did not occur at all.

2.2 Assay of S-Adenosyl-L-Methionine:Anthranilic Acid N-Methyltransferase (NMT) Activity

Enzyme extraction and assay were performed in a modified version after Baumert et al. (1983).

2.3 Identification and Determination of Secondary Metabolites

The content of acridone epoxides was determined microbiologically using the TLC-strip assay as described by Wolters and Eilert (1982). Furoquinolines and furanocoumarins were determined using capillary gas chromatographic separation techniques. The compounds were identified by comparison with the GLC-MS data of authentic material. GLC separation was performed under similar conditions as given above and mass spectra were recorded with a Finnigan MAT 4515 (quadrupole) instrument.

2.4 Elicitor Preparation and Elicitor Experiments

The preparation of *Rhodotorula* elicitor homogenate was performed as described previously (Eilert et al. 1984). Amounts as indicated were added to the cultured cells at times as indicated in the respective experiment. Cells and medium were harvested at intervals given in the figures.

3 Results

3.1 Acridone Epoxide Formation

In R-MS cultures a rapid 10- to 15-fold increase of NMT activity, which peaks approx. 12 to 16 h after elicitation, was observed. Then the activity declines again. In cultures of different age the time course of NMT activity is similar, however, younger cultures reach higher levels and show a much slower decline.

When subjected to elicitation in the newly established cell line (RN) NMT activity takes a similar course (Fig. 1b) as observed previously in R-MS cultures, and the same maximum activity of 15 μkat kg^{-1} was determined. Constitutive levels of acridone epoxides were somewhat higher than in the R-MS line, but similarly in response to elicitation, the OH-RE level multiplied more than the RE level. In contrast to the other culture, the onset of elicitor-induced accumulation was delayed and both compounds reached a maximum level about 72 h after elicitation (Fig. 1a).

The shoot teratoma culture (RAT) of *R. graveolens* is characterized by vigorous growth and it is almost free of dedifferentiated calluslike cell masses. During 3 years of subculturing root formation was never observed. Characteristic for shoot tissue is a low constitutive amount of acridone epoxides (Table 1a). Constitutive NMT

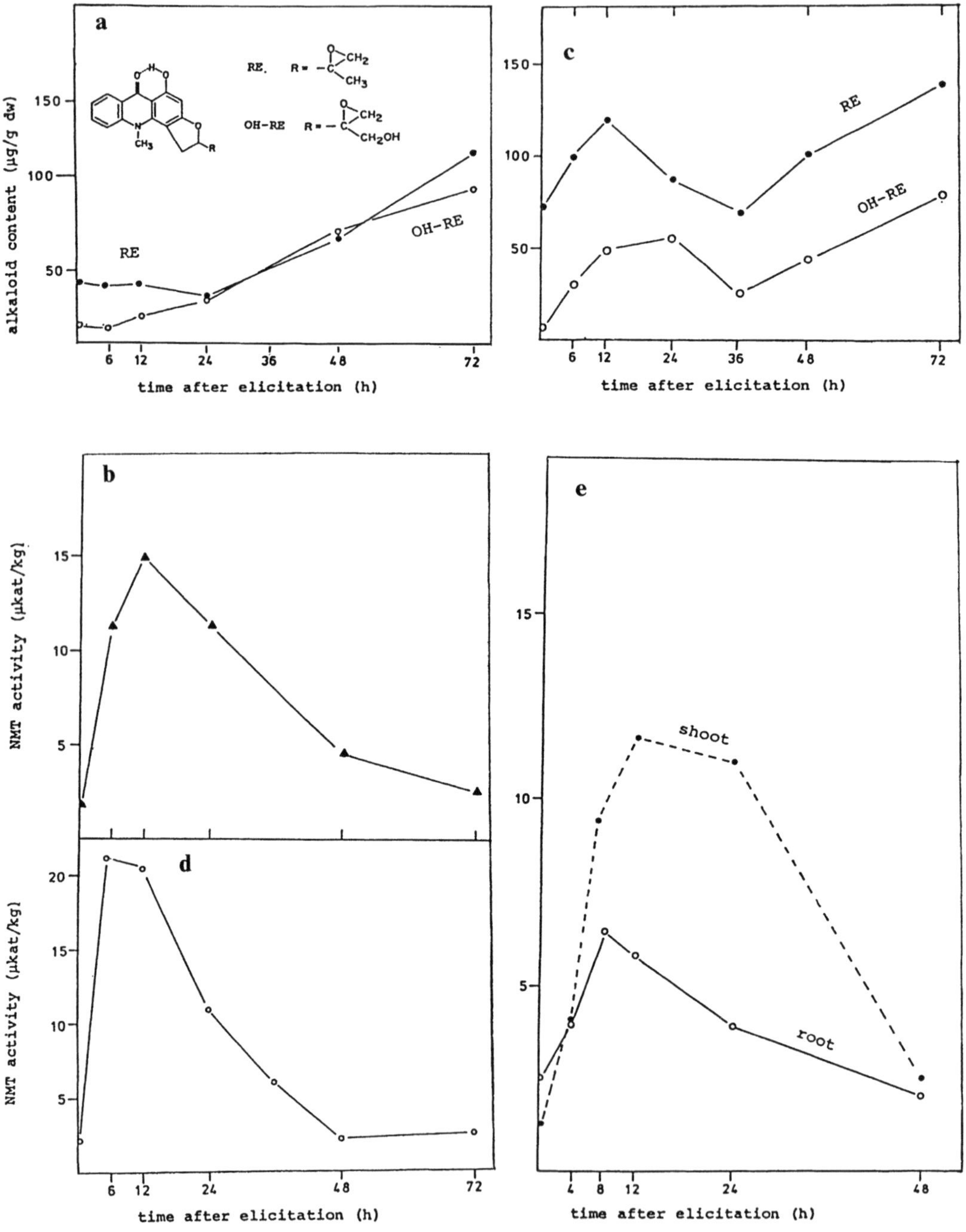

Fig. 1a-e. Time courses of elicitor-induced NMT activity and acridone epoxide accumulation in different in vitro systems of *Ruta graveolens*. **a, b** RN suspension, 10-day culturing period, 1B5 medium, 0.5 ml elicitor/100 ml culture. **c, d** RAT teratoma culture, 12-day culturing period, 0B5 medium, 0.3 ml elicitor/100 ml culture. **e** R-OH hydroponically grown plants, 28-day culturing period, 0B5 medium, 1 ml elicitor/100 ml culture

Table 1. Constitutive and maximum inducible levels of secondary metabolites in different in vitro culture systems of *Ruta graveolens* at (**a**) the constitutive level and (**b**) maximum of elicitor-induced levels

Table 1a

R. graveolens in vitro system		Acridone alkaloid[a] ($\mu g\ g^{-1}$ dry wt.) RE	OH-RE
R-MS suspension	a)	10– 50	1– 20
(MS medium)	b)	30–100	150–300
RN suspension	a)	10– 50	1– 10
(1B5 medium)	b)	30–100	20– 70
RAT culture	a)	50–300	5– 50
(OB5 medium)	b)	200–500	50–150
R-OH hydroponics	a)	100–200	10– 20
(OB5 medium)	b)	300–500	20– 50

[a] RE, rutacridone epoxide; OH-RE, hydroxy-rutacridone epoxide.

Table 1b

R. graveolens in vitro system		Furoquinoline alkaloids[a] ($\mu g\ g^{-1}$ dry wt.) D	F	S	K	E
R-MS suspension	a)	Trace	0– 20	Trace	Trace	150– 200
(MS medium)	b)	100–200	30–100	Unaltered	Unaltered	300– 400
RN suspension	a)	20– 40	80–120	200–300	80–200	400– 700
(1B5 medium)	b)	80–100	150–250	Unaltered	Unaltered	Unaltered
RAT culture	a)	10– 40	80–200	200–500	40–100	1500–3000
(OB5 medium)	b)	100–200	150–500	Unaltered	Unaltered	2000–4000
R-OH hydroponics	a)	1– 10	200–500	300–700	50–100	2000–6000
(OB5 medium)	b)	Unaltered	Unaltered	Unaltered	Unaltered	Unaltered

[a] D, dictamnine; F, γ-fagarine; S, skimmianine; K, kokusaginine; E, edulinine.

Table 1c

R. graveolens in vitro system		Furanocoumarins[a] ($\mu g\ g^{-1}$ dry wt.) P	X	B
R-MS suspension	a)	0– 50	25– 100	75– 150
(MS medium)	b)	200– 400	Unaltered	Unaltered
RN suspension	a)	20– 100	200– 600	600–1000
(1B5 medium)	b)	500– 700	Unaltered	Unaltered
RAT culture	a)	150– 250	3000–6000	4000–8000
(OB5 medium)	b)	600–1000	Unaltered	Unaltered
R-OH hydroponics	a)	2000–3000	8000–10000	1000–2000
(OB5 medium)	b)	Unaltered	Unaltered	Unaltered

[a] P, psoralen; X, xanthotoxin; B, bergapten.

activity was as low as in dedifferentiated cultures. Elicitor treatment resulted in the same rapid activity increase as in R-MS and RN cultures, reaching an even higher maximum activity with about 20 μkat kg^{-1} followed by a decline of activity in the well-known mode (Fig. 1d). Acridone epoxide accumulation was also induced. The time course (Fig. 1c) resembles the one of acridone epoxide accumulation in elicitor-treated R-MS cultures (Eilert et al. 1984). As in the other systems, in relation to the initial concentrations, the OH-RE level was much more raised than the RE level. Comparing the absolute values, however, the contents of both compounds increased by similar amounts.

Aseptically grown plants constitute an excellent system to study whether elicitor-induced changes of secondary metabolism play any important role in the intact plant. Any previous microbial contact which might cause a change in the qualitative and quantitative pattern of constitutive compounds is excluded. At the end of experiments roots and shoots were harvested separately. Figure 1e shows the time courses of NMT activity in roots and shoots in response to elicitor treatment. The initial activity in roots, the main site of constitutive acridone epoxide accumulation, was twice as high as in shoot tissue. In both cases elicitor treatment resulted in an increase of NMT activity which followed the usual pattern. The maximum level of NMT activity in shoots compares very well with the ones of the dedifferentiated cell cultures and does not reach the activity as induced in shoot teratoma cultures. Interestingly, the roots only respond with a comparatively weak increase of NMT activity. As the hydroponically grown plants produce only comparatively small amounts of root biomass, acridone epoxide analysis remained restricted to the shoot tissue. Here, the constitutive levels of RE and OH-RE varied considerably from flask to flask. To eliminate this variation control samples were taken from each flask prior to elicitation. Comparison of the data clearly showed, on the average, a two-fold increase in RE and a two- to six-fold increase of OH-RE content 24 to 48 h after elicitation (Table 1a).

3.2 Furoquinoline Accumulation

In contrast to acridone epoxides, which are like all rutacridone-type compounds predominantly located in the roots, furoquinolines are prevailingly found in the above ground parts of *R. graveolens*. The furoquinolines dictamnine, τ-fagarine, skimmianine and kokusagine, differing only in their methoxylation pattern, and edulinine, a 2(1*H*)-quinolone, are the main compounds.

In the R-MS cells edulinine constitutes the main compound of this group. Figure 2(a, b and c) shows the time courses of the three compounds edulinine, dictamnine and τ-fagarine after treatment of cultured R-MS cells with different amounts of elicitor. Dictamnine rapidly accumulates upon elicitation. Maximum levels were obtained after addition of 1 ml elicitor per 100 ml culture and 24 h after elicitation. While the medium of controls was free of alkaloids, 40 to 60% of the total dictamnine of a culture (cells and medium) was present in the medium at that time point. Accumulation of τ-fagarine, a methoxylated metabolite of dictamnine, starts later and lasts at least over 72 h after elicitation. Again about 50% of the total alkaloid was found in the medium. Edulinine content increased to a maximum of

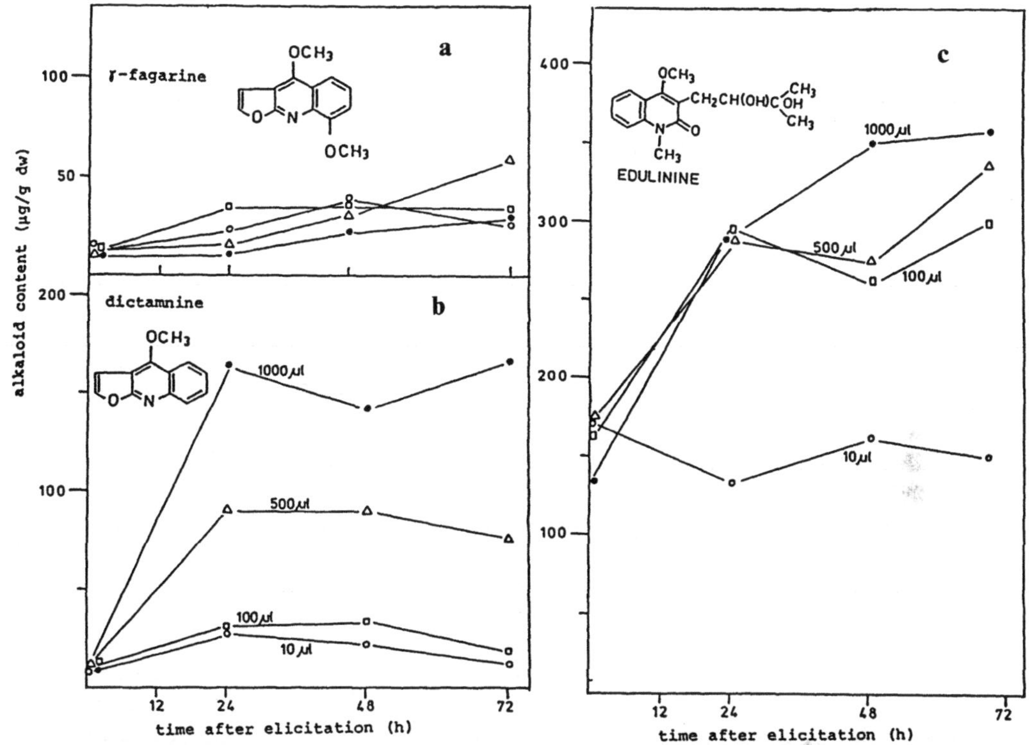

Fig. 2a-c. Time course of elicitor-induced furoquinoline alkaloid accumulation and effect of elicitor concentration. In vitro system: R-MS suspension culture, 8-day culturing period, MS medium

500 µg g^{-1} dry weight 72 h elicitation with only traces being found in the medium. In contrast to dictamnine, but as with acridone epoxides, maximum amounts were already inducible with 100 µl elicitor per 100 ml culture.

The constitutive furoquinoline pattern in cells of the RN line deviated significantly from the one of the R-MS line (Table 1b). In this cell line with the bimethoxylated τ-fagarine and the trimethoxylated compounds, skimmianine and kokusagine, the biosynthetically further advanced metabolites are present in similar amounts as in the teratoma or the shoots (Table 1b). Upon elicitor treatment, however, only dictamnine and τ-fagarine levels were increased (Table 1b). The RAT shoot teratoma responds to elicitation only with a five- to six-fold increase of the cellular dictamnine level (Table 1b), but 60 to 80% of the total dictamnine may be found in the culture medium. The absolute levels accumulated are similar to the ones in the other systems. Hydroponically grown shoots, in contrast, contain only traces of dictamnine and elicitor-induced accumulation was neither noted in the cells (Table 1b) nor in the medium.

3.3 Furanocoumarin Accumulation

The constitutive furanocoumarin spectrum of the R-MS culture is fairly simple and the amounts present are only a fraction of the content of the differentiated teratoma cultures and hydroponics. The two isomers, xanthotoxin and bergapten, are the main compounds (Table 1c). No furanocoumarins were present in the medium. Treatment with 0.5 to 1 ml elicitor for an optimum response resulted in a dramatic accumulation of psoralen, a precursor both of xanthotoxin and bergapten. The levels of the other furanocoumarins remained unaffected. Accumulation reached its maximum 24 h after elicitation with 50 to 80% of the total psoralen present in the medium. After 48 to 72 h the psoralen level dropped back to its initial concentration.

The constitutive furanocoumarin level of the RN culture proved to be five to ten times higher as compared to the R-MS culture (Table 1c) with the pattern much more complex. Isopimpinellin present at about 500 $\mu g\ g^{-1}$ dry weight was a major compound besides xanthotoxin and bergapten. In addition, chalepensin showed up at a concentration of about 100 $\mu g\ g^{-1}$ dry wt. and psoralen was present with about 50 $\mu g\ g^{-1}$ dry wt. Similarly, elicitation stimulated a rapid psoralen accumulation, reaching already its maximum level (600 $\mu g\ g^{-1}$ dry wt.) 12 h later. In contrast to the R-MS culture, this level was maintained over a period of 72 h. The level of other coumarins remained unaffected.

Teratoma cultures and hydroponically grown plants possess similar qualitative furanocoumarin patterns but deviate somewhat in the ratio of the compounds. In RAT cultures bergapten is the major compound while in ROH cultures xanthotoxin takes this position. Psoralen levels of RAT cultures are only 10% of the amount found in hydroponics. Upon elicitor treatment the psoralen content of RAT cultures rapidly increases about five-fold but still does not meet the level constitutively present in ROH cultures. Here, elicitor treatment does not lead to any increase of furanocoumarin accumulation.

4 Discussion

The investigation proves that elicitor treatment induces an extensive transient redirection of secondary metabolite accumulation in *Ruta graveolens* cell cultures. Elicitation, thus, is a multicomponent process. Figure 3 shows the pathways involved and at the same time points out some similarities of the biochemical reactions involved.

Comparing the effect of elicitor treatment on acridone biosynthesis in the in vitro systems in all cases, an induction of NMT activity was observed. The roots, however, possessed a measurable constitutive NMT activity and showed only a weak increase in response to elicitation. Acridone epoxide accumulation also resulted in the systems tested but the amounts were not correlated to the NMT activity. The constitutive levels of acridone epoxides also increased with increasing differentiation. The results strongly support the concept of acridone epoxides having a true phytoalexin character at least in the aerial part of the plant.

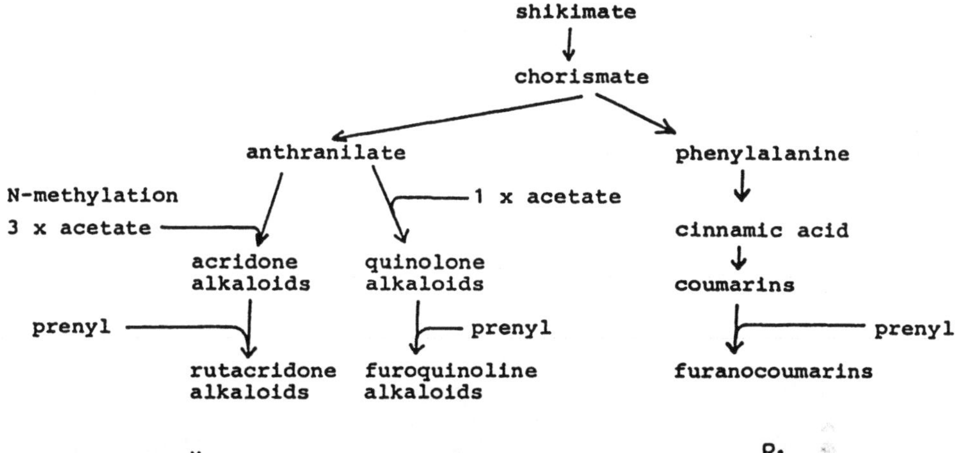

Fig. 3. Biosynthetic pathways of elicitor-inducible secondary compounds of *Ruta graveolens*

Elicitor-induced accumulation of furoquinoline alkaloids is reported for the first time. The experiments show their induction in dedifferentiated cultures as well as in the teratoma line. The lack of any induction in the hydroponics strongly suggests that they are not true phytoalexins.

A comparison shows that different elicitor concentrations are necessary for optimum accumulation of the three groups of compounds. Maximum accumulation of acridone epoxides is induced at low elicitor concentrations. At five to ten times this amount, furoquinolines and furanocoumarins are additionally induced. At present the question cannot be answered whether the same elicitor compound is inducing different pathways in a concentration-dependent mode or whether three different elicitor compounds in the crude homogenate are responsible for the differential elicitation of the pathways.

Elicitor induction of furanocoumarins was reported for parsley, celery and *Ammi majus* (Matern et al. 1988) and these compounds can be considered phytoalexins for these Apiaceae. In *Ruta* the constitutive levels in the plant are high and are not increased further by stress.

In cell cultures, however, with constitutive furanocoumarin accumulation being poor, microbial stress stimulates rapid psoralen accumulation. Here, the term "phytoalexin" does not apply to the furanocoumarins. These data show that in the related families *Rutaceae* and *Apiaceae* the same inducible response mechanism to microbial stress is present, at least genetically. In addition, in the well-differentiated shoot tissue of *Ruta*, evolution selected for a constitutive expression which seems to meet already the physiological limits of the cells. Thus, the inducible mechanism remains silent or hidden.

In conclusion, the experiments show that a transient gradient exists between constitutive and elicitor-inducible expression of secondary metabolism in *Ruta graveolens*. With an increasing degree of differentiation a shift towards constitutive expression is noted. An elicitor-inducible pathway in an in vitro system does not necessarily reflect a plant physiological relevant process; the respective compounds should not be referred to as phytoalexins but rather a new term, elicitation metabolite, is suggested.

Acknowledgements. I thank Dr. L. Witte for recording the GLC-MS spectra and his assistance with the interpretation. This work was supported by the DFG.

References

Baumert A, Hieke M, Gröger D (1983) N-methylation of anthranilic acid to N-methylanthranilic acid by cell-free extracts from *Ruta graveolens* tissue cultures. Planta Med 48:258–262

Eilert U (1987) Elicitation: methodology and aspects of application. In: Constabel F, Vasil IK (eds) Cell culture and somatic cell genetics of plants, vol 4. Cell culture in phytochemistry. Academic Press, New York London, pp 153–196

Eilert U, Wolters B (1987) Elicitor induction of S-adenosylmethionin:anthranilic acid N-methyltransferase activity in suspension cultures of *Ruta graveolens*. Pharm Weekbl 9:233 (Abstr)

Eilert U, Ehmke A, Wolters B (1984) Elicitor-induced accumulation of acridone alkaloid epoxides in *Ruta graveolens* suspension cultures. Planta Med 6:459–532

Eilert U, Kurz WGW, Constabel F (1985) Stimulation of sanguinarine accumulation in *Papaver somniferum* cell cultures by fungal elicitors. J Plant Physiol 119:65–76

Eilert U, Constabel F, Kurz WGW (1986) Elicitor-stimulation of monoterpene indole alkaloid formation in suspension cultures of *Catharanthus roseus*. J Plant Physiol 126:11–22

Eilert U, DeLuca V, Kurz WGW, Constabel F (1987) Alkaloid formation by habituated and tumorous cell suspension cultures of *Catharanthus roseus*. Plant Cell Rep 6:271–274

Gamborg OL, Miller RA, Ojima K (1968) Nutrient requirements of suspension cultures of soybean root cell. Exp Cell Res 50:151–158

Hart JH (1981) Role of phytostilbenes in decay and disease resistance. Annu Rev Phytopathol 19:437–458

Kuc J (1972) Phytoalexins. Annu Rev Phytopathol 10:207–232

Matern U, Strasser H, Wendorff H, Hamerski D (1988) Coumarins and furanocoumarins. In: Constabel F, Vasil IK (eds) Cell culture and somatic cell genetics of plants, vol 5: Phytochemicals in plant cell cultures. Academic Press, New York London, pp 3–23

Murashige T, Skoog F (1962) A revised medium for rapid growth and bioassays with tobacco tissue cultures. Physiol Plant 15:473–497

Reinhard E, Corduan G, Volk OH (1968) Über Gewebekulturen von *Ruta graveolens*. Planta Med 16:8

Wolters B, Eilert U (1981) Antimicrobial substances in callus cultures of *Ruta graveolens*. Planta Med 43:166–174

Wolters B, Eilert U (1982) Acridonepoxidgehalte in Kalluskulturen von *Ruta graveolens* und ihre Steigerung durch Mischkultur mit Pilzen. Z Naturforsch C Biosci 37:575–583

Wolters B, Eilert U (1984) Elicitor-induzierte Akkumulation "konstitutiver" antimikrobieller Pflanzenstoffe. Farm Tijdschr Belg 61:283 (Abstr)

Stimulation of Phytoalexin Accumulation in Fungus-Infected Roots and Elicitor-Treated Cell Cultures of Soybean (*Glycine max* L.)

J. Ebel, E.G. Cosio, D. Grab, and H. Habereder[1]

1 Introduction

Plants possess a variety of defence mechanisms which render them resistant to pathogen attack. One such mechanism is the inducible accumulation of low-molecular-weight antimicrobial compounds known as phytoalexins at the site of attempted infection. The production of isoflavonoid phytoalexins and their role in defence have been studied in some detail in soybean (*Glycine max* L. Merrill) [1,2]. Two experimental systems have been used to investigate the mechanisms underlying phytoalexin formation in this plant. One utilizes the inoculation of soybean primary roots with zoospores of the fungal pathogen *Phytophthora megasperma* f.sp. *glycinea*, the causal agent of root and stem rot, and mimics the natural infection process. Fungus-induced phytoalexin synthesis in roots has been analyzed in relation to fungal race:plant cultivar-specific resistance and to the regulation of enzymes of phytoalexin biosynthesis. The second experimental system uses cell-suspension cultures and a β-glucan elicitor isolated from *P. megasperma* cell walls and represents a system of reduced complexity. The analysis of the elicitor-cell culture interaction may provide valuable information on the molecular basis of the initial recognition event between fungus-derived components and the plant cell surface and the subsequent steps involved in the transduction of the primary signal. Different soybean tissues have been used to characterize plasma membrane-associated β-glucan binding proteins which could be related to putative elicitor "receptors". These studies may, thus, help to understand the nature of the hypothetical "recognition" event between the whole plant and the live pathogen preceding the activation of plant defence responses.

2 Regulation of Phytoalexin Accumulation

Soybean cultivars are differentially resistant to several *P. megasperma* races [3]. A detailed analysis of a single soybean cultivar has shown marked differences in dynamics and sites of accumulation of glyceollin I, the major phytoalexin of soybean (Fig. 1), following infection of roots with zoospores of either a virulent (compatible interaction, plant susceptible) or an avirulent race (incompatible interaction, plant resistant) of *P. megasperma* [4]. Only resistant soybean plants

[1]Lehrstuhl für Biochemie der Pflanzen am Biologischen Institut II der Universität Freiburg i. Br., Schänzlestraße 1, D-7800 Freiburg i. Br., FRG

Primary and Secondary Metabolism of Plant Cell Cultures, ed. by W.G.W. Kurz

rapidly accumulate glyceollin I at infection sites and fungal growth is arrested early in the immediate vicinity of the inoculation site. Similar race:cultivar-specific differences have also been observed for the increases in the activities of several enzymes involved in glyceollin biosynthesis (Fig. 1) during early stages of infection (2-8 h) [5, 6]. Inducible changes at the mRNA level have recently been investigated for two enzymes of phytoalexin biosynthesis, phenylalanine ammonia-lyase and chalcone synthase [7]. Cloned cDNA [8, 9] has been used to demonstrate that in the incompatible interaction rapid increases in the mRNA amounts of both enzymes occur at the time of penetration of fungal germ tubes into the epidermal cell layers (1 to 2 h after inoculation; [10]), reaching highest levels at about 7 h (Fig. 2). In the compatible interaction, only a slight, early enhancement of the mRNA levels has been found and no further increase occurs until about 9 h after inoculation (Fig. 2). For one of the enzymes, chalcone synthase, mRNA activity changes have also been analyzed. The kinetics for mRNA activity shows again major differences between the incompatible and compatible interactions [7]. The rapidity of mRNA induction

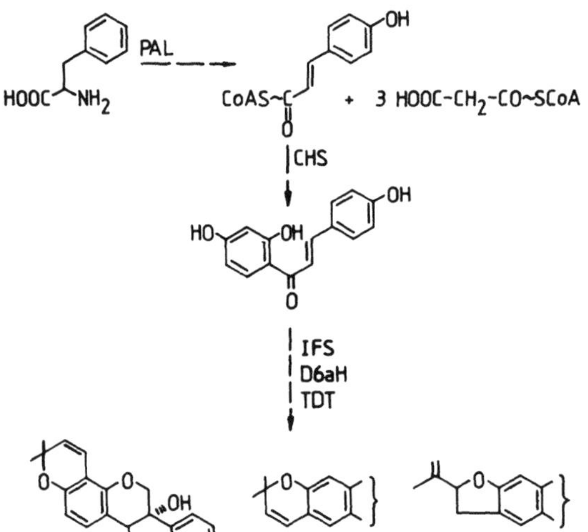

Fig. 1. Proposed biosynthetic pathway of the pterocarpan phytoalexins, glyceollin I, II, and III in soybean. The enzymes marked by abbreviations are: phenylalanine ammonia-lyase (*PAL*), chalcone synthase (*CHS*), "isoflavone synthase" (*IFS*), 3,9-dihydroxypterocarpan 6a-hydroxylase (*D6aH*), 3,6a,9-trihydroxypterocarpan dimethylallyltransferase (*TDT*)

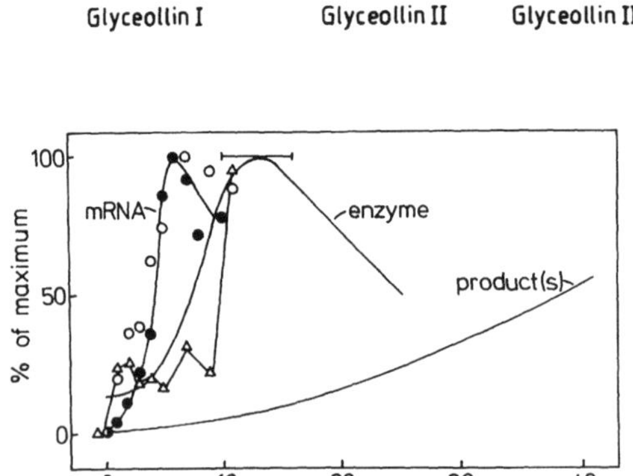

Fig. 2. Time course of changes associated with phytoalexin accumulation either in soybean cell cultures following elicitor treatment or in soybean roots after inoculation with zoospores of *P. megasperma*. Product(s): total amount of glyceollins in cell cultures; enzyme: chalcone synthase activity in cell cultures; mRNA: chalcone synthase mRNA amount following elicitor treatment of cell cultures (●) and infection of roots with the incompatible race 1 (○) or the compatible race 3 (△) of *P. megasperma*

in the incompatible interaction, as compared with the timing of fungal invasion of the soybean roots, might imply that at least some of the components involved in the perception and transduction of the signal produced by the invading fungus in the root tissue are available within a short period after inoculation. Another interesting observation is that the kinetics of mRNA induction in the incompatible root-fungus interaction is very similar to that measured after elicitor treatment of soybean cell-suspension cultures (Fig. 2). Thus, in the soybean root not only phytoalexin accumulation itself but also the induction of phytoalexin biosynthetic enzymes and their mRNAs reflect the known differences in the physiological interaction between *P. megasperma* and the soybean plant [1].

3 Characteristics of Fungal β-Glucan Binding Sites

A necessary step during initiation of the phytoalexin response is the perception of an appropriate signal by the plant cell. It is assumed that the initial event in the elicitor-mediated stimulation of phytoalexin synthesis is the interaction of the elicitor with primary target sites or receptors. Candidates for the location of such sites are the cell membranes, including the plasma membrane and the intracellular membranes. However, soluble cellular components as primary target sites cannot be excluded.

The components of *P. megasperma* mycelial wall extracts which show phytoalexin elicitor activity in soybean are β-1,3-, β-1,6-linked glucans [2, 11]. These glucans are prominent structural polysaccharides of the oomycetes. Different *Phytophthora* spp. β-glucans have been used as ligands to demonstrate specific glucan binding sites in soybean membranes [12, 13] and to analyze ligand specificity [13, 14]. Schmidt and Ebel [13] have prepared a [^{3}H]-labelled glucan fraction from *P. megasperma* consisting entirely of glucose and having a glucosyl-linkage pattern typical of the *P. megasperma* cell-wall branched glucans. With the [^{3}H]glucan fraction, binding is of high affinity, saturable, and reversible [13, 14]. The binding sites are present in membrane fractions from soybean roots, hypocotyls, cotyledons, and cell cultures and they follow the distribution of a plasma-membrane marker, 1,3-β-glucan synthase, in sucrose density gradients. Localization of the binding sites in the plasma membrane has been further supported by the result that displaceable β-glucan binding occurs on the surface of intact soybean protoplasts derived from cell cultures [14]. This latter observation allows the possibility of studying properties of the binding sites in vivo, including their putative role in mediating the elicitor stimulus in the cell cultures.

Competition studies with a variety of poly- and oligosaccharides have been carried out to investigate the specificity of the glucan binding [13, 14]. The ability of these other ligands to compete for the binding sites with the [^{3}H]glucan, is dependent on both their structure and their size. Only β-1,3-glucan-type poly/oligosaccharides have proved to be competitors. Most interestingly, a strong positive correlation has been found between the ability to compete for [^{3}H]glucan binding sites and the activity of the ligands as elicitors of glyceollin synthesis in a cotyledon bioassay system.

Further investigations on the nature of the glucan binding sites and their possible role in the activation of defence responses in soybean require their isolation and characterization. To achieve this goal, a modified β-glucan ligand with considerably higher specific radioactivity than that of the [^{3}H]glucan used previously has been synthesized [14], an assay for measuring soluble binding activity has been introduced (Table 1) and several detergents have been tested for their ability to solubilize the glucan binding protein(s) from soybean membranes. A [^{125}I]glucan (about 100 Ci mmol^{-1}) has been obtained by reductive amination of a β-glucan fraction with tyramine, followed by oxidative incorporation of [^{125}I] into the tyramine moiety (Fig. 3). This modification of the β-glucan at the reducing end neither affected binding affinity nor biological activity as measured in a cotyledon bioassay [14].

Ionic and non-ionic detergents have been found to solubilize glucan-binding activity to different degrees (Table 1). Zwittergent 3–12 has been chosen for further studies because of its capability to preserve binding activity and its suitability for protein purification. Glucan-binding affinity of solubilized binding sites is slightly higher (apparent K_d = 11 nM) than that of the membrane-associated sites (apparent K_d = 37 nM; [14]). Detergent-solubilized binding activity can be fractionated by conventional methods of protein purification, including gel-permeation and ion-exchange chromatography (Fig. 4). The behaviour of the solubilized-β-glucan binding activity during fractionation and its sensitivity to heat and proteolytic digestion provide strong evidence for the protein nature of the plasma membrane-localized glucan binding sites from soybean.

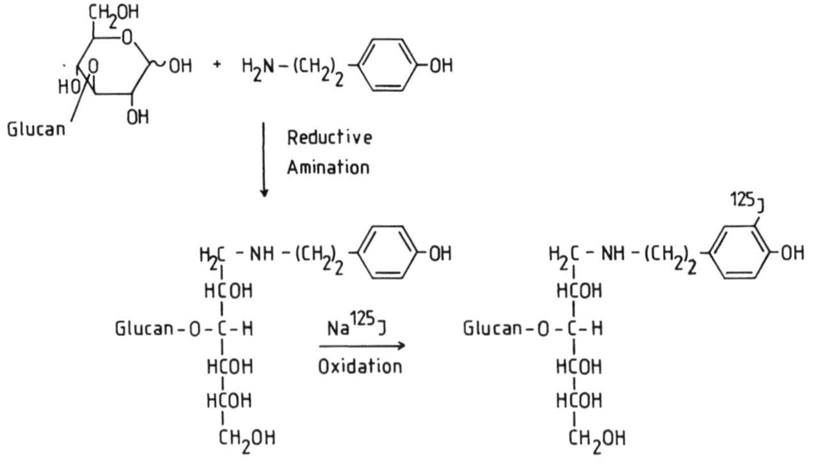

Fig. 3. Synthesis of a [^{125}I]glucan ligand by reductive amination of a *P. megasperma* β-glucan fraction with tyramine followed by oxidative iodination

Table 1. Solubilization of ß-glucan binding proteins from soybean microsomal preparations with various detergents. Soybean membranes were suspended and stirred (60 min at 4°C) in 25 mM Tris-HCl, pH 7, 0.1 M NaCl, 2 mM DTT, and various detergents. After centrifugation at 100 0000 × g for 1 h, specific binding and protein were measured in supernatants and resuspended pellets. Binding activity was determined by rapid filtration of the assay mixture (14) through glass fiber filters impregnated with polyethyleneimine (ZW: Zwittergent)

Detergent	Concentration (%)	Specific binding Soluble (%)	Pellet (%)	Protein solubilized (%)
None	–	0	100	0
Lubrol PX	0.5 (v/v)	6	104	38
Triton X-100	0.5 (v/v)	20	76	43
Octyl glucoside	2.0 (w/v)	0	43	82
Dodecyl maltoside	2.0 (w/v)	54	61	47
CHAPS	2.0 (w/v)	4	105	52
ZW 3-14	0.2 (w/v)	20	14	33
ZW 3-12	1.0 (w/v)	40	14	77
Cholate	1.0 (w/v)	14	62	77

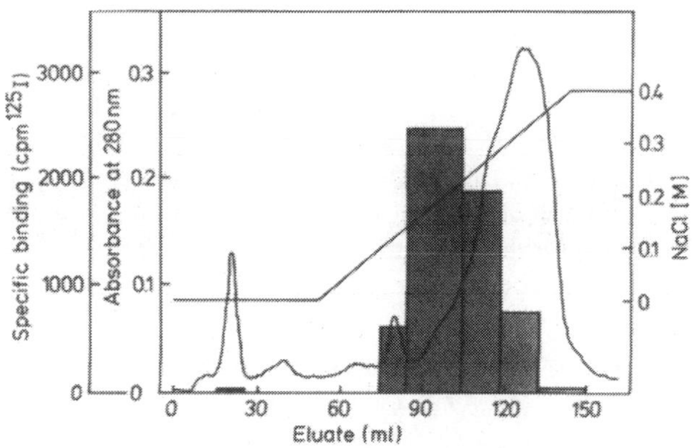

Fig. 4. Ion-exchange chromatography of detergent-solubilized β-glucan binding activity on DEAE-Sepharose. Adsorbed proteins were eluted with a linear gradient of 0 to 0.4 M NaCl. Specific binding (*shaded area*) was measured as outlined in Table 1

4 Effects Associated with Elicitation in Cell-Suspension Cultures

The existence of plasma membrane-localized binding proteins allows one to investigate the hypothesis whether binding of β-glucan fragments from *P. megasperma* cell walls to these sites initiates the signal resulting in enhanced transcription of the genes for the glyceollin biosynthetic enzymes. In soybean cell cultures, the glucan elicitor increases the activities of enzymes of phytoalexin biosynthesis and stimulates phytoalexin accumulation when the culture medium contains 1.2 mM Ca^{2+} [15]. However, if one depletes the exchangeable external Ca^{2+} with EGTA (residual Ca^{2+} concentration about 10 μM) or adds low concentrations of the Ca^{2+} antagonist La^{3+} to the medium containing a normal Ca^{2+}

concentration, the elicitor-mediated phytoalexin response is effectively abolished. These results and the observation that amino acid uptake is rapidly inhibited [15] suggest the possibility that the initial response to elicitor binding to soybean cells is a number of substrate (ion) flux changes across the plasma membrane. These fluxes might include or affect Ca^{2+} fluxes. This interpretation is supported by the observation that in the presence of Ca^{2+} in the medium, but not in its absence, the Ca^{2+} ionophore A 23187 stimulates the phytoalexin response in a dose-dependent manner [15]. In addition, the Ca^{2+} channel blocker verapamil, which has been shown to inhibit Ca^{2+} influx into carrot protoplasts [16], causes inhibition of the elicitor-mediated phytoalexin response at concentrations higher than 0.1 mM [15]. Further refinements of these techniques could establish the relationship between the external Ca^{2+} pool and putative Ca^{2+} as well as other ion flux changes in stimulus-response coupling during elicitor action.

Addition of the β-glucan elicitor to soybean cell-suspension cultures also results in rapid changes in the phosphate turnover of several phosphoproteins (Grab D and Ebel J, unpublished results). The effect of the elicitor on protein phosphorylation has been tested after labelling of the cells with [^{32}P]orthophosphate. As revealed by gel electrophoretic analysis and autoradiography (Fig. 5), decreases as well as increases in the labelling of several phosphoproteins occur rapidly, being detectable within 5 min after elicitor application, and persist for at least 15 min. As judged by their relative molecular masses, M_r, and isoelectric points, pI, a number of proteins which are radioactively labelled in vivo are also phosphorylated in vitro by endogenous protein kinase activity in the presence of Ca^{2+}. The most pronounced effect has been observed with a protein substrate with an M_r = 69 000 and pI = 5.7 whose phosphate labelling decreases in response to elicitor treatment in vivo (Fig. 5). Phosphorylation in vitro of the same or a closely similar protein with an M_r = 69 000 is strongly enhanced by a phosphorylation-stimulating factor (effector) derived from soybean cell cultures (Fig. 6). The effector possesses a low apparent M_r, is negatively charged at pH 7.3,

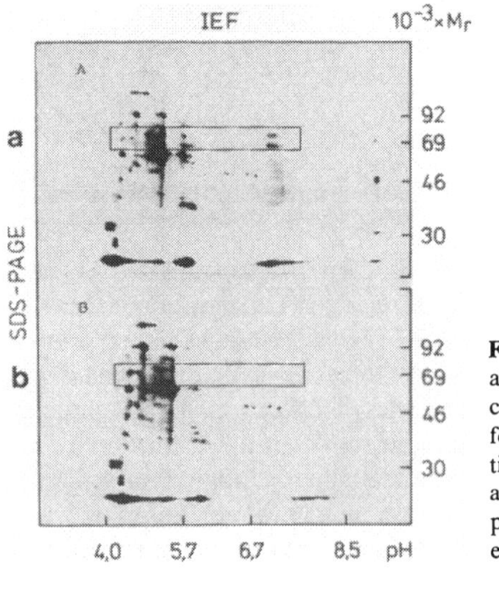

Fig. 5a,b. Two-dimensional gel electrophoretic analysis of phosphoproteins isolated from soybean cell cultures which had been treated with glucan elicitor for 10 min (**b**) and from untreated cells (**a**). Radioactively labelled phosphoproteins were visualized by autoradiography. The *boxes* denote several phosphoproteins whose labelling was affected by the elicitor treatment

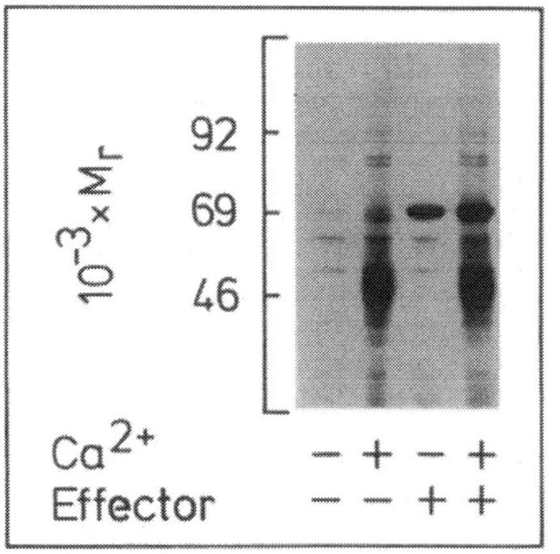

Fig. 6. Effect of Ca^{2+} (3 μM) and a phosphorylation-stimulating factor (effector) on endogenous protein phosphorylation in extracts from soybean cell cultures. A cell-free extract was incubated with [γ-^{32}P]ATP and the indicated additions for 7 min and then subjected to gel electrophoresis in the presence of sodium dodecyl sulphate and to autoradiography. A protein with $M_r = 69\,000$ was strongly phosphorylated in the presence of the effector

and is relatively heat-stable. Phosphorylation of the protein with $M_r = 69\,000$ is only slightly stimulated by Ca^{2+} (Fig. 6), but is not enhanced by several other substances known as activators of protein kinases in various systems. Further characterization of the components involved in phosphorylation/dephosphorylation of endogenous soybean phosphoproteins will help to understand the regulation of elicitor-triggered post-translational metabolic processes.

Acknowledgements. This work was supported by the Deutsche Forschungsgemeinschaft (SFB 206), Fonds der Chemischen Industrie, the Alexander von Humboldt Foundation (E.G.C.), and Studienstiftung des Deutschen Volkes (D.G.).

References

1. Ebel J, Grisebach H (1988) Defense strategies of soybean against the fungus *Phytophthora megasperma* f.sp. *glycinea*: a molecular analysis. Trends Biochem Sci 13:23–27
2. Ebel J (1986) Phytoalexin synthesis: the biochemical analysis of the induction process. Annu Rev Phytopathol 24:235–264
3. Compendium of soybean diseases (1975) Am Phytopathol Soc, St. Paul, MN, USA
4. Hahn MG, Bonhoff A, Grisebach H (1985) Quantitative localization of the phytoalexin glyceollin I in relation to fungal hyphae in soybean roots infected with *Phytophthora megasperma* f.sp. *glycinea*. Plant Physiol 77:591–601
5. Bonhoff A, Loyal R, Ebel J, Grisebach H (1986) Race: cultivar-specific induction of enzymes related to phytoalexin biosynthesis in soybean roots following infection with *Phytophthora megasperma* f.sp. *glycinea*. Arch Biochem Biophys 246: 149–154
6. Bonhoff A, Loyal R, Feller K, Ebel J, Grisebach H (1986) Further investigations of race: cultivar-specific induction of enzymes related to phytoalexin biosynthesis in soybean roots following infection with *Phytophthora megasperma* f.sp. *glycinea*. Biol Chem Hoppe-Seyler 367:797–802
7. Habereder H, Schröder G, Ebel J (1989) Rapid induction of phenylalanine ammonia-lyase and chalcone synthase mRNAs during fungus infection of soybean (*Glycine max* L.) roots or elicitor treatment of soybean cell cultures at the onset of phytoalexin synthesis. Planta 177:58–65
8. Edwards K, Cramer CL, Bolwell GP, Dixon RA, Schuch W, Lamb CJ (1985) Rapid transient induction of phenylalanine ammonia-lyase mRNA in elicitor-treated bean cells. Proc Natl Acad Sci USA 82:6731–6735

9. Ryder TB, Cramer CL, Bell JN, Robbins MP, Dixon RA, Lamb CJ (1984) Elicitor rapidly induces chalcone synthase mRNA in *Phaseolus vulgaris* cells at the onset of the phytoalexin defense response. Proc Natl Acad Sci USA 81:5724–5728
10. Beagle-Ristaino JE, Rissler JF (1983) Histopathology of susceptible and resistant soybean roots inoculated with zoospores of *Phytophthora megasperma* f.sp. *glycinea*. Phytopathology 73:590–595
11. Darvill AG, Albersheim P (1984) Phytoalexins and their elicitors – a defense against microbial infection in plants. Annu Rev Plant Physiol 35:243–275
12. Yoshikawa M, Keen NT, Wang M-C (1983) A receptor on soybean membranes for a fungal elicitor of phytoalexin accumulation. Plant Physiol 73:497–506
13. Schmidt WE, Ebel J (1987) Specific binding of a fungal glucan phytoalexin elicitor to membrane fractions from soybean *Glycine max*. Proc Natl Acad Sci USA 84:4117–4121
14. Cosio EG, Pöpperl H, Schmidt WE, Ebel J (1988) High-affinity binding of fungal ß-glucan fragments to soybean (*Glycine max* L.) microsomal fractions and protoplasts. Eur J Biochem 175:309–315
15. Stäb MR, Ebel J (1987) Effects of Ca^{2+} on phytoalexin induction by fungal elicitor in soybean cells. Arch Biochem Biophys 257:416–423
16. Graziana A, Fosset M, Ranjeva R, Hetherington AM, Lazdunski M (1988) Ca^{2+} channel inhibitors that bind to plant cell membranes block Ca^{2+} entry into protoplasts. Biochemistry 27:764–768

Part IV
Gene Expression in Transformed and Stressed Plant Cells

Genes of Secondary Metabolism: Differential Expression in Plants and In Vitro Cultures and Functional Expression in Genetically Transformed Microorganisms

M. Wink[1,2]

1 Introduction

Plants had to develop strategies for defence against microorganisms and herbivores during evolution. The evolution of chemical defence, i.e. the production of the so-called secondary products can be regarded as one of the most important means in this context (Swain 1977, Rosenthal and Janzen 1979, Wink 1988). Since secondary metabolites evolved as biologically active compounds, it is not surprising that man has found a use for many of them, e.g. as medicinal agents, flavours, spices, colours, stimulants, hallucinogens, or pesticides. As a consequence, quite a number of secondary plant products are of economical importance.

Therefore, the challenge exists today of how to produce valuable natural products economically. Traditionally, the cultivation of the respective plants and the subsequent isolation of the products is the method of choice. Alternatively, chemical synthesis or biotechnology provide another option.

Plant cell cultures have been favoured by plant biotechnologists for the production of valuable products for more than two decades (Berlin 1984). In general, plant cell cultures turned out to be extremely useful for biochemical and physiological studies, but as far as the production of economically important compounds is concerned, progress is rather slow and difficult to predict (Drapeau et al. 1987). As discussed below, a major problem involved is that plant secondary metabolism is regulated in an organ-, tissue-, or development-specific process in the intact plant (Wiermann 1981; Wink 1987c). The correct expression of both the genes for biosynthesis and for storage (which are necessary for the production of a compound) seems to be difficult to control in the undifferentiated state of plant cell suspension or callus cultures. To overcome these problems, many researchers have turned to organized tissue cultures, such as root or shoot cultures, which were facilitated by the use of transformation of plant tissue with *Agrobacterium rhizogenes* (Hamill et al. 1987).

In order to overcome the limitations of differential gene expression in plants it would be plausible to take advantage of the rapid advance of genetic engineering. In theory it should be possible to isolate the plants' genetic information for a biosynthetic pathway and to transfer it to microorganisms. Expression of the respective genes could either lead to the production of the enzymes in vitro and

[1]Genzentrum der Universität München, Pharmazeutische Biologie, Karlstr. 29, D-8000 München 2, FRG

[2]Present address: Universität Mainz, Institut für Pharmazie, Saarstr. 21 D-6500 MAINZ, FRG

Primary and Secondary Metabolism of Plant Cell Cultures, ed. by W.G.W. Kurz

their use in "enzyme reactors" or to the functional expression of the genes and their products directly in the microorganism, e.g. *E. coli*.

In this communication our results on alkaloid formation in cell and organ cultures of *Lupinus* are summarized as an example for tissue-specific control of alkaloid production in plants (Wink 1987a, b). Secondly, we report on the successful transfer and expression of a gene of plant secondary metabolism in *Escherichia coli*, i.e. that of phenylalanine ammonium lyase (which catalyzes the deamination of phenylalanine to cinnamic acid, a precursor of many secondary products), indicating that the genetic approach could indeed be feasible in the long run for the production of plant secondary metabolites in genetically transformed microorganisms.

2 Differential Gene Expression in Cell and Organ Cultures and Intact Plants of *Lupinus polyphyllus*

2.1 Cell Suspension Cultures

Cell suspension cultures were established from more than ten quinolizidine alkaloid (QA) producing plant species of the genus *Lupinus, Baptisia, Cytisus, Genista, Laburnum,* and *Sophora.* Whereas the alkaloid pattern of the respective differentiated plant was complex and diverse, all cell cultures produced lupanine as the major alkaloid (Wink et al. 1983). The amounts of alkaloids produced were in the range of 0.01 to 5 $\mu g\ g^{-1}$ fresh weight, i.e. orders of magnitude lower than those of the differentiated plant (Table 1). This indicates that the genes for QA formation are expressed at a very low level in cell suspension cultures only.

Alkaloid formation in cell suspension cultures was positively correlated with greening and followed a light-mediated diurnal rhythm (Wink and Hartmann 1980, 1982a; Wink et al. 1981). This phenomenon can be explained by the finding that QA biosynthesis occurs in the chloroplast (Wink and Hartmann 1982b).

Table 1. Expression of quinolizidine alkaloid biosynthesis in cell-, organ cultures and intact plants

Cultivation	Alkaloid content ($\mu g\ g^{-1}$ fresh weight)		
	L. polyphyllus	*Cytisus scoparius*	*Laburnum anagyroides*
Cell cultures			
Callus culture	0.01–1	0.01–5	0.1–1[a]
Suspension culture	0.01–5	0.8–12	0.1–1[a]
Organ cultures			
Root cultures	4	0	0
Shoot cultures	76	64	21
Intact plant			
Leaves	500–5000	100–5500	100–900
Seeds	35000	210	10000

[a] *Laburnum alpinum*.

QA are not stable end products in cell cultures but undergo extensive degradation. Some cultures could even grow for 6 months on a medium which contained the QA sparteine as the sole nitrogen source (Wink and Witte 1985). The degradation can be observed both inside the cells and in the medium (into which part of the alkaloids are released) and is catalyzed by vacuolar and extracellular (secreted) enzymes (Wink 1984, 1985a). The molecular biology of these enzymes is presently being studied in our laboratory (Perrey et al. 1989).

QA formation can be stimulated by a number of chemical or biological "elicitors" (Wink 1985b), as is the case with many other secondary metabolites (so-called phytoalexins; for review, Eilert 1987). Inducing agents include alkaloids, polyamines, cAMP, HCl, and bacterial cell walls (Wink 1985b, and unpublished observations). Preliminary evidence indicates that the activation involves an induction of the respective alkaloid genes (Wink unpublished). The increase of alkaloid formation is transient and after 3–5 days the alkaloids that were formed disappear due to degradation (Wink 1985b).

2.2 Organ Cultures

In order to establish organ cultures of QA-producing plants, lupins and other legumes were infected with *Agrobacterium rhizogenes*. However, hairy roots could be induced only in *Spartium junceum* and *Lupinus mutabilis* (Wink and Witte 1987). Since the method failed with *L. polyphyllus*, we established root and shoot cultures from "normal" seedlings: aseptically grown seedlings of *Lupinus polyphyllus, L. succulentus, L. subcarnosus, Cytisus scoparius,* and *Laburnum anagyroides* were dissected into roots, shoots, and cotyledons and cultured in liquid medium (Wink 1987d). Alkaloid analysis by capillary GLC revealed that root and cotyledon cultures did not produce alkaloids, whereas the shoot cultures did (Table 1). In contrast to the finding with the cell suspension cultures (see above), the alkaloid patterns of the shoot cultures resembled closely those of the differentiated plants (Wink 1987d). Feeding experiments with the biogenetic precursor cadaverine confirmed the inactivity of roots and cotyledons towards QA biosynthesis (Wink 1987d). We can conclude that the genes of QA biosynthesis are correctly expressed in differentiated shoot cultures but are silent in roots.

2.3 Intact Plants

In order to understand the physiology of QA in cell culture, their metabolism and biochemistry were studied in detail in intact plants (for review, Wink 1987b). The findings are briefly outlined here: QA biosynthesis takes place in the chloroplast of leaf mesophyll cells and is regulated by light. After synthesis the alkaloids are translocated all over the plant via the phloem (Wink and Witte 1984). Accumulation and storage of QA can be observed predominantly in epidermal and subepidermal cells (Wink 1986). A subcellular compartment of QA storage is the vacuole. Strong evidence could be presented that QA pass the tonoplast via a specific carrier system and that the energy for accumulating the alkaloids against a concentration gradient is provided by an ATP-dependent proton-antiport mechanism (Mende and Wink 1987; Wink and Mende 1987). Another storage organ for QA are the seeds of QA are partly remobilized during germination (Wink and Witte 1985). Besides minor metabolic functions (e.g. nitrogen transport and nitrogen storage), the main biological objective of QA seem to be that of chemical defence (Wink 1985c, 1987e, 1988) against microorganisms and herbivores.

2.4 Conclusions

The advances of molecular biology (Darnell et al. 1987) during the last decade make it possible to speculate on the regulation of the genes encoding secondary metabolism. Because the biochemical mechanisms which govern secondary metabolism are almost identical to those of primary metabolism, it seems adequate to extrapolate from "ordinary" eukaryotic genes on genes of secondary metabolism.

Since prokaryotic gene control is now very well understood and eukaryotic gene expression still relatively mysterious, we shall make use of the detailed knowledge of prokaryotic gene regulation to speculate on the possible mechanisms of eukaryotic gene control.

As illustrated in Fig. 1A a typical bacterial gene consists of the DNA which codes for a protein and of DNA for the regulatory elements such as promotor and operator. It is common in bacteria (but uncommon in eukaryotes) for one promotor to serve a series of clustered genes, which produce enzymes that are active in a single metabolic pathway (Fig. 1B), the so-called operon. The coding portions of the respective polycistronic mRNA are independently translated, i.e. there a multiple starts and stops. Not all the potential promotors in a bacterial cell are available at every instant in the lifetime of the cell; they are controlled by the nutritional content of the culture medium. The control mechanism needs regulatory proteins, which are either negative-acting proteins ("repressor") (as in the lac-operon) or positive-acting proteins ("activators") (as in the ara-operon). Regulatory proteins bind to DNA at or near the promotors (the "operator site") and modulate the efficiency with which the RNA polymerase binds to the promotor. Repressor and activator can combine with small molecules ("effectors") which greatly affects the binding affinity of these regulatory proteins for their operator site (Fig. 1C).

As a model for the regulation of eukaryotic genes, the genes that encode the enzymes for arginine biosynthesis in *E. coli* are chosen, because these genes are not organized in operons but scattered over a number of non-adjacent sites on the *E. coli* chromosome (Fig. 1D), similar to the situation in eukaryotes. When arginine is in plentiful supply, transcription of all these genes is repressed by a single repressor molecule. The presumed operator sequences around the RNA initiation sites are very similar, which may explain why one repressor recognizes these distant sites in DNA.

A simplified eukaryotic gene is shown in Fig. 1E, which in addition to the promotor site shows additional enhancer (and/or silencer) sites, which are probably important for the tissue-specific control of gene activation (Dunn and Gough 1984; Ptashne 1986; Darnell et al. 1987).

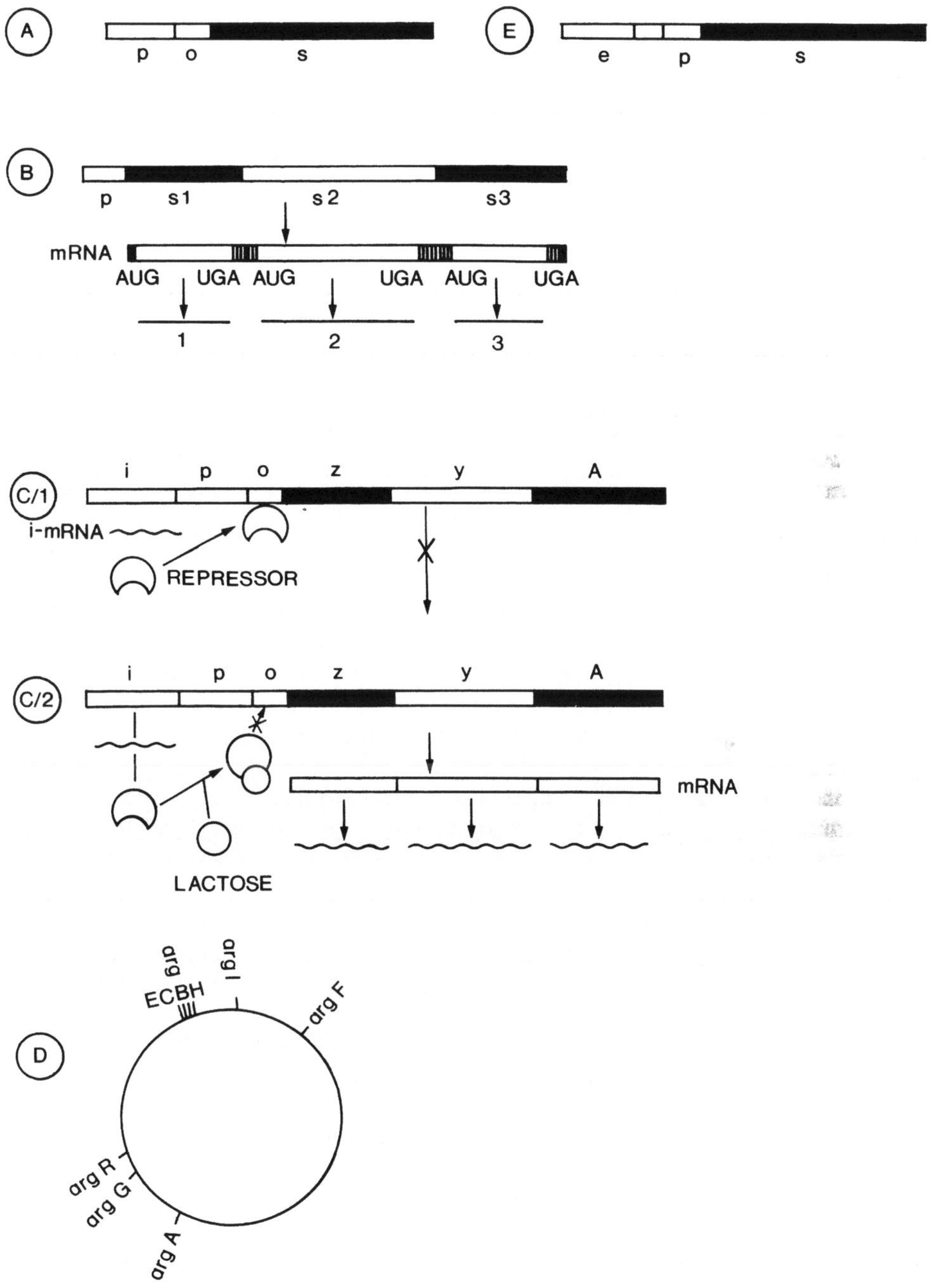

Fig. 1A-E. Schematic illustration of gene structure in pro- and eukaryotes. **A** Prokaryotic gene; **B** prokaryotic operon; **C** regulation of the lac operon. **C/1** Repressor binds to operator and thus inhibits transcription; **C/2** lactose binds to repressor, which then cannot bind to operator. As a result the lac operon is expressed. **D** Organization of the arginine genes in *E. coli*; **E** eukaryotic gene. *p* promotor; *o* operator; *s* structural gene; *e* enhancer; *i* repressor gene

It was recently established that plant genes, similar to the better studied vertebrate genes, are in general regulated in a tissue-, organ-, or development-specific fashion (Sengupta-Gopalan et al. 1985; Goldberg 1986; Rosahl et al. 1986; Stockhaus et al. 1987), although the regulating factors involved are not fully understood. In a differentiated organism the right gene is only turned on at the right time and the right site (cell, tissue, or organ). This is achieved through signal molecules (e.g. hormones) which bind to respective regulatory DNA-binding proteins (such as repressors or activators) or to regulatory proteins which are themselves expressed only in a tissue- or organ-specific way. Similar to the situation with the bacterial arginine regulon (Fig. 1D), these specific transcription factors will recognize their target genes, even if they are scattered all over the genome.

This concept can be applied also for secondary metabolism. In most cases studied (with intact plants) it was found that the biosynthesis of secondary metabolites does not occur in all cells of a plant but is organ- or even tissue-specific. In addition, the site of biosynthesis is not necessarily the site of accumulation which itself is often also tissue- or organ-specific (see above; for review, Wiermann 1981; Wink 1987c).

For the production of a secondary metabolite, e.g. QA such as lupanine, we need the correct expression of a set of genes. First, the genes encoding the biosynthetic enzymes and second, those genes which encode the transport proteins, which are necessary to facilitate membrane passage of lupanine and which determine the correct accumulation of QA, e.g. in the epidermis. We assume that the genes for QA biosynthesis and transport are not organized in operons, but scattered all over the genome (which would be typical of eukaryotes; Darnell et al. 1987). Since the biosynthesis genes must be turned on concomitantly we can assume that each single gene is organized as shown in Fig. 1E. They are only turned on coordinately (Fig. 1D) when the regulatory genes for mesophyll cells (because QA biosynthesis was only found to be expressed in green tissue) are active and the alkaloid genes receive the correct molecular signal. The "transport genes" should be activated by other regulatory genes that are active in epidermal cells.

Let us now consider the situation in plant cell cultures which grow as morphologically more or less undifferentiated cells. Which genes are turned on or off in this situation? It is likely that only those genes are turned on which are necessary for primary metabolism, i.e. which are active in all tissues. However, those specialized genes which are normally only expressed in a differentiated organ or tissue are very likely to be turned off because the undifferentiated cells do not receive the correct molecular signal in vitro. This would explain why the formation of secondary metabolites which are usually formed in leaves of plants (such as cardenolides, terpenoids, some alkaloids, etc.; Table 2) are not produced to a substantial degree in heterotrophic cell suspension cultures. The situation would be even worse in those instances in which the production (i.e. biosynthesis and accumulation) takes place in different tissues (such as mesophyll and epidermis in the case of QA) and therefore would demand a complex set of gene expressions.

On the other hand, when these cells differentiate into organized tissue, such as the situation in organ cultures, the cells must have received the correct molecular signal before. As a consequence, now the corresponding regulatory genes are

Table 2. Influence of differentiation on the production of secondary metabolites (Data from Wink 1987c and Wink unpublished)

	Level of organization				
	Callus	Suspension	Embryoid Cultures[a]	Root	Shoot
Digitalis					
Cardenolides	–	–	+	?	++
Atropa/Hyoscyamus					
Tropane alkaloids	–	(+)	?	+++	–
Papaver					
Morphinane alkaloids	–	–	(+)	?	++
Lupinus					
Lupanine	(+)	(+)	?	–	++

[a](–) 0; (+) trace amounts; (+, ++, +++) small, medium, high amounts; (?) data not available.

activated which in turn switch on all their target genes (and also those of secondary metabolism) which are typical of that organ or tissue. In effect this is just the situation which has been determined experimentally (Table 2). Whereas there is no substantial formation of morphine, hyoscyamine, digitalis glycosides or lupanine (for review, Wink 1987c) in cell suspension cultures, the production sets in differentiated organ cultures.

In view of these considerations it is surprising that a small number of cell cultures indeed produce remarkable amounts of secondary metabolites, such as shikonine in *Lithospermum*, berberine in *Coptis* or *Thalictrum*, or betalaine in *Beta*. What is different in these cases? First, in plants the formation of these compounds takes place in the roots (Wink 1987c). Studying ethanol formation in cell suspension cultures and the isoenzyme pattern of peroxidase (Wink 1985a; Perrey et al. 1989) we concluded that cell suspension cultures of *L. polyphyllus* (at least) are biochemically not totally undifferentiated, but resemble root cells. If this is also the case in other cell cultures, then the spontaneous expression of root-dependent pathways should take place at least in some cells, because the molecular signal "root" is obviously expressed. Second, these compounds are usually stored in the same cell in which they are produced. Therefore, no long-distance transport needs to take place (which would be an obstacle in cell cultures). Third, all compounds involved are coloured, natural products, which facilitate the selection of those cell clones which by chance show the correct gene expression.

We conclude that the mass production of secondary metabolites could be achieved in those cases which are not complicated by long-distance transport, i.e. where the site of synthesis is also the site of accumulation and those, in which the biosynthesis either takes place in all cells or in the roots.

However, if more specialized tissues and long-distance transport are involved, the situation is certainly more difficult. The application of genetic engineering and the use of microorganisms for the production of the respective compounds might be a long-term solution of this problem, since most of the restrictions present in plants would not exist in these systems.

3 Expression of Plant Genes Which Code for Enzymes of Secondary Metabolism in Microorganisms

3.1 Cloning Strategies

How can we employ microorganisms to produce plant secondary metabolites? First, we need the genetic information from plants, which might be obtained from a cDNA library (Fig. 2). If the amino acid sequence of an enzyme is known, an alternative would be the chemical synthesis of the respective gene. The latter approach has the advantage that the codon composition could be adjusted in a way that is optimal for the respective microorganism into which the gene is to be cloned and expressed later. Subsequently, the cDNA clone or the synthetic gene has to be ligated into an expression vector, e.g. a plasmid. The recombinant gene can be either expressed as a fusion protein with another plasmid-derived protein, e.g. β-galactosidase, or as a free protein. In both cases it must be assured that the gene is inserted into the plasmid in the correct reading frame. For the expression of a native protein, it is also necessary to add the correct promotor, start and termination sequences, which have to be in the correct reading frame. Then *E. coli* or another microorganism is to be transformed with the expression vector and recombinant bacteria can be selected by use of a resistance gene to antibiotics which is additionally encoded on the plasmid. If the construct and transformation was successful and the inducer added, *E. coli* will then start to produce the desired protein. At this stage we have the option either to produce the enzyme in *E. coli* or to extract the enzyme and use it in an enzyme reactor, where it catalyzes the conversion of substrate A into product B. Alternatively, substrate A could be administered to the living bacteria, in the hope that a biotransformation would take place inside the cells from substrate A to product B.

Unfortunately, the biosynthesis of most economically interesting compounds may involve not only one but several enzymic conversions, i.e. between two and ten. It is evident that this feature potentiates the problems, which already exist in cloning and expressing one single gene. Theoretically two routes seem to be possible (Fig. 2): (1) to clone and to express all the genes independently which are

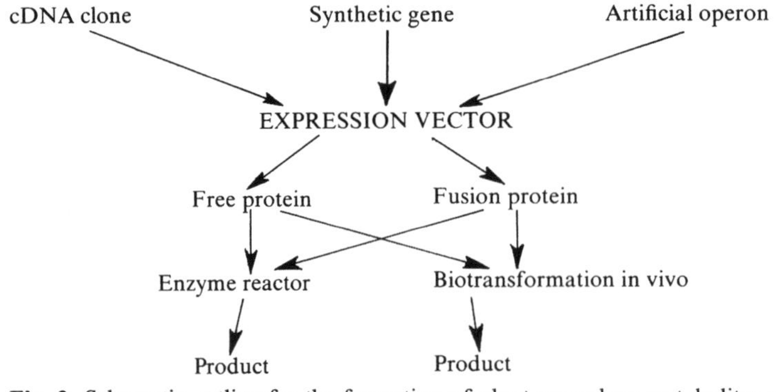

Fig. 2. Schematic outline for the formation of plant secondary metabolites with the aid of genetic engineering

necessary for the biosynthesis, then to isolate the recombinant enzymes and use them sequentially in an enzyme reactor. (2) More fascinating are the prospects of cloning the whole pathway and letting the bacteria make the end product, e.g. hyoscyamine or morphine. To achieve this goal it will be necessary to construct artificial operons (Fig. 1B) which code for all the enzymes of a pathway since in eukaryotes the genes that contribute to one pathway are not clustered into operons (Darnell et al. 1987), whereas in most prokaryotes genes are often organized in such a way. For example, the genes which encode the biosynthesis of antibiotics in *Streptomyces* are already clustered in nature, which facilitates their expression to a substantial degree (Martin and Gil 1984).

It is evident from this outlook that progress in this new field will be slow and that many basic aspects need to be solved before we can produce plant secondary metabolites in microorganisms.

3.2 Functional Expression of the PAL Gene in *E. coli*

For a number of secondary metabolites the amino acid phenylalanine serves as a precursor (Fig. 3). In many instances phe is deaminated in the first step to yield cinnamic acid by the enzyme phenylalanine ammonium lyase (PAL). PAL is one of the best-studied enzymes of secondary metabolism and cDNA clones have been isolated (Fritzemeier et al. 1987).

We have chosen PAL to learn how to express a plant gene functionally in *E. coli* with the goal to achieve the direct biotransformation of phe to cinnamic acid, a reaction which is normally not performed by *E. coli.*

We obtained a cDNA clone of PAL, isolated from *Petroselinum* from Prof. Dr. K. Hahlbrock (Köln), cloned into the Pst site of pBR322. Plasmid DNA was obtained through a large-scale preparation (Maniatis et al. 1982). The PAL-DNA was cleaved with the restriction enzyme Pst I and separated from the plasmid DNA by agarose gel electrophoresis. The PAL-DNA band (1.7 kb) was cut out of the gel and the DNA was recovered from the agarose electrophoretically in an Elu-trap chamber (Schleicher and Schüll) (Fig. 4). The purified PAL-DNA was ligated into the Pst I site of the expression plasmids pUC 8, 9 and 19 (Fig. 5). This plasmid is available with a number of slightly different polylinkers. This polylinker inserts in the gene which encodes β-galactosidase. Therefore, all constructs cloned in pUC vectors will lead to a fusion protein of β-galactosidase with the desired protein. Insertion of DNA into the Pst I site of pUC8, pUC9 and pUC19 makes sure that at least one construct has the correct reading frame (Fig. 5).

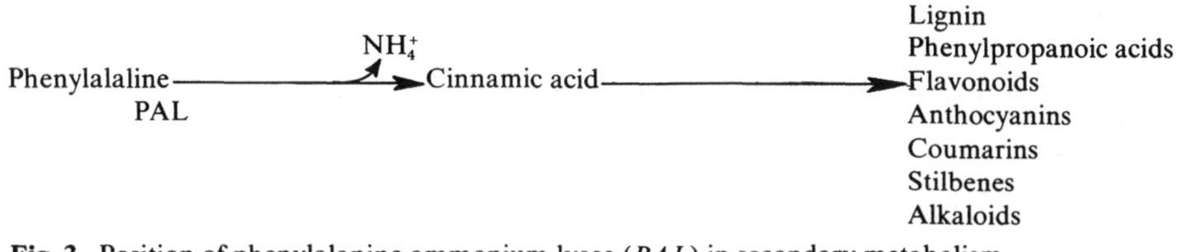

Fig. 3. Position of phenylalanine ammonium lyase (*PAL*) in secondary metabolism

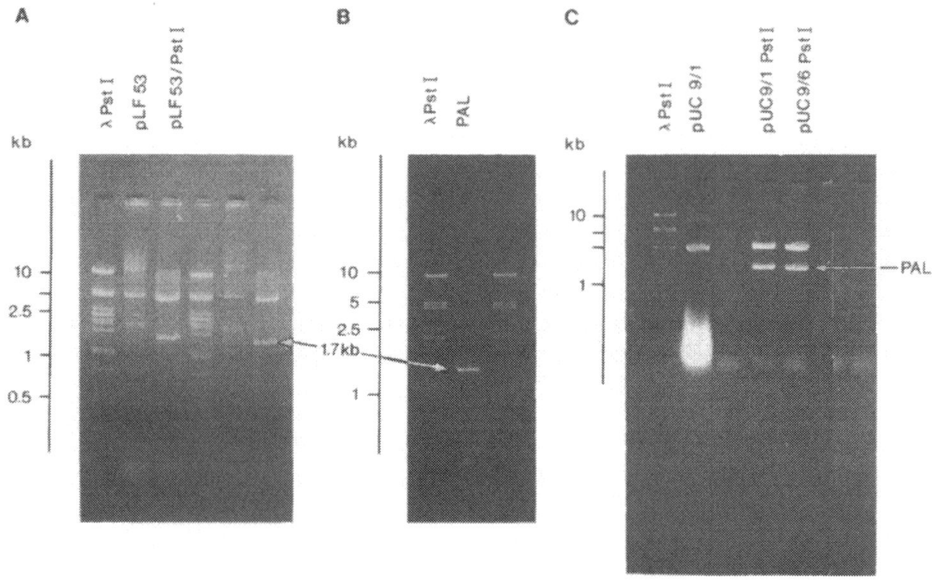

Fig. 4A-C. Isolation and subcloning of PAL-cDNA. **A** Isolation of pLF53 (PAL-cDNA in pBR322) and restriction analysis with Pst I. Agarose gel electrophoresis (1% agarose); DNA stained with ethidium bromide and visualized under UV light. **B** Isolation of PAL-DNA and purification by electrophoresis (Elu trap). **C** Subcloning of PAL gene in pUC9 and restriction analysis with Pst I

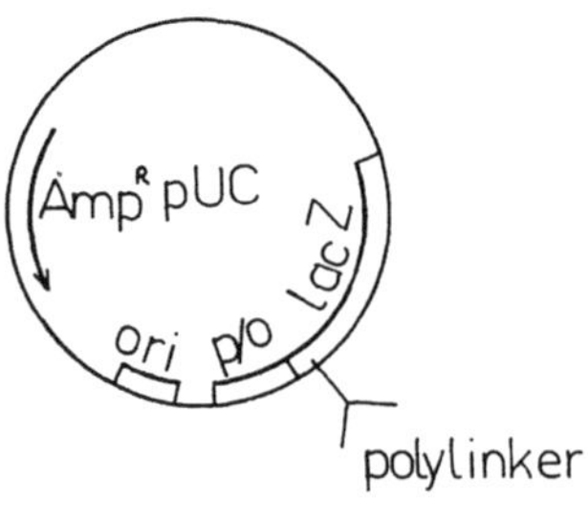

Fig. 5. Nucleotide sequence of the polylinker region of pUC8, pU69, and pUC19. The Pst I site is *underlined*

pUC8 ATG ACC ATG ATT ACG AAT TCC CGG GGA TCC GTC GAC CTG CAG CCA AGC TTG GCA CTG GCC

pUC9 ATG ACC ATG ATT ACG GCA TTG GCT GCA GGT CGA CGG ATC CCC GGG AAT TCA CTG GCC GCC

pUC19 ATG ACC ATG ATT AGG CCA AGC TTG CAT GCC TGC AGG TCG ACT CTA GAG GAT CCC CGG GTA CCG

About 15 recombinant clones were selected for each construct and the presence of the PAL gene was analyzed by agarose gel electrophoresis (Fig. 4). Each clone was cultured in 500 ml LB medium (Maniatis et al. 1982) containing ampenicillin (only transformed bacteria possess the pUC plasmid which confers resistance to ampenicillin) and IPTG, a strong inducer of the lac operon (Winnacker 1987) to achieve the expression of the fusion protein. Cells were harvested after 24 h of incubation at 37°C and suspended in fresh minimal medium

(M9) which contained 10 mM phenylalanine. After another 24–48 h at 37°C, cells and media were made acidic with acetic acid and extracted with ethylacetate employing Chem Elut columns (Analytichem; ICT, Frankfurt). After evaporation of the solvents the extracts were derivatized with methanol/HCl to produce methylesters of cinnamic acid. These extracts were analyzed by capillary gas-liquid chromatography (Fig. 6) and GLC-mass spectrometry. In most cultures we detected small amounts of phenyllactate as a transformation product but no cinnamic acid. Only one clone which contained the PAL gene (pUC9–1a) was obviously expressing the PAL as an active enzyme, since we were able to detect a

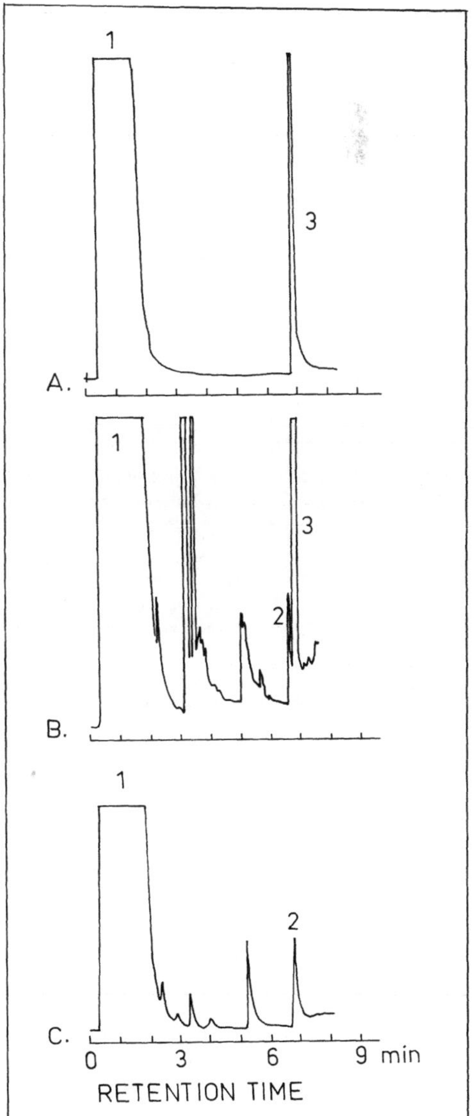

Fig. 6A-C. Analysis of the biotransformation products of *E. coli* after application of phenylalanine by capillary GLC. **A** Authentic cinnamic acid; **B** DH5 alpha with pUC9–1a (PAL gene in correct reading frame); **C** non-transformed culture *E. coli* DH5 alpha. GLC conditions: Perkin Elmer GLC 8500; capillary column DB1 (30 m × 0.3 mm); flame ionization detection; carrier gas: helium, 90 kPa; split injection, 1:20; injector temperature: 250°C; detector temperature: 320°C; oven: 100–300°C, 20°C min^{-1}. 1 = solvent peak; 2 = phenyllactate; 3 = cinnamic acid large amount of cinnamic acid in the culture (Fig. 6). This is a surprising result, because PAL is present as a tetramer in nature and here it was active as a fusion protein with β-galactosidase. Unfortunately, we were not able to study this clone in detail, because it died very soon during subculture. Probably, this transformation is more or less lethal for *E. coli* unless we can control the expression of PAL in a very stringent way. This problem needs to be solved, since many secondary products have antimicrobial properties which would interfere with production to a high degree.

Our results show that it is possible to express PAL and probably also other genes of plant secondary metabolism in *E. coli* but also that we need more basic research before the process can be used for the production of plant secondary metabolites in microorganisms.

Acknowledgements. Our research was supported by grants of the Deutsche Forschungsgemeinschaft and a Heisenberg fellowship. Financial support of the Genzentrum München and the Fonds der chemischen Industrie is gratefully acknowledged. I thank Mrs. B. Österreicher, U. Schade, and M. Weyerer for technical assistance and Dr. L. Witte (Technische Universität Braunschweig) for GLC-MS measurements.

References

Berlin J (1984) Plant cell cultures – a future source of natural products? Endeavour 8:5–8
Darnell J, Lodish H, Baltimore D (1987) Molecular cell biology. Sci Am Books, WH Freeman and Company, New York, NY, USA
Drapeau D, Blanch HW, Wilke CR (1987) Economic assessment of plant cell culture for the production of ajmalicine. Biotech Bioeng 30:946–953
Dunn AR, Gough N (1984) Tissue-specific enhancer. TIBS 9:81–82
Eilert U (1987) Elicitation: methodology and aspects of application. In: Constabel F, Vasil IK (eds) Cell culture and somatic cell genetics of plants, vol 4. Academic Press, New York London, pp 153–196
Fritzemeier K-H, Cretin C, Kombrink E, Rohwer F, Taylor J, Scheel D, Hahlbrock K (1987) Transient induction of phenylalanine ammonia lyase and 4-coumarate: CoA ligase in potato leaves infected with virulent or avirulent races of *Phytophthora infestans*. Plant Physiol 85:34–41
Goldberg RB (1986) Regulation of plant gene expression. Philos Trans R Soc London Ser B 314:343–353
Hamill JD, Parr AJ, Rhodes MJC, Robins RJ, Walton NJ (1987) New routes to plant secondary products. Bio/Technology 5:800–804
Heidecker G, Messing J (1986) Structural analysis of plant genes. Annu Rev Plant Physiol 37: 439–466
Maniatis T, Fritsch EF, Sambrock J (1982) Molecular cloning. A laboratory manual. Cold Spring Harbour Lab, NY
Martin JF, Gil JA (1984) Cloning and expression of antibiotic production genes. Bio/Technology 2:63–72
Mende P, Wink M (1987) Uptake of the quinolizidine alkaloid lupanine by protoplasts and vacuoles of *Lupinus polyphyllus* cell suspension cultures. Diffusion or carrier-mediated transport? J Plant Physiol 129:229–242
Perrey R, Hauser M-Th, Wink M (1989) Cellular and subcellular localization of peroxidase and its isoenzymes in cell cultures and plants of *Lupinus polyphyllus*. (submitted)
Ptashne M (1986) Gene regulation by proteins acting nearby and at a distance. Nature (London) 322:697–71
Rosahl S, Eckes P, Schell J, Willmitzer L (1986) Organ-specific gene expression in potato. Isolation and characterization of tuber-specific cDNA sequences. Mol Gen Genet 202:368–373
Rosenthal GA, Janzen DH (1979) Herbivores: their interaction with secondary metabolites. Academic Press, New York London

Sengupta-Gopalan C, Reichert NA, Barber RF, Hall TC, Kemp JD (1985) Developmentally regulated expression of the bean β-phaseoline gene in tobacco seeds. Proc Natl Acad Sci USA 82:3320–3324
Stockhaus J, Eckes P, Rocha-Sosa M, Schell J, Willmitzer L (1987) Analysis of cis-active sequences involved in the leaf-specific expression of a potato gene in transgenic plants. Proc Natl Acad Sci USA 84:7943–7947
Swain T (1977) Secondary compounds as protective agents. Annu Rev Plant Physiol 28:479–501
Wiermann R (1981) Secondary products and cell and tissue differentiation. In: Conn EE (ed) The biochemistry of plants, vol 7. Academic Press, New York London, pp 85–115
Wink M (1984) Evidence for an extracellular lytic compartment of plant cell suspension cultures. The cell culture medium. Naturwissenschaften 71:635–636
Wink M (1985a) Composition of the spent cell culture medium. 1. Time-course of ethanol formation and the excretion of hydrolytic enzymes into the culture medium of suspension-cultured cells of *Lupinus polyphyllus*. J Plant Physiol 121:287–293
Wink M (1985b) Metabolism of quinolizidine alkaloids in plants and cell suspension cultures: induction and degradation. In: Neumann KH, Barz W, Reinhard E (eds) Primary and secondary metabolism of plant cell cultures. Springer, Berlin Heidelberg New York Tokyo, pp 107–116
Wink M (1985c) Chemische Verteidigung der Lupinen: Zur biologischen Bedeutung der Chinolizidinalkaloide. Plant Syst Evol 150:65–81
Wink M (1986) Storage of quinolizidine alkaloids in epidermal tissues. Z Naturforsch C Biosci 41:375–380
Wink M (1987a) Why do lupin cell cultures fail to produce alkaloids in large quantities? Plant Cell Tissue Org Cult 8:103–111
Wink M (1987b) Quinolizidine alkaloids: biochemistry, metabolism, and function in plants and cell suspension cultures. Planta Med 53:509–514
Wink M (1987c) Physiology of the accumulation of secondary metabolites with special reference to alkaloids. In: Constabel F (ed) Cell culture and somatic cell genetics of plants, vol 4. Academic Press, New York London, pp 17–42
Wink M (1987d) Site of lupanine and sparteine biosynthesis in intact plants and in vitro organ cultures. Z Naturforsch C Biosci 42:868–872
Wink M (1987e) Chemical ecology of quinolizidine alkaloids. ACS Symp Ser 330:524–533
Wink M (1988) Plant breeding: importance of plant secondary metabolites for protection against pathogens and herbivores. Theor Appl Genet 75:225–233
Wink M, Hartmann T (1980) Production of quinolizidine alkaloids by photomixotrophic cell suspension cultures: biochemical and biogenetic aspects. Planta Med 40:149–155
Wink M, Hartmann T (1982a) Diurnal fluctuations of quinolizidine alkaloid accumulation in legumes and cell suspension cultures. Z Naturforsch C Biosci 37:369–375
Wink M, Hartmann T (1982b) Localization of the enzymes of quinolizidine alkaloid biosynthesis in leaf chloroplasts of *Lupinus polyphyllus*. Plant Physiol 70:74–77
Wink M, Mende P (1987) Uptake of lupanine by alkaloid-storing epidermal cells of *Lupinus polyphyllus*. Planta med 53:465–469
Wink M, Witte L (1984) Turnover and transport of quinolizidine alkaloids: diurnal variation of lupanine in the phloem sap, leaves and fruits of *Lupinus albus* L. Planta 161:519–524
Wink M, Witte L (1985) Quinolizidine alkaloids as nitrogen source for lupin seedlings and cell suspension cultures. Z Naturforsch C Biosci 40:767–775
Wink M, Witte L (1987) Alkaloids in stem roots of *Nicotiana tabacum* and *Spartium junceum* transformed by *Agrobacterium* rhizogenes. Z Naturforsch C Biosci 42:69–72
Wink M, Witte L, Hartmann T (1981) Quinolizidine alkaloids of cell suspension cultures and plants of *Sarothamnus scoparius* and of its root parasite *Orobanche rapum-genistae*. Planta Med 43:342–352
Wink M, Witte L, Hartmann T, Theuring C, Volz V (1983) Accumulation of quinolizidine alkaloids in plants and cell suspension cultures: genera *Lupinus, Cytsisus, Baptisia, Genista, Laburnum*, and *Sophora*. Planta Med 48:253–257
Winnacker E-L (1987) From genes to clones. Chemie, Weinheim

Microsequencing of Plant Proteins and Cloning of the Corresponding Genes by Oligonucleotide Probing

D. Inzé, J. Vandekerckhove, and M. van Montagu[1]

1 Introduction

The production of secondary metabolites in plants requires the coordinate and developmentally regulated expression of a set of genes. The understanding of the molecular mechanisms underlying the spatial expression of plant genes such as these involved in secondary product biosynthesis is emerging as one of the most exciting fields in plant molecular biology research. Essential for such an analysis is the ability to clone the appropriate genes. In this chapter, we report the development of a powerful new approach for isolating plant genes whose the expression is correlated with certain phenotypes or environmental conditions. This approach is based on the ability of obtaining N-terminal amino-acid sequences from proteins which are immobilized on glass-fiber sheets coated with a polybase. Subsequently, oligonucleotides matching the amino-acid sequences are used to clone corresponding cDNA clones.

2 Two-Dimensional Polyacrylamide Gel Electrophoresis and Microsequence Analysis of Immobilized Proteins

Two-dimensional polyacrylamide gel electrophoresis (2D-PAGE) of (plant)-proteins is a powerful tool for analyzing the effect of external conditions on tissues (Bravo 1984). An example of such an analysis is shown in Fig. 1. *Nicotiana plumbaginifolia* cell cultures grown for 14 days in medium with auxin (0.1 mg l^{-1} NAA) as sole plant hormone were subdivided into two identical cultures. One culture was further grown with only auxin as growth factor while the other subculture was supplied with both an auxin (0.1 mg l^{-1} NAA) and a cytokinin (1 mg l^{-1} BAP). Seventy-two hours after subculture, both cell suspensions were analyzed by 2D-PAGE as described by Bauw et al. (1987). As can be seen in Fig. 1, a number of proteins are induced or repressed by the addition of cytokinin to the culture.

In order to characterize hormonally regulated proteins, a sensitive microsequencing method has been developed. The method was originally described by Vandekerckhove et al. (1985) and its application to the analysis of *Nicotiana*

[1]Laboratory of Genetics, State University Ghent, K.L. Ledeganckstraat 35, B-9000 Gent, Belgium

Primary and Secondary Metabolism of Plant Cell Cultures, ed. by W.G.W. Kurz

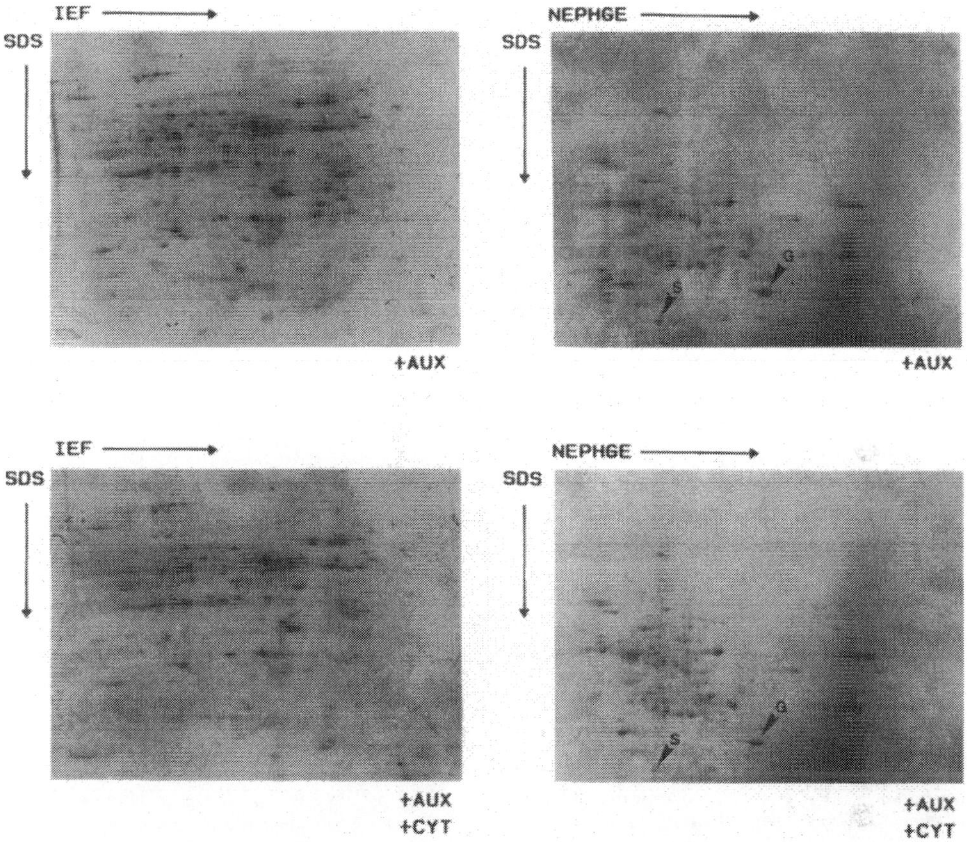

Fig. 1. *Nicotiana plumbaginifolia* proteins separated by 2D gel electrophoresis. Total protein extracts of cell suspensions grown for 2 weeks with only 0.1 mg l^{-1} naphthalene acetic acid (+aux) or subsequently supplemented for 72 h with 1 mg l^{-1} benzylamino purine (+aux; +cyt) were separated by isoelectrofocussing (*IEF*) or non-equilibrium pH gel electrophoresis (*NEPHGE*) in the horizontal direction and by $NaDodSO_4$/PAGE in the vertical direction. The position of the microsequenced β-glucanase (*G*) and manganese superoxide dismutase (*S*) is indicated

plumbaginifolia cell cultures has been described by Bauw et al. (1987). In brief, proteins which are separated by 2D-PAGE are electroblotted onto a glass-fiber membrane which is coated with a polybase such as poly-(4-vinyl-N-methylpyridinium iodide) (P4VMP). The high density of positive charges at the glass-fiber surface is necessary to attract and bind the negatively charged $NaDodSO_4$ protein complexes, which would otherwise pass through the relatively large pores of the glass fiber. The immobilized proteins on the glass-fiber membrane can be visualized by a short staining with fluorescamine. Since this dye reacts with protein amino groups, only very diluted concentrations have to be used to avoid excessive blocking of the NH_2 groups. Subsequently, the desired protein spots are excised and directly applied to a gas-phase sequenator for NH_2-terminal amino-acid sequence analysis. By comparing the obtained sequences to protein β-Glucanase

Nicotiana plumbaginifolia	E S I G V C Y G M L G N N L P N H W E V I
Nicotiana tabacum	L G N N L P N H W E V I
Barley	(ES)I G V C Y G M S A N N L P A A S T V W

Mn-superoxide dismutase

Nicotiana plumbaginifolia	L Q T F S L P D L P Y D X G A L E P A I G D
Human	K H S L P D L P Y D Y G A L E P H I N A Q
S. cerevisiae	K V T L P D L K W D F G A L E P Y I S G Q

Fig. 2. Comparison of two *Nicotiana plumbaginifolia* protein sequences with homologous enzymes identified in other eukaryotes. Amino-acid residues are given by the one-letter code. *X* denotes an unidentified post-translationally modified residue. Regions showing complete homology are shown in *boxes*

sequence data bases some of the proteins can be readily identified. In this way, one of the proteins which was repressed by cytokinin addition to the *Nicotiana plumbaginifolia* cell cultures was found to be a β-glucanase. One of the abundant, "constitutive" proteins present in the cell suspensions was shown to have homology to the manganese superoxide dismutase of *Saccharomyces cerevisiae* and humans (Bauw et al. 1987). The extent of homology is shown in Fig. 2.

3 Cloning of Genes Corresponding to the Microsequenced Proteins

The N-terminal or internal amino-acid sequences obtained by microsequencing are usually of sufficient length allowing the synthesis of one or more specific oligonucleotide probes. However, due to the degeneration of the genetic code, the correctly matching oligonucleotide would only represent one species of a large number of possible probes. For this reason, we have used deoxyinosine at ambiguous codon positions of the oligonucleotide probes. Deoxyinosine was shown to base pair with A, C and T (Ohtsuka et al. 1985) and seems neither to seriously disturb nor to destabilize the DNA duplex. Last, but not least, single-chain probes carrying deoxyinosine can be synthesized more easily than mixed deoxyoligonucleotides whose exact base contents at specific positions are difficult to determine. The use of this approach to isolate cDNA clones corresponding to β-glucanase and Mn-superoxide dismutase is described below.

3.1 β-Glucanase

The sequence of the oligonucleotide probe used to screen a cDNA library is outlined in Fig. 3. This 27-mer contains five deoxyinosines and was synthesized as a mixture of eight different oligonucleotides. For the serine, a thymidine was used at the first codon position.

A cDNA library was constructed from poly(A)$^+$ RNA of *Nicotiana plumbaginifolia* cell suspensions depleted of cytokinins for 14 days. The screening of this cDNA library with the end-labelled mixture of oligonucleotides yielded several positive clones of which one was chosen for further analysis. The clone contained a plasmid, designated pBEG, with a cDNA insert of 1294 bp. The complete nucleotide sequence of the cDNA insert is shown in Fig. 4. There is one continuous, open reading frame of 362 amino acids, which is incomplete at the amino-terminal end (De Loose et al. 1988). The reading frame ends with two consecutive stop codons (TGA and TAA). The 3′ end of the cDNA is flanked by ten deoxyadenosines, which are presumably part of the poly(A) tail. A possible polyadenylation sequence (AATAAT) is located 28 bp upstream from this poly(A) tail. To our surprise, the N-terminal amino-acid sequence as determined from the cDNA was found to be different from the amino-acid sequence obtained by microsequencing the protein. The cDNA contains two different bases with respect to the oligo sequence, thereby resulting in two amino-acid changes. The Glu at the first position of the microsequenced mature protein is replaced by a Gln, while the Ile at the third position is replaced by a Val. It must be assumed that *Nicotiana plumbaginifolia* has different isoforms for the β-glucanase and that by chance a distinct isoform has been cloned as compared to the one originally microsequenced by Bauw et al. (1987).

```
  E S I G V C Y G M L G N N L P N H W E V I   N-terminal amino-acid sequence

  Glu  Ser  Ile  Gly  Val  Cys  Tyr  Gly  Met

    A                        T    T
5' GA   TC(I) A T(I) GG(I) GT(I) TG   TA   GG(I) ATG  3'  Mixed oligonucleotide probe
    G                        C    C
```

Fig. 3. Nucleotide sequence of the oligonucleotide corresponding to β-glucanase. Deoxyinosine residues are *encircled*

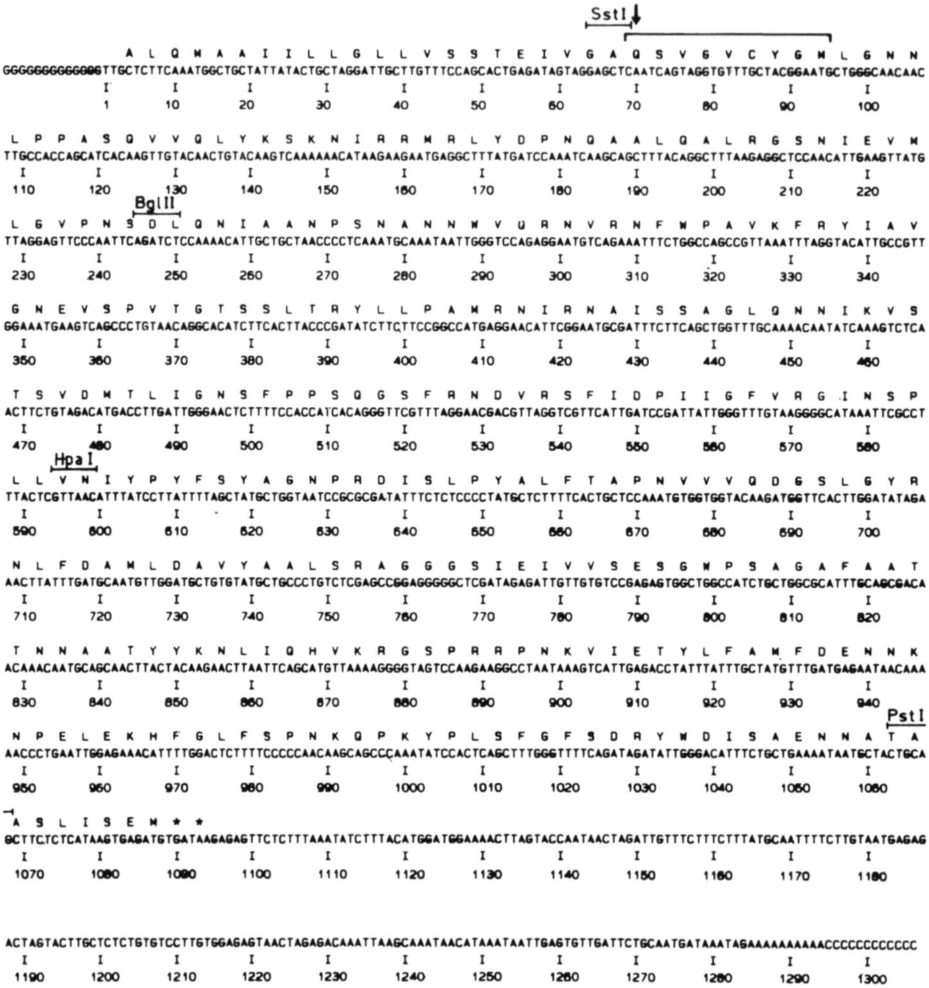

Fig. 4. cDNA and amino-acid sequence of a *Nicotiana plumbaginifolia* β-glucanase. The extent of the used oligonucleotide is indicated by a *line above*. The first amino acid of the mature protein is indicated by an *arrow*. The positions of the *Sst*I, *Bgl*II, *Hpa*I and *Pst*I sites in the cDNA are shown

3.2 Mn-Superoxide Dismutase (MnSOD)

Part of the N-terminal amino-acid sequence of an MnSOD from *Nicotiana plumbaginifolia* was chosen to design a corresponding oligonucleotide probe (Fig. 5). The 29-mer contains deoxyinosines at ambiguous codon position (nine in total). Thus, we were able to test the application of deoxyinosine at its limits. Screening of the cDNA library from *Nicotiana plumbaginifolia* cell suspension RNA yielded a full length cDNA clone. The entire sequence of the 996 bp cDNA insert is shown in Fig. 6. It contains one continuous, open reading frame, corresponding to a protein with 228 amino acids. The presumed initiation codon (ATG) at nucleotide

L Q T F S L P D L P Y D Y G A L E P A I ? G D N-terminal amino-acid sequence

Pro Tyr Asp Tyr Gly Ala Leu Glu Pro Ala

5' CC(I) TA(I) GA(I) TA(I) GG(I) GC(I) CT(I) GA(I) CC(I) GC 3' Oligonucleotide probe

Fig. 5. Nucleotide sequence of the oligonucleotide corresponding to Mn-superoxide dismutase. Deoxyinosine residues are *encircled*

position 99 is preceded by two in-frame stop codons. The sequence starting from amino acid 25 is identical to the previously determined N-terminal amino-acid sequence of the mature protein. The molecular weight of the mature protein, calculated from the cloned sequence, is 22.8 kD, which compares well with our previous estimate of 24 kD (Bauw et al. 1987). The sequence of 24 amino acids upstream from the mature protein (Fig. 6) has features typical of a signal peptide for translocation to the mitochondrial matrix (Schatz 1987). It is characterized by five arginines distributed among uncharged amino acids, and is devoid of acidic residues but rich in the hydroxylated amino acids serine and threonine.

4 Perspectives for the Cloning of Genes Involved in the Biosynthesis of Secondary Metabolites

The biosynthesis of most secondary metabolites in plants requires the expression of a large number of genes. The isolation of these genes will allow a detailed study of their expression in relation to secondary metabolite biosynthesis. Although an extensive analysis of the enzymes involved in secondary product formation remains a prerequisite, their further characterization can be simplified by the technical approach described here. Several advantages are emerging. First, proteins can be sequenced from a total or only partially purified cellular extract. The only requirement is that the enzyme of interest can be localized as a clearly distinguishable band or spot on a one- or two-dimensional polyacrylamide gel respectively. The association of enzymatic activities with specific protein spots on a gel will be facilitated by the development of specific enzyme-mediated staining procedures which can be carried out on the membrane-immobilized enzymes. Second, relatively short segments of amino-acid sequences (containing between 10–20 residues) can be readily used to clone corresponding cDNAs which in their turn will allow the prediction of the complete amino-acid sequence of the protein involved. Finally, knowledge of the NH_2-terminal sequence of the mature protein (by microsequencing) will provide more information on NH_2-terminal processing during protein maturation.

```
1     GGGGGGGGGG GGGGGGCTGG CCTCTCTGGG CATGACCTGC AACTATAAAA GGACACCATA GAGTTAACAG
                                                                         Hpa I

                                      M   A   L   R   T   L   V   S   R   R   T   L
71    CTAGAAAGCA TTTAGGAATA TCTCAAAA ATG GCA CTA CGA ACC CTA GTG AGC AGA CGG ACC TTA
                                                                          10
                                                        ↓
       A   T   G   L   G   F   R   Q   Q   L   R   G   L   Q   T   F   S   L   P
135   GCA ACA GGG CTA GGG TTC CGC CAG CAA CTC CGC GGC TTG CAG ACC TTT TCG CTC CCC
                              20                                      30

       D   L   P   Y   D   Y   G   A   L   E   P   A   I   S   G   D   I   M   Q
193   GAT CTC CCC TAC GAC TAT GGA GCA CTG GAG CCG GCA ATT AGC GGT GAC ATA ATG CAG
                                      40                                      50

       L   H   H   Q   N   H   H   Q   T   Y   V   T   N   Y   N   K   A   L   E
249   CTC CAC CAC CAG AAT CAC CAT CAG ACT TAC GTC ACC AAT TAC AAT AAA GCC CTT GAA
                                          60

       Q   L   H   D   A   I   S   K   G   D   A   P   T   V   A   K   L   H   S
305   CAG CTA CAT GAC GCC ATT TCC AAA GGA GAT GCT CCT ACC GTC GCC AAA TTG CAT AGC
       70                                  80

       A   I   K   F   N   G   G   G   H   I   N   H   S   I   F   W   K   N   L
363   GCT ATC AAA TTC AAC GGC GGA GGT CAC ATT AAC CAC TCG ATT TTC TGG AAG AAT CTT
           90                                      100

       A   P   V   R   E   G   G   G   E   P   P   K   G   S   L   G   W   A   I
420   GCC CCT GTC CGC GAG GGT GGT GGT GAG CCT CCA AAG GGT TCT CTT GGT TGG GCT ATC
               110                                     120

       D   T   N   F   G   S   L   E   A   L   V   Q   K   M   N   A   E   G   A
477   GAC ACT AAC TTT GGC TCC CTA GAA GCT TTA GTT CAA AAG ATG AAT GCA GAA GGT GCT
                  130              Hind III                140

       A   L   Q   G   S   G   W   V   W   L   G   V   D   K   E   L   K   R   L
534   GCT TTA CAG GGC TCT GGC TGG GTG TGG CTT GGT GTG GAC AAA GAG CTT AAG CGC CTG
                      150                                     160

       V   I   E   T   T   A   N   Q   D   P   L   V   S   K   G   A   N   L   V
591   GTG ATT GAA ACC ACT GCT AAT CAG GAC CCT TTG GTT TCT AAA GGA GCA AAT TTG GTT
                          170                                     180

       P   L   L   G   I   D   V   W   E   H   A   Y   Y   L   Q   Y   K   N   V
645   CCT CTT CTG GGA ATA GAC GTT TGG GAA CAT GCA TAC TAC TTG CAG TAC AAA AAT GTA
                              190                                     200

       R   P   D   Y   L   K   N   I   W   K   V   M   N   W   K   Y   A   N   E
706   AGA CCT GAT TAT CTG AAG AAC ATA TGG AAA GTT ATG AAC TGG AAA TAT GCA AAT GAA
                                  210                                     220

       V   Y   E   K   E   C   P   *
762   GTT TAT GAG AAA GAA TGT CCT TGAACAGGGA TATTTGATGT TGTTTTGAGG ACGTCTGTAA

823   AACTTTTTGA TGGGAAATAA GGCTGAGTGA CATGAGCAGG TGTCCTGTTT TTCTTGCATG TAGTCGCTGG

893   CTGATGTACT TGATGTATTT CTGGAAAAGG TTGATGTATG TACTTGATAT ATGGAGCCTA AATAAAACTA

963   CTCTATCGTT TGAGCGCAAA CCCCCCCCCC CCCC
```

Acknowledgements. This work was supported by grants from the A.S.L.K.-Kankerfonds, the Fonds voor Wetenschappelijk Geneeskundig Onderzoek (FGWO 3.0001.82) and the Services of the Prime Minister (O.O.A. 12.0561.84 and U.I.A.P. 120C0187).

D. Inzé is a Senior Research Assistant and J. Vandekerckhove a Research Associate of the National Fund for Scientific Research (Belgium).

References

Bauw G, De Loose M, Inzé D, Van Montagu M, Vandekerckhove J (1987) Alterations in the phenotype of plant cells studied by NH_2-terminal amino acid-sequence analysis of proteins electroblotted from two-dimensional gel-separated total extracts. Proc Natl Acad Sci USA 84:4806–4810

Bravo R (1984) Two-dimensional gel electrophoresis: a guide for the beginner. In: Celis JE, Bravo R (eds) Two-dimensional gel electrophoresis of proteins. Academic Press, New York London, pp 3–36

De Loose M, Alliotte T, Gheysen G, Genetello C, Gielen J, Soetaert P, Van Montagu M, Inzé D (1988) Primary structure of a hormonally regulated β-glucanase of *Nicotiana plumbaginifolia*. Gene pp 70:13–23

Ohtsuka E, Matsuki S, Ikehara M, Takahashi Y, Matsubara K (1985) An alternative approach to deoxyoligonucleotides as hybridization probes by insertion of deoxyinosine at ambiguous codon positions. J Biol Chem 260:2605–2608

Schatz G (1987) Signals guiding proteins to their correct locations in mitochondria. Eur J Biochem 165:1–6

Vandekerckhove J, Bauw G, Puype M, Van Damme J, Van Montagu M (1985) Protein-blotting on polybrene-coated glass-fiber sheets. A basis for acid hydrolysis and gas-phase sequencing of picomole quantities of protein previously separated on sodium dodecyl sulfate/polyacrylamide gel. Eur J Biochem 152:9–19

Fig. 6. Sequence of a manganese superoxide dismutase (MnSOD) of *Nicotiana plumbaginifolia*. Complete nucleotide sequence of the pSOD1 cDNA insert of 996 base pairs with its flanking G/C homopolymer tails added during the cloning procedure. The sequence homologous to the oligonucleotide probe is *underlined*. The potential polyadenylation site AATAAA (position 953) is indicated by *boxing*. The predicted coding sequence starting at position 99 (start codon ATG) to position 783 (stop codon TGA marked with an *asterisk*) has been translated into the corresponding amino-acid sequence, written above the nucleotide sequence in the one-letter code. The *arrow* on amino acid 25 points to the N-terminus of the mature protein (Bauw et al. 1987)

Switching on Production of Thiarubrines and Other Polyines by Transformation of Plant Tissue Cultures

G.H.N. Towers[1]

1 Introduction

About 6 years ago we planned to extend our research on the photobiology and the photobiochemistry of the characteristic polyines and related sulphur derivatives of the Asteraceae to studies on the biogenesis of these compounds and, in particular, the enzymes involved in these processes; no enzymes are known for thiophene, thiarubrine or polyine biosynthesis. For this purpose we decided to use cell and callus cultures of selected plant species. Unfortunately, we were not well prepared for the now well-known phenomenon of the switching off of secondary metabolism in such systems.

2 Callus and Cell Cultures

Our first attempts were with callus and cell suspension cultures of the pan-tropical weed, *Bidens alba* (Norton 1984). The leaves of this species contain high concentrations of the phototoxin, 1-phenylhepta-1,3,5-triyne (PHT) in addition to minor amounts of other related toxic polyines. In spite of varying culture media, light regimes, temperature and hormonal levels we were unsuccessful in inducing synthesis of these compounds.

We were equally unsuccessful with other composites which are rich in polyines and their sulphur derivatives such as *Eriophyllum lanatum* and *Chaenactis douglasii*. We were able to obtain in one instance very low concentrations of thiophenes in long-term callus cultures of *Tagetes patula* grown in SH medium with 4 mg l^{-1} NAA. This was a callus with small roots and the spectrum of chemical products was essentially the same as that of normal roots (Norton et al. 1985). Other composite species such as *Bidens cervicata* and *B. hillebrandiana* also failed to yield polyines. There are, however, reports on the production of very low amounts of polyines in short-term callus cultures of species in the Asteraceae (Jente 1971; Ichihara and Noda 1977) and more recently of thiophenes in *Tagetes* (Breteler, this symposium).

Polyines in *Bidens*, as in other Asteraceae, occur in all parts of the plant and typically the spectrum of compounds is organ-specific. Those that occur in *Bidens* have typical multiple absorption bands in the ultraviolet, a characteristic of com-

[1]Botany Department, the University of British Columbia, Vancouver, B.C. V6T 2B1, Canada

Primary and Secondary Metabolism of Plant
Cell Cultures, ed. by W.G.W. Kurz

pounds with conjugated double and triple bonds. Screening of large numbers of samples by spectrophotometric analysis of pet ether fractions often is therefore relatively easy. The polyines occur in resin canals which are distributed throughout the plant and they accumulate extracellularly, i.e. in the lumen of the resin canal.

3 Transgenic Cultures

Our failures to induce polyine production in cell and tissue cultures by manipulating growth parameters ended when we discovered that callus obtained from crown galls of *Bidens*, obtained with *Agrobacterium tumefaciens*, yielded PHT and other polyines in culture. Earlier studies of secondary metabolism (Klein and Link 1955, Kovacs et al. 1964; Kado 1976) had been carried out with crown gall tumours attached to the plant or with tissues in prolonged culture in media without hormones and it was speculated that the production of secondary metabolites in these systems was a carryover from the tissues where the tumour originated.

Figure 1 lists the polyines as well as some related sulphur-containing congeners which we have obtained from crown-gall tumour cultures. (Tumours were induced in greenhouse-grown plants or were induced in vitro in surface-sterilized stem or

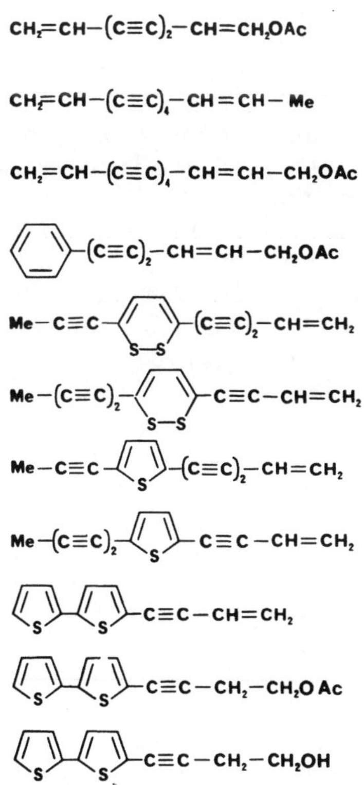

I EDE-OAc
II ETE
III ETE-OAc
IV PDE-OAc
V Thiarubrine A
VI Thiarubrine B
VII Thiophene A
VIII Thiophene B
IX BBT
X BBT-OAc
XI BBT-OH

Fig. 1. Major polyines and sulphur derivatives identified in callus and root cultures of transgenic tissues petiole segments.) These lines were polyclonal in origin and were maintained over 3 years.

Of the compounds we have studied perhaps the red dithiane heterocycles, which I have called thiarubrines, are the most interesting. Because of their colour they allow for rapid visual screening of cultures. They occur in about ten genera of the Asteraceae including the species *Chaenactis douglasii* and *Aspilia mossambicensis*. These compounds have been of interest to us because of their possible biogenetic relationship to the thiophenes which we have been studying and particularly because of their intense antifungal activity (Towers et al. 1985). Selective transfer of transgenic tissue active in their accumulation gave a line with yields equal to those of plant roots (Cosio et al. 1986).

These cultures provided evidence for a correlation between differentiation and polyine production, an observation in agreement with results of others (Staba 1980; Lindsey and Yeoman 1983). Organized xylem was clearly evident in sections of transgenic callus stained with phloroglucinol-HCl and the thiarubrines were seen to occur in intercellular spaces close to tracheids. Addition of auxins to cultures, to suppress xylogenesis, resulted in the disappearance of the thiarubrines. Similarly, we found that cell suspension cultures prepared from transgenic, thiarubrine-producing callus also failed to yield these compounds. Perhaps these two correlated effects arise from factors controlling the transcription of the auxin and cytokinin genes in T-DNA which has been incorporated into the plant genome.

A second useful discovery with transgenic callus was that the cultures bleached rapidly when exposed to light and thiarubrine synthesis ceased. Thiarubrines are sensitive to ultraviolet or blue light being readily converted to corresponding thiophenes which are colourless.

Tumours were induced, using *A. tumefaciens*, in the stems of greenhouse-grown *Chaenactis douglasii* and the callus derived from these was found to produce higher levels of thiarubrines A and B as well as their corresponding thiophenes than the levels occurring in roots. The production of "root" compounds in stem-derived tissues was expected and contradicts previous notions that crown-gall tumour cultures only produce secondary compounds characteristic of the tissue of origin (Teuscher 1973). With *Tagetes* and *Bidens*, patterns of polyines resembled those of stems, where the tumours originated (Norton et al. 1985). Dark-grown transgenic callus, when transferred to culture conditions for 4 weeks under cool-white fluorescent lamps lost their thiarubrines and started synthesizing chlorophylls and anthocyanins. On return of the cultures to dark thiarubrine production resumed. Thiarubrines accumulated in nodes which, in section, resembled root cross-sections (Cosio et al. 1986). The compounds accumulated in intercellular cavities surrounding the "stele" rather than in resin canals. Yields were decreased when callus was grown in the presence of either IAA, NAA or 2,4-D.

Endogenous growth regulator levels were analyzed in three crown-gall tumour cultures of *Chaenactis* originating in stems transformed by *A. tumefaciens* strain A277. These cultures displayed different tumour morphologies and only one line, CD-5, accumulated polyines. Polyine production was correlated with xylogenesis. Analyses of auxin and cytokinin levels in these transgenic cultures revealed levels of isopentenyladenosine (iPA) in one of the inactive lines that was eight-fold higher than what was found in the other two lines. The other non-producing line exhibited levels of bound IAA that were nine-fold higher than the levels in the other lines (Cosio et al.1987). These results were difficult to interpret and the studies did not shed much light on the problem except to show clearly once more that thiarubrine production is correlated with xylogenesis.

4 Root Cultures

We now realized that in view of the fact that these anatomically "rootlike" transgenic cultures could produce thiarubrines, we should be growing root cultures of *Chaenactis* and perhaps even transgenic root cultures obtained with either *A. tumefaciens* or with *A. rhizogenes*. This is, in fact, what we turned our attentions to.

Our initial studies were with root cultures of *Bidens alba* (Norton and Towers 1986). Cultured roots were similar to intact roots in the types of polyines produced but contained significantly less amounts of compounds. High carbohydrate/ nitrogen ratios in the medium favoured greater production, whereas light, on the other hand, reduced levels drastically. Ichihara and Noda (1977), however, demonstrated an increase of polyines in illuminated callus cultures of safflower. Root cultures obtained from regenerated plantlets of *Bidens* were also studied (Norton 1984). The plantlets were regenerated from crown-gall tumour cultures. These transgenic roots differed in their increased sensitivity to external hormonal levels and in their propensity to form callus in older cultures. Polyine production was similar to that of normal root cultures.

Similarly, normal root cultures of *Chaenactis* were found to produce thiarubrines and in concentrations similar to those found in intact plant roots (Cosio et al. 1986).

5 Hairy Root Cultures

As an approach to obtaining even higher-yielding thiarubrine-producing cultures we transformed *Chaenactis* with *Agrobacterium rhizogenes* strain TR7 (Constabel and Towers 1988a). The hairy root cultures obtained produced the highest yields of the thiarubrines of any plant tissue culture system reported so far. These results are similar to those of many other groups studying secondary compound production in roots (Flores and Filnex 1985; Kamada et al. 1986; Hamill et al. 1986; Jung and Tepfer 1987).

Hairy root cultures of *Chaenactis* showed growth and biosynthetic characteristics which could not be duplicated in non-transformed root cultures through the addition of exogenous NAA. Thiarubrine biosynthesis was found to parallel growth in batch cultures of hairy roots and maximum accumulation occurred at the beginning of the stationary phase when maximum biomass was achieved. This parallel pattern suggests that the biosynthetic pathway is expressed constitutively throughout the culture period and is not turned on at specific times. We were thus finally in a position to initiate studies of the biogenesis of these very interesting antibiotics.

6 Biosynthesis Studies

Normal roots and both normal and hairy root cultures of *Chaenactis* contain, in addition to thiarubrines A and B, the corresponding thiophenes 3 and 4 as minor constituents. These thiophenes are known from almost 100 species of Asteraceae and have usually been reported as occurring without the thiarubrines (Bohlmann and Zdero 1985). We found that the patterns of incorporation of S^{35} into the thiarubrines and thiophenes in hairy root cultures were similar. The label appeared in both types of compound at the same time and once incorporated, remained at the same level in every case (Constabel and Towers 1988). This is inconsistent with a hypothesis that these two types of sulphur-containing polyines are synthesized sequentially as part of the same pathway with either thiarubrines giving rise to thiophenes or vice versa. Thiarubrines are readily converted to corresponding thiophenes on exposure to light. The similar specific activities, which were determined for both types of compound, as well as the fact that contraction of the dithiane ring to yield a thiophene ring during workup of samples occurs, suggest that in *Chaenactis* a part of the thiophenes detected are artefacts. Bohlmann and Bresinsky (1967) had previously suggested that thiophene biosynthesis *in plant* via a dithiacyclohexadiene was unlikely. Our data also indicated that there was no turnover of thiarubrines during late logarithmic and stationary growth phases suggesting that these compounds, which are fungicidal and nematocidal, play a protective rather than a physiological function in the roots.

Acknowledgements. The work reported here is based on studies by Dr. R.A. Norton, Dr. E. Cosio and Peter Constabel and was supported by the Natural Sciences and Engineering Research Council of Canada.

References

Bohlmann F, Bresinsky E (1967) Synthese einiger di-und triacetate natürlich vokammendex Polyine. Chem Ber 100:1209–1212
Bohlmann F, Zdero C (1985) Thiophene and its derivatives, Pt 1. In: Weissberger A (vol. ed) The chemistry of heterocyclic compounds, vol 44. Gronowitz S (vol. ed) John Wiley and Sons, New York, pp 261–324
Constabel CP, Towers GHN (1988a) Thiarubrine accumulation in hairy root cultures of *Chaenactis douglasii*. J Plant Physiol 133:67–70
Constabel CP, Towers GHN (1988b) Incorporation of S into dithiacyclohexadiene and thiophene polyines in hairy root cultures of *Chaenactis douglasii*. Phytochemistry (in press)
Cosio EG, Norton RA, Towers E, Finlayson AJ, Rodriguez E, Towers GHN (1986) Production of antibiotic thiarubrines by a crown gall tumor line of *Chaenactis douglasii*. J Plant Physiol 124:155–164
Cosio EG, Towers GHN, McPherson J (1987) Endogenous growth regulator levels and polyacetylene accumulation in crown gall tumor lines of *Chaenactis douglasii*. J Plant Physiol 129:1–11
Flores HE, Filner P (1985) Hairy roots of Solanaceae as a source of alkaloids. Plant Physiol 77S:12
Hamill JD, Parr AJ, Robins PJ, Rhodes MJC (1986) Secondary product formation by cultures of *Beta vulgaris* and *Nicotiana rustica* transformed with *Agrobacterium rhizogenes*. Plant Cell Rep 5:111–114
Ichihara KI, Noda M (1977) Distribution and metabolism of polyacetylenes in safflower. Biochem Biophys Acta 487:269–270

Jente R (1971) Polyacetylenverbindungen in Gewebekulturen von *Centaurea ruthenica* Lam. Tetrahedron Lett 27:4077–4083

Jung G, Tepfer D (1987) Use of genetic transformation by the Ri T-DNA of *Agrobacterium rhizogenes* to stimulate biomass and tropane alkaloid production in *Atropa belladonna* and *Calystegia sepium* roots grown in vitro. Plant Sci 50:145–151

Kado DI (1976) The tumor-inducing substances of *Agrobacterium tumefaciens*. Annu Rev Phytopathol 14:265–308

Kamada HN, Okamura M, Sataka H, Harada H, Shimomura K (1986) Alkaloid production by hairy root cultures in *Atropa belladonna*. Plant Cell Rep 5:239–242

Klein RM, Link GKK (1955) The etiology of crown gall. Q Rev Biol 30:207–277

Kovacs BA, Wakkary JA, Goodfriend L, Rose B (1964) Isolation of an antihistamine principle resembling tomatine from crown gall tumors. Science 144:295–296

Lindsey K, Yeoman MM (1983) The relationship between growth rate, differentiation and alkaloid accumulation in cell cultures. J Exp Bot 34:1055–1065

Norton RA (1984) Studies on polyacetylene production in normal and transformed tissue cultures of *Bidens alba*. PhD Thesis, Univ British Columbia

Norton RA, Towers GHN (1985) Synthesis of polyacetylenes in tumor callus of *Bidens alba*. J Plant Physiol 120:273–283

Norton RA, Towers GHN (1986) Factors affecting synthesis of polyacetylenes in root cultures of *Bidens alba*. J Plant Physiol 122:41–53

Norton RA, Finlayson RJ, Towers GHN (1985) Thiophene production by crown galls and callus tissues of *Tagetes patula*. Phytochemistry 24:719–722

Staba EJ (1980) Plant tissue culture as a source of biochemicals. CRC, Boca Raton, Fl

Teuscher E (1973) Probleme der Produktion sekundarer Pflanzenstoffe mit Hilfe von Zellkulturen. Pharmazie 28:6–18

Towers GHN, Abramowski Z, Finlayson AJ, Zucconi A (1985) Antibiotic properties of thiarubrine A, a naturally occurring dithiacyclohexadiene polyine. Planta Med 3:225–229

Regulation of Gene Expression in Biologically Stressed Bean Cell Cultures

R.A. Dixon[1], E.R. Blyden[2], M. Dron[3], M.J. Harrison[1], C.J. Lamb[3], M.A. Lawton[3], and M. Mavandad[1]

1 Introduction

For many years, an important aim of plant biotechnology has been to manipulate cultured plant cells in order to increase the metabolic flux into specific pathways of secondary product formation. In most cases this has involved alteration of physical and chemical components of the culture environment, the most common of which has been the systematic variation of the concentration of growth regulators in the culture medium. This approach has led to the design of protocols for the preliminary optimization of culture conditions for secondary product formation in specific plant species [1], although the commercial exploitation of cell cultures as sources of valuable chemicals has been disappointing [2].

Several groups of workers have investigated the possibility of using biotic elicitors to increase and/or modify secondary product synthesis in cultured plant cells. Biotic elicitors are molecules of microbial (or host plant) origin first identified on the basis of their ability to induce defense responses, in particular the accumulation of antimicrobial phytoalexins, in plants. The original work on phytoalexin accumulation in response to elicitors indicated that cultures of plant cells such as bean and soybean responded to polysaccharide components from the cell walls and culture fluids of plant pathogenic fungi by the increased synthesis and accumulation of the same isoflavonoid derivatives as had previously been shown to accumulate in the intact plant under natural conditions of infection [3, 4]. Subsequently, many plant species have been shown to respond to elicitors by increased accumulation of their respective phytoalexins [5, 6, 7]. More recently, elicitor treatment has been shown in some cases to result in the formation of novel metabolites or compounds apparently not associated with defense against pathogens [7, 8, 9]. This has in part re-kindled interest in the use of cell cultures as a source of secondary products [10].

Work in our laboratories has focussed on the molecular mechanisms underlying the elicitor-mediated accumulation of isoflavonoid phytoalexins and other defense response products in cell suspension cultures of the French bean, *Phaseolus vulgaris.* This system is highly amenable to studies of molecular control

[1]Plant Biology Division, The Samuel Roberts Noble Foundation, P.O. Box 2180, 2510 Highway 199 East, Ardmore, Oklahoma 73402, USA
[2]Department of Biochemistry, Royal Holloway and Bedford New College, University of London, Egham Hill, Egham, Surrey TW20 OEX, United Kingdom
[3]Plant Biology Laboratory, Salk Institute for Biological Studies, P.O. Box 85800, San Diego, California 92138, USA

Primary and Secondary Metabolism of Plant Cell Cultures, ed. by W.G.W. Kurz

mechanisms, as it is characterized by rapid but transient enzyme induction. We review progress on understanding the factors which might control overall biosynthetic capacity in the isoflavonoid pathway. These include mechanisms for gene activation by *cis*- and *trans*-acting regulatory elements, the role of multigene families and corresponding isoenzymes for the key regulatory enzymes phenylalanine ammonia-lyase (PAL) and chalcone synthase (CHS), and the possible role of pathway intermediates as signals for the down-regulation of the pathway at the transcriptional and post-translational levels. Elicitor induction of other defense response genes is also briefly considered.

2 Molecular Responses of Cultured Bean Cells to Biotic Elicitors

The isoflavonoid phytoalexins phaseollin and kievitone are the major low M_r defense-related products to accumulate in cultured bean cells in response to a crude elicitor preparation heat-released from cell walls of the fungal bean pathogen *Colletotrichum lindemuthianum*. Their accumulation is preceded by increased extractable activities of a number of enzymes involved in their synthesis; these include PAL, cinnamic acid 4-hydroxylase, 4-coumarate: CoA ligase, CHS and chalcone isomerase (reviewed in [6]). At the same time, some of the flux through the phenylpropanoid pathway is directed towards the synthesis of wall-bound phenolics (lignin-like material); this is preceded by rapid induction of the lignin-specific marker enzyme cinnamyl alcohol dehydrogenase (CAD) [10a].

Hydroxyproline-rich glycoproteins (HRGPs) accumulate in the cell walls of elicitor-treated bean cells [11]. Prior to their deposition, the apo-proteins are post-translationally modified in the endoplasmic reticulum by hydroxylation of proline residues and subsequent arabinosylation; this is closely associated with elicitor-mediated induction of the extractable activities of prolyl hydroxylase (PH) and protein: arabinosyl transferase [11]. Other elicitor-induced enzymes include chitinase [12] and 1, 3-β-D-glucanase [13], which exhibit antifungal activity by degrading fungal cell wall polymers. In addition, a number of as yet unidentified elicitor-induced proteins have been observed following two-dimensional electrophoretic analysis of bean polypeptides synthesized either in vivo (in pulse labelling experiments) or in vitro from mRNA [14, 15].

The kinetics of the appearance of defense-related enzymes/proteins in elicitor-treated bean cells are summarized in Table 1.

The above studies have considered the multicomponent response of bean cells to a single elicitor preparation from cell walls of a fungal bean pathogen. Broadly similar responses have been recorded using a variety of other elicitors. These include various fractions isolated from *Colletotrichum* culture fluids [15, 16] or host-derived elicitor components from autoclaved bean hypocotyl extracts [17]. The latter probably contain elicitor-active oligogalacturonides originating from the pectic material of the host-cell wall; pure oligogalacturonide preparations from commercially available pectic material are also active as elicitors in the bean system [17]. Most of these elicitors induce the de novo synthesis of PAL, CHS and CHI. However, in the case of the *Colletotrichum* culture fluid elicitors, this is not activity. These data are summarized in Table 1. The rapidity of transcriptional activation of PAL and CHS clearly indicates a direct signal transduction pathway from binding of elicitor to its putative receptor and the nuclear events which determine the selective activation of defense response genes. The delayed activation of other genes (e.g. these encoding the HRGPs) suggests the operation of more than one signal transduction pathway, and the different times required for attainment of maximum transcript levels may reflect different rates of mRNA processing, stability or transcriptional shutoff.

5 *Cis*-Acting Elements and *Trans*-Acting Factors for Elicitor-Induced Gene Expression

5′-flanking regions of bean PAL and CHS genes have been sequenced [29, 30], and putative regulatory elements are currently under analysis. In order to study the early events in gene activation following exposure to elicitor, a chimeric gene was constructed comprising the 5′-flanking region of a CHS gene fused to a bacterial chloramphenicol acetyltransferase (CAT) gene. The chimeric gene, when electroporated into soybean protoplasts, was rapidly and transiently expressed in response to *Colletotrichum* elicitor or glutathione (a potent and selective elicitor of defense response genes in cultured bean cells [35]). Functional analysis of 5′ deletions showed that promoter activity was determined by an elicitor-regulated activator located between the TATA box and −170 and an upstream silencer between −170 and −326 [36].

Binding of nuclear proteins to the silencer region in vitro has been detected by gel retardation and DNase I protection footprinting [37]. Footprinting indentified binding to three discrete sites in the element. Binding was specifically competed by CHS promoter sequences and by the promoter of a coordinately induced defense response gene (PAL), but not by unrelated sequences, e.g. CHS coding sequences. A synthetic oligonucleotide homologous to the major protected site in the silencer region did not compete binding of nuclear protein to the promoter as a single monomeric sequence. However, competition was obtained with a trimeric oligonucleotide. Electroporation of the silencer element enhanced *in trans* the activity of the CHS promoter presumably by competition for binding of the *trans*-acting repressors. These data suggest approaches for engineering the expression of a set of coordinately regulated biosynthetic enzymes in cultured plant cells.

6 Down-Regulation of Phenylpropanoid Biosynthesis

The induction of several of the defense response enzymes in elicitor-treated bean cells is not only rapid but also highly transient. A number of studies have provided evidence to suggest that the immediate product of the PAL reaction, *trans*-cinnamic acid, might act as a signal to regulate the expression of the phenylpropanoid result from post-translational modifications. The tetrameric, active charge isoforms of PAL differ in their K_m values for phenylalanine, and the lower K_m forms are preferentially induced by elicitor [26]. Chromatofocussing likewise reveals differences in the patterns of active PAL isoenzymes after induction by different stimuli (wounding, light, fungal infection) in intact plant tissues, reflecting differential activation of the PAL small multigene family [15]. Further work is needed to define the functional significance of the differential expression of PAL genes. Are, for example, the different forms of the enzyme differentially feedback inhibited, involved in the synthesis of different phenylpropanoid end products or localized in different subcellular compartments? If specific PAL genes do indeed encode enzyme subunits which function in a specific branch of phenylpropanoid metabolism, there may be potential here for fine manipulation of secondary product synthesis via genetic engineering.

CHS, the first enzyme specific for the flavonoid/isoflavonoid branch of phenylpropanoid synthesis, is encoded by at least six genes in bean [20]. Levels of transcripts from all six genes increase in suspension cultured cells exposed to fungal elicitor [20]. The coding regions of the six CHS genes are more homologous to one another than in the case of the PAL family, although 3′ and 5′ untranslated regions are highly divergent and the extent of induction of individual members of the family differs in response to light, wounding or fungal infection [20]. In bean, two distinct CHS activities have been proposed in order to account for the synthesis of the 5-hydroxylated series of isoflavonoids (leading to kievitone) and the 5-deoxy series leading to phaseollin [27]. Phaseollin and kievitone accumulate at different rates in bean cell cultures (with kievitone preceding phaseollin by at least 12 h) and differential timing of attainment of maximum transcript levels is also observed for the CHS genes [28]. However, it has not yet been possible to ascribe functional differences to the different CHS gene products.

An elicitor-induced cDNA complementary to the lignin-specific enzyme CAD has recently been described [22]. It is not yet fully clear whether a second gene may be involved specifically in lignification during xylogenesis.

4 Rapid Activation of Gene Transcription Underlies Elicitor-Induced Secondary Product Synthesis

Post-translational modifications may generate increased subunit polymorphism for PAL [26] and the activation of pre-existing inactive enzyme may represent a portion of the elicitor-mediated increase in CHI activity [24]. However, the general picture to emerge from studies using a variety of in vivo labelling and in vitro techniques (reviewed in [6, 29, 30]) is one of increased de novo synthesis, arising from increased steady state transcript levels, underlying the elevated activities of PAL, CHS, CHI, CAD, chitinase, HRGPs and PH. Significant differences do, however, occur with respect to the timing of attainment of maximum rates of transcription (determined by nuclear run-off experiments with isolated nuclei), steady state transcript levels (measured by Northern hybridization), translatable mRNA activities (measured by in vitro translation) and extractable enzyme activity. These data are summarized in Table 1. The rapidity of transcriptional activation of PAL and CHS clearly indicates a direct signal transduction pathway from binding of elicitor to its putative receptor and the nuclear events which determine the selective activation of defense response genes. The delayed activation of other genes (e.g. these encoding the HRGPs) suggests the operation of more than one signal transduction pathway, and the different times required for attainment of maximum transcript levels may reflect different rates of mRNA processing, stability or transcriptional shutoff.

5 *Cis*-Acting Elements and *Trans*-Acting Factors for Elicitor-Induced Gene Expression

5′-flanking regions of bean PAL and CHS genes have been sequenced [29, 30], and putative regulatory elements are currently under analysis. In order to study the early events in gene activation following exposure to elicitor, a chimeric gene was constructed comprising the 5′-flanking region of a CHS gene fused to a bacterial chloramphenicol acetyltransferase (CAT) gene. The chimeric gene, when electroporated into soybean protoplasts, was rapidly and transiently expressed in response to *Colletotrichum* elicitor or glutathione (a potent and selective elicitor of defense response genes in cultured bean cells [35]). Functional analysis of 5′ deletions showed that promoter activity was determined by an elicitor-regulated activator located between the TATA box and -170 and an upstream silencer between -170 and -326 [36].

Binding of nuclear proteins to the silencer region in vitro has been detected by gel retardation and DNase I protection footprinting [37]. Footprinting indentified binding to three discrete sites in the element. Binding was specifically competed by CHS promoter sequences and by the promoter of a coordinately induced defense response gene (PAL), but not by unrelated sequences, e.g. CHS coding sequences. A synthetic oligonucleotide homologous to the major protected site in the silencer region did not compete binding of nuclear protein to the promoter as a single monomeric sequence. However, competition was obtained with a trimeric oligonucleotide. Electroporation of the silencer element enhanced *in trans* the activity of the CHS promoter presumably by competition for binding of the *trans*-acting repressors. These data suggest approaches for engineering the expression of a set of coordinately regulated biosynthetic enzymes in cultured plant cells.

6 Down-Regulation of Phenylpropanoid Biosynthesis

The induction of several of the defense response enzymes in elicitor-treated bean cells is not only rapid but also highly transient. A number of studies have provided evidence to suggest that the immediate product of the PAL reaction, *trans*-cinnamic acid, might act as a signal to regulate the expression of the phenylpropanoid pathway (reviewed in [29]). Exogenously applied cinnamic acid inhibits the transcription of PAL and CHS genes, thereby preventing further synthesis of the enzyme in elicitor-treated cells [38]. While thus inhibiting expression of the phenylpropanoid pathway's biosynthetic machinery, cinnamic acid also appears to induce the synthesis of a number of characteristic polypeptides [38], including one which may be involved in the irreversible inactivation of PAL via effects at the enzyme's active site [39]. The effects of cinnamic acid on CHS gene transcription in cultured cells are mimicked in protoplasts electroporated with a CHS promoter – CAT construct. Work is now in progress aimed at identifying (1) cinnamic acid-induced transcripts and their possible role in the regulation of PAL activity and (2) the mechanisms by which cinnamic acid inhibits gene transcription. One strategy for maximizing the biosynthetic potential of the phenylpropanoid pathway in cultured cells could be to control the processes which underlie the cessation of gene transcription/enzyme synthesis and the turnover of enzyme activity. Regulation here could be extremely subtle in view of the potential for differential inhibition of transcription of PAL or CHS isogenes or differential inactivation of PAL isoforms.

7 Implications for the Manipulation of Secondary Product Synthesis

This chapter has reviewed our current knowledge of the molecular mechanisms which regulate the expression of the phenylpropanoid pathway in elicitor-treated bean cell cultures. There are still extensive gaps in this knowledge, in particular with respect to the mode of elicitor recognition at the cell surface and subsequent initial stages of signal transduction. A basis, however, is now clear for the manipulation of this and other secondary product pathways by genetic engineering. Elicitor-inducible promoters can be used to drive the expression of chimeric genes in transgenic plants; such genes could feasibly encode enzymes which modify the plants' natural phenylpropanoid metabolism (e.g. by hydroxylation, methylation, isoprenylation, etc.), thereby producing a new spectrum of secondary products in a given plant species. Overall synthetic capacity could perhaps be increased by engineering lack of responsiveness to endogenous, negative, regulatory effectors (e.g. cinnamic acid). Alternatively, sequences which bind and therefore sequester *trans*-acting silencer factors could be expressed in plants, or silencer transcripts could be nullified by expression of antisense RNA in vivo. In the case of the isoflavonoid pathway in legumes, experiments along these lines may also lead to novel strategies for the manipulation of disease resistance in the intact plant.

References

1. Morris P, Scragg AM, Smart NJ, Stafford A (1985) Secondary product formation by cell suspension cultures. In: Dixon RA (ed) Plant cell culture. A practical approach. IRL Oxford, pp 127–167
2. Fuller KW, Bartlett DJ (1985) The chemosynthetic potential of plants and its realization by immobilized systems. Annu Proc Phytochem Soc Eur 26:229–247
3. Ebel J, Ayers AR, Albersheim P (1976) Host-pathogen interactions. XII. Response of suspension cultured soybean cells to the elicitor isolated from *Phytophthora megasperma* var. *sojae*, a fungal pathogen of soybeans. Plant Physiol 57:775–779
4. Dixon RA, Fuller KW (1977) Characterization of components from culture filtrates of *Botrytis cinerea* which stimulate phaseollin biosynthesis in *Phaseolus vulgaris* cell suspension cultures. Physiol Plant Pathol 11:287–296
5. Dixon RA, Dey PM, Lamb CJ (1983) Phytoalexins: enzymology and molecular biology. Adv Enzymol Rel Areas Mol Biol 53:1–136
6. Dixon RA (1986) The phytoalexin response: elicitation, signalling and control of host gene expression. Biol Rev 61:239–291
7. Kessmann H, Barz W (1987) Accumulation of isoflavones and pterocarpan phytoalexins in cell suspension cultures of different cultivars of chickpea (*Cicer arietinum*). Plant Cell Rep 6:55–59
8. Ayabe S-I, Iida K, Furuya T (1986) Stress-induced formation of echinatin and a metabolite, 5′-prenyl-licodione, in cultured *Glycyrrhiza echinata* cells. Phytochemistry 25:2803–2806
9. Gugler K, Funk C, Brodelius P (1988) Elicitor-induced tyrosine decarboxylase in berberine-synthesizing suspension cultures of *Thalictrum rugosum*. Eur J Biochem 170:661–666
10. DiCosmo F, Tallevi SG (1985) Plant cell cultures and microbial insult: interactions with biotechnological potential. Trends Biotechnol 3:110–111

10a. Grand C, Sarni F, Lamb CJ (1987) Rapid induction by fungal elicitor of the synthesis of cinnamyl alcohol dehydrogenase, a specific enzyme of lignin synthesis. Eur J Biochem 169:73–77

11. Bolwell GP, Robbins MP, Dixon RA (1985) Metabolic changes in elicitor-treated bean cells. Enzymic responses associated with rapid changes in cell wall components. Eur J Biochem 148:571–578
12. Hedrick SA, Bell JN, Boller T, Lamb CJ (1988) Chitinase cDNA cloning and mRNA induction by fungal elicitor, wounding and infection. Plant Physiol 86:182–186
13. Kombrink E, Hahlbrock K (1985) Dependence of the level of phytoalexin and enzyme induction by fungal elicitor on the growth stage of *Petroselinum crispum* cell cultures. Plant Cell Rep 4:277–280
14. Cramer CL, Ryder TB, Bell JN, Lamb CJ (1985) Rapid switching of plant gene expression induced by fungal elicitor. Science 227:1240–1243
15. Hamdan MAMS, Dixon RA (1987) Differential patterns of protein synthesis in bean cells exposed to elicitor fractions from *Colletotrichum lindemuthianum*. Physiol Mol Plant Pathol 31:105–121
16. Hamdan MAMS, Dixon RA (1987) Fractionation and properties of elicitors of the phenylpropanoid pathway from culture filtrates of *Colletotrichum lindemuthianum*. Physiol Mol Plant Pathol 31:91–103
17. Dixon RA, Jennings AC, Davies LA, Gerrish C, Murphy DL (1989) Elicitor-active components from French bean hypocotyls. Physiol Mol Plant Pathol 34:99–115
18. Edwards K, Cramer CL, Bolwell G, Dixon RA, Schuch W, Lamb CJ (1985) Rapid transient induction of phenylalanine ammonia-lyase mRNA in elicitor-treated bean cells. Proc Natl Acad Sci USA 82:6731–6735
19. Cramer CL, Edwards K, Dron M, Liang X, Dildine SL, Bolwell GP, Dixon RA, Lamb CJ, Schuch W (1989) Phenylalanine ammonia-lyase gene organization and structure. Plant Mol Biol (in press)
20. Ryder TB, Hedrick SA, Bell JN, Liang X, Clouse SD, Lamb CJ (1987) Organization and differential activation of a gene family encoding the plant defense enzyme chalcone synthase in *Phaseolus vulgaris*. Mol Gen Genet 210:219–233
21. Mehdy MC, Lamb CJ (1987) Chalcone isomerase cDNA cloning and mRNA induction by fungal elicitor, wounding, and infection. EMBO J 6:1527–1533

22. Walter MH, Grima-Pettenati J, Grand C, Boudet A, Lamb CJ (1988) Cinnamyl alcohol dehydrogenase, a molecular marker specific for lignin synthesis: cDNA cloning and mRNA induction by fungal elicitor. Proc Natl Acad Sci (USA) 85(15):5546–5550
23. Corbin DR, Sauer N, Lamb CJ (1987) Differential regulation of a hydroxyproline-rich glycoprotein gene family in wounded and infected plants. Mol Cell Bol 7:6337–6344
24. Robbins MP, Dixon RA (1984) Induction of chalcone isomerase in elicitor-treated bean cells. Comparison of rates of synthesis and appearance of immunodetectable enzyme. Eur J Biochem 145:195–202
25. Liang X, Dron M, Cramer CL, Dixon RA, Lamb CJ (1989) Differential regulation of phenylalanine ammonia-lyase genes during plant development and by environmental cues. J. Biol Chem (in press)
26. Bolwell GP, Bell JN, Cramer CL, Schuch W, Lamb CJ, Dixon RA (1985) L-phenylalanine ammonia-lyase from *Phaseolus vulgaris.* Characterization and differential induction of multiple forms from elicitor-treated cell suspension cultures. Eur J Biochem 149:411–419
27. Dewick PM, Steele MJ, Dixon RA, Whitehead IM (1982) Biosynthesis of isoflavonoid phytoalexins: incorporation of sodium [1,2-$^{13}C_2$] acetate into phaseollin and kievitone. Z Naturforsch 37c:363–368
28. Ellis J, Blyden R, Jennings A, Dixon R (1988) Gene expression in elicitor-treated bean cells. J Cell Biochem Suppl 12c:261
29. Dixon RA, Bolwell GP (1986) Modulation of the phenylpropanoid pathway in bean (*Phaseolus vulgaris*) cell suspension cultures. In: Morris P, Scragg AJ, Stafford A, Fowler MW (eds) Secondary metabolism in plant cell cultures. Univ Press, Cambridge, pp 89–102
30. Lamb CJ, Bell JN, Corbin DR, Lawton MA, Mehdy MC, Ryder TB, Sauer N, Walter MH (1987) Activation of defense genes in response to elicitor and infection. In: Arntzen CJ, Ryan C (eds) Molecular strategies for crop protection. Liss, Boca Raton, pp 49–58
31. Lawton MA, Lamb CJ (1987) Transcriptional activation of plant defense genes by fungal elicitor, wounding and infection. Mol Cell Biol 7:335–341
32. Ryder TB, Cramer CL, Bell JN, Robbins MP, Dixon RA, Lamb CJ (1984) Elicitor rapidly induces chalcone synthase mRNA in *Phaseolus vulgaris* cells at the onset of the phytoalexin response. Proc Natl Acad Sci USA 81:5726–5728
33. Dixon RA, Blyden ER, Robbins MP, Van Tunen AJ, Mol JNM (1988) Comparative biochemistry of chalcone isomerases. Phytochemistry 27:2801–2808
34. Bolwell GP, Dixon RA (1986) Membrane-bound hydroxylases in elicitor-treated bean cells. Rapid induction of the synthesis of prolyl hydroxylase and a putative cytochrome P450. Eur J Biochem 159:163–169
35. Wingate VPM, Lawton MA, Lamb CJ (1988) Glutathione causes a massive and selective induction of plant defense genes. Plant Physiol 87:206–210
36. Dron M, Clouse SD, Dixon RA, Lawton MA, Lamb CJ (1988) Glutathione and fungal elicitor regulation of a plant defense gene promoter in electroporated protoplasts. Proc Natl Acad Sci USA 85:6738–6742
37. Lawton MA, Kragh K, Jenkins SM, Dron M, Clouse SD, Dixon RA, Lamb CJ (1988) Nuclear protein binding to a *cis*-acting silencer in a defense gene promoter. J Cell Biochem Suppl 12c:222
38. Bolwell GP, Mavandad M, Millar DJ, Edwards K, Schuch W, Dixon RA (1988) Inhibition of mRNA levels and activities by *trans*-cinnamic acid in elicitor-induced bean cells. Phytochemistry 27:2109–2117
39. Bolwell GP, Cramer CL, Lamb CJ, Schuch W, Dixon RA (1986) L-Phenylalanine ammonia-lyase from *Phaseolus vulgaris*: modulation of the levels of active enzyme by *trans*-cinnamic acid. Planta 169:97–107

Regulated Expression of *lacZ* Gene Fusions in Tobacco

T.H. Teeri[1], H. Lehväslaiho[1,3], M. Franck[1], J. Uotila[1], P. Heino[1,4], E.T. Palva[1,4], M. van Montagu[2], and L. Herrera-Estrella[2,5]

Analysis of gene regulation with fusions to reporter genes has proved to be successful in many biological systems including both prokaryotes and eukaryotes. The regulatory regions of the genes under study are joined to the coding sequence of the reporter gene and subsequent to transfer of the gene fusion into the host cell, the activity of the reporter gene product will reflect the activity of the regulatory regions. A good choice of the reporter gene allows convenient, standard and quantitative measurement of gene expression and, in some cases, detection of reporter gene activity directly in the tissue. Genes such as *nptII, cat, luc* and *uidA* have been used to monitor the activity of plant promoters controlling their expression in transgenic plants (Bevan et al. 1983; Herrera-Estrella et al. 1983; Ow et al. 1986; Jefferson et al. 1987). These genes code for the enzymes neomycin phosphotransferase II, chloramphenicol acetyltransferase, the firefly luciferase and β-glucuronidase (GUS) respectively, which are relatively easy to assay.

We have used the *E. coli* gene *lacZ* as a reporter gene for plant gene expression studies. The gene codes for a soluble β-galactosidase (LacZ). Both the gene and its product have been extensively studied since the 1960s (Ullman et al. 1965; Brickman et al. 1979; Kalnins et al. 1983). The 27 N-terminal amino acids of LacZ are dispensable and they can be removed or replaced by foreign polypeptides without loss of enzymatic activity (Casadaban et al. 1983), allowing the effective use of fusion genetics. The wide application of *lacZ* as a reporter gene in microbial and animal systems (Bassford et al. 1978; Casadaban and Cohen 1980; Hiromi et al. 1985) has led to the development of a variety of chromogenic substrates which are commercially available.

We fused 5′ to *lacZ*, the regulatory regions and amino terminus coding sequences of various genes active in plant cells (Fig. 1). In *ss-lacZ*, the 5′ sequences originate from the pea gene *ss3.6* (Cashmore 1983) coding for a ribulose bisphosphate carboxylase small subunit, in *cab-lacZ* they come from the gene *AB80* (Cashmore 1984) also from pea and coding for a chlorophyll a/b binding protein, in *nos-lacZ* from the nopaline synthase gene (Depicker et al. 1982) and in *ocs-lacZ*

[1]Molecular Genetics Laboratory, Department of Genetics, University of Helsinki, Arkadiankatu 7, SF-00100 Helsinki, Finland
[2]Laboratorium voor Genetica, Rijksuniversiteit Gent, B-9000 Gent, Belgium
[3]Present address: Transplantation Laboratory and Department of Virology, University of Helsinki, SF-00590 Helsinki, Finland
[4]Present address: Department of Molecular Genetics, Swedish University of Agricultural Sciences, S-75007 Uppsala, Sweden
[5]Present address: Department of Plant Genetic Engineering, CINVESTAV, Unidad Irapuato, Apartado Postal 629, 26500 Irapuato, GTO Mexico

Primary and Secondary Metabolism of Plant Cell Cultures, ed. by W.G.W. Kurz

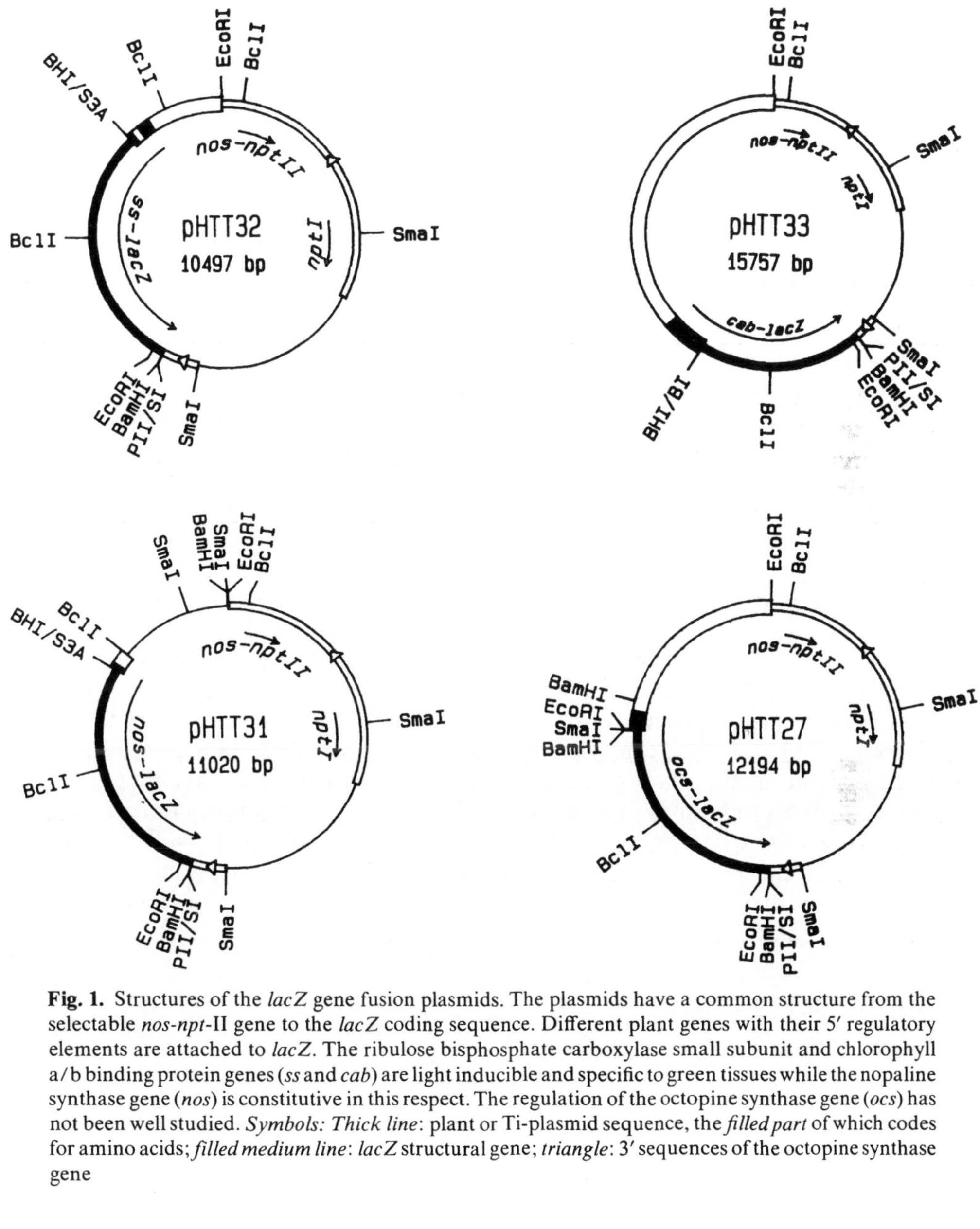

Fig. 1. Structures of the *lacZ* gene fusion plasmids. The plasmids have a common structure from the selectable *nos-npt*-II gene to the *lacZ* coding sequence. Different plant genes with their 5′ regulatory elements are attached to *lacZ*. The ribulose bisphosphate carboxylase small subunit and chlorophyll a/b binding protein genes (*ss* and *cab*) are light inducible and specific to green tissues while the nopaline synthase gene (*nos*) is constitutive in this respect. The regulation of the octopine synthase gene (*ocs*) has not been well studied. *Symbols: Thick line*: plant or Ti-plasmid sequence, the *filled part* of which codes for amino acids; *filled medium line*: *lacZ* structural gene; *triangle*: 3′ sequences of the octopine synthase gene

from the octopine synthase gene (De Greve et al. 1982). Both *nos* and *ocs* originate from the *Agrobacterium* T-DNA. In addition, the 35S transcript promoter of the cauliflower mosaic virus (Odell et al. 1985) and *lac Z* were fused. Downstream of the *lac Z* gene fusions, transcriptional terminator and polyadenylation signals of T-DNA origin were attached. All of the fusion genes code for LacZ polypeptides lacking the 8 N-terminal amino acids. In *ss-*, *cab-*, *nos-* and *ocs-lacZ* they are replaced by 81, 247, 16 and 90 N-terminal amino acids respectively, originating from the plant gene product. The 35S-*lacZ* fusion does not code for amino acids originating from the plant gene. The constructions were stably transferred into tobacco cells with an *Agrobacterium*-mediated system (Horsch et al. 1985).

Essential for the effective use of a reporter gene is that its product can be assayed without interference of substances in the plant cell. The presence of endogenous β-galactosidases active at neutral pH (Fig. 2) has been the major slowdown for the otherwise very attractive *lacZ* system in plants (Helmer et al. 1984; Matsumoto et al. 1988). Animals, which like plants have β-galactosidase activities with acidic pH optima, seem to lack the neutral activity (Nielsen et al. 1983). The LacZ enzyme is very resistant to inactivation by heat, detergents, proteases or urea (Ullman and Perrin 1970; Langley and Zabin 1976). However, attempts to inactivate the endogenous enzyme in solution and under conditions in which LacZ is still active have not succeeded (data not shown).

Fortunately, the endogenous β-galactosidase activity at pH 7 is not very high. Thus *lacZ* gene fusions, expressing relatively well in tobacco, could be studied by measuring the enzyme activity directly in plant extracts. In cases where LacZ activity could not be measured because of the background, we separated the two activities from each other electrophoretically in an SDS-polyacrylamide gel. The β-galactosidase bands in the gel can be visualized with a fluorogenic substrate.

The β-galactosidase activities in leaf, stem and root tissues of plants harboring the 35S- and *cab-lacZ* gene fusions are shown in Table 1. The regulatory elements activate the *lacZ* gene in a very similar pattern as other investigators have found them to activate other reporter genes (Odell et al. 1985; Simpson et al. 1986). In addition to being specifically expressed in green tissues, the *ss* and *cab* genes are also regulated by light. Probably due to the low turnover rate of the very stable LacZ enzyme in the plant cells, we could not demonstrate clearly that *ss*- or *cab-lacZ* plants would have much reduced LacZ levels after keeping the plants in dark. However, by using the self-pollinated offspring of these plants and by germinating the seeds in light or complete darkness, we could show the light regulation of these *lacZ* fusions (Fig. 3). In the *cab-lacZ* plants, the regulation is

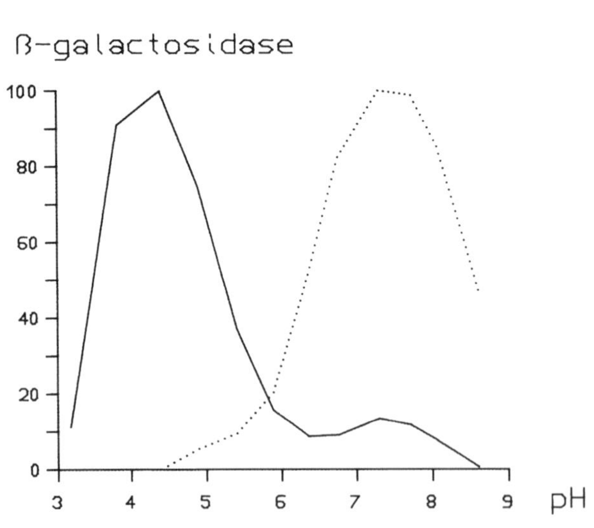

Fig. 2. The pH dependence of the tobacco endogenous β-galactosidases (*solid line*) and the *E. coli* LacZ β-galactosidase (*dotted line*) determined by the ONPG hydrolysis assay (Miller 1972). The values are relative to the maximal activity in both organisms

Table 1. β-Galactosidase activities (U mg[-1] protein) in tissue extracts of the control SR1 plant and transgenic plants harboring 35S- and *cab-lacZ* gene fusions[a]

		Leaf	Stem	Root
	SR1	5	16	18
35S-*lacZ*,	Plant 1	95	100	235
	Plant 2	100	112	306
	Plant 3	98	115	222
cab-lacZ,	Plant 1	81	51	23
	Plant 2	448	173	21
	Plant 3	146	117	23

[a] The enzyme was measured by ONPG hydrolysis (Miller 1972) and one unit is defined as the amount of enzyme hydrolyzing 1 nmol of ONPG per min at 28°C. The 35S-*lacZ* is expressed at high levels in all tissues and best in the root. Although the individual *cab-lacZ* plants show variation, the gene fusion is expressed always at the highest levels in the leaf, intermediate in the stem and not above background in the root tissue.

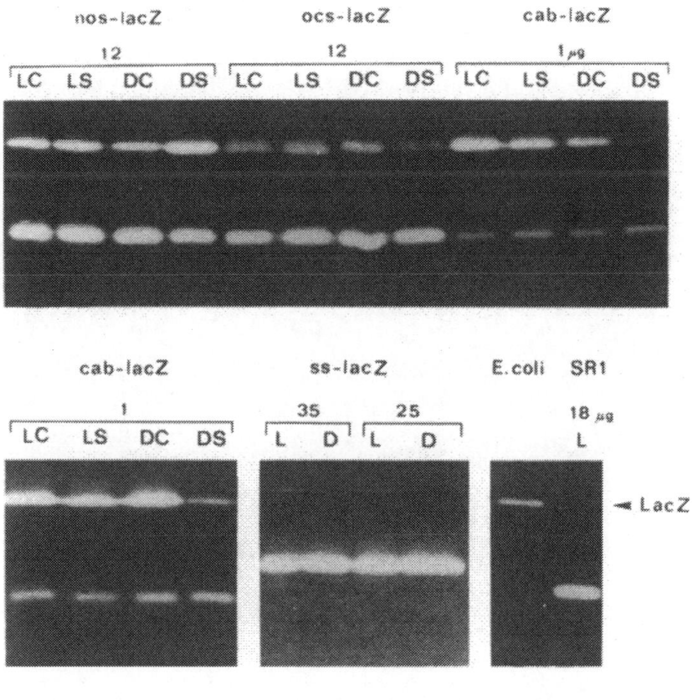

Fig. 3. Light-inducible expression of the *lacZ* fusions. Seeds from self-pollinated transgenic plants were germinated in light (*L*) and dark (*D*). Except for the seedlings harboring the *ss-lacZ* gene, cotyledons (*C*) and stems (*S*) were separated and similar amounts of protein as indicated within each construction were analyzed by SDS-polyacrylamide gel electrophoresis and staining with the fluorogenic substrate 4-MUG (Teeri et al., in preparation). The endogenous, neutral β-galactosidase migrates faster than LacZ. *ss-lacZ* is clearly inducible by light and the *cab-lacZ* gene product has a notably reduced level in stems of dark-grown seedlings as does the *ocs-lacZ* gene product. *SR1* is the non-transformed control plant

clearly detectable only in the stem fraction of the seedlings. The *nos* regulatory element, attached to the *nptII* gene, has been shown earlier not to be regulated by light (Simpson et al. 1985) and similarly, the *nos-lacZ* fusion is not either. However, the much less studied *ocs* gene, fused to *lacZ*, suggestively shows reduced expression in the stems of dark-grown plantlets.

The study of gene expression in different tissues becomes tedious and at some point impossible if one wants to increase spatial resolution and is limited to the types of enzymatic assays described above. The LacZ β-galactosidase can be detected in the plant tissues in situ by using a histochemical stain leading to indigo precipitation at sites of enzymatic activity. The enzyme can be effectively fixed in the tissue by glutaraldehyde which also reduces the background to undetectable levels. We stained LacZ in seedlings with this method and could study light induction and tissue specificity together with high resolution. Figure 4 principally repeats what we analyzed with the gel assay in Fig. 3. The *nos-lacZ* is not reacting, while the *cab-lacZ* reacts clearly only in the stem. We studied the expression of these fusions also in the presence of the herbicide Norfluorazon, which in light halts chloroplast development and leads to bleaching of the plants (Bartels and McCullough 1972). In this condition, the leaves of the *cab-lacZ* seedlings are void of LacZ activity, but the *nos-lacZ* is actually stimulated. Interestingly, the *ocs-lacZ* is inactive in the Norfluorazon-treated seedlings.

A close look at the roots of the histochemically stained material reveals that the *cab-lacZ* plantlets stain also in the vascular cylinder of the root. This activity in the root fraction is hidden by the background when the endogenous β-galactosidase is not inactivated (Table 1) but is detectable with the more sensitive gel assay (not shown). In the *ocs-lacZ* plants, the tip of the root is particularly intense in blue color. We embedded stained *ocs-lacZ* root tips in paraffin and could detect in thin sections that the LacZ activity is present only in the epidermal cell layer (Fig. 5).

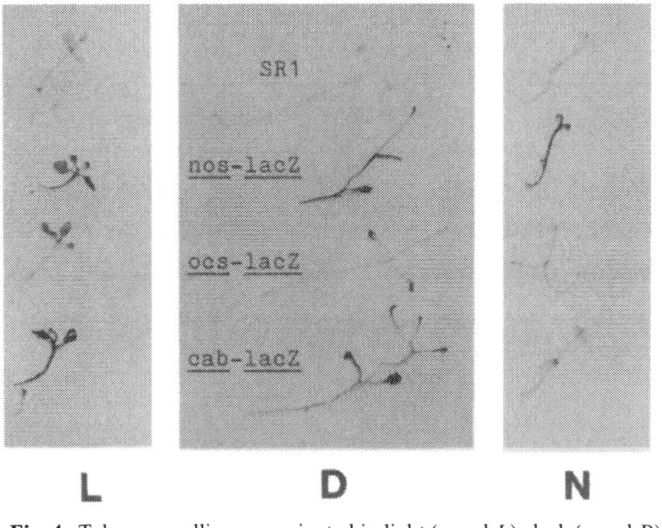

Fig. 4. Tobacco seedlings, germinated in light (panel *L*), dark (panel *D*) or in light in the presence of the herbicide Norfluorazon (panel *N*), were fixed with glutaraldehyde and stained for β-galactosidases with XGal (Teeri et al., in preparation). Norfluorazon arrests chloroplast development and reduces both the *ocs*- and *cab-lacZ* expression levels but not the *nos-lacZ* expression level

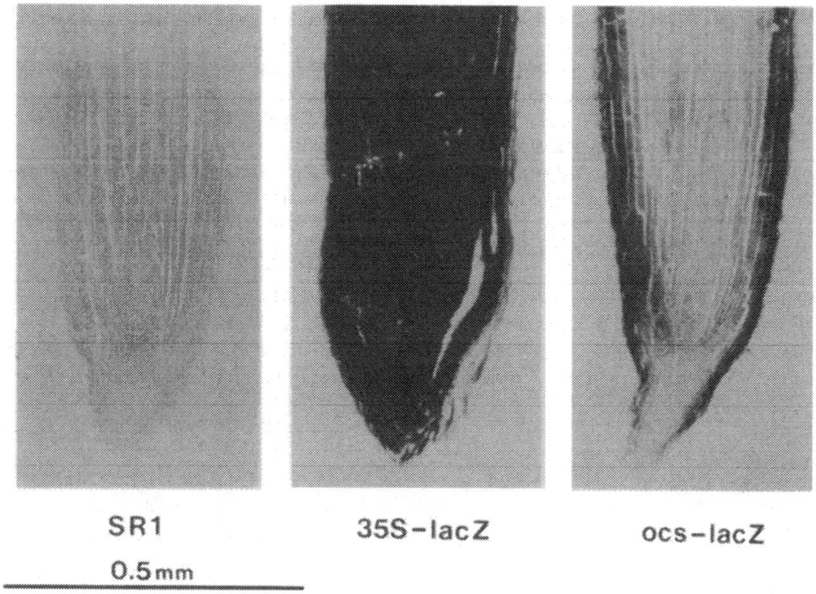

Fig. 5. Thin-sectioned root tips of XGal-stained plants. The endogenous background is inactivated by fixation (*SR1*) and the constitutively expressed 35S-*lacZ* fusion stains well throughout the tissue. The *ocs-lacZ* is active only in the epidermis of the root tip

In conclusion, we have shown that *lacZ* can be conveniently used as a reporter gene to study gene expression in plant cells. The complications of the endogenous background can be solved by electrophoretic separation of the LacZ and background activities prior to enzymatic detection. The most useful property of *lacZ* is its amenability to histochemical analysis. This is due to the fact that LacZ can be immobilized in the tissue by glutaraldehyde fixation, which simultaneously inactivates the endogenous β-galactosidases but not LacZ. The histochemical analysis immediately reveals new properties of the plant regulatory elements attached to *lacZ*.

Acknowledgements. Thanks are due to Pirjo Rahkola and Eija Holma for excellent technical assistance and to Barbara von Schoultz for preparing the thin sections. This research was supported by the Finnish National Fund for Research and Development (SITRA) – Project 17005.1 YP/B and the Academy of Finland. TT had part of the time an Emil Aaltonen Foundation fellowship.

References

Bartels PG, McCullough C (1972) A new inhibitor of carotenoid synthesis in higher plants: 4-chloro-5-(dimethylamino)-2-alpha, alpha, alpha, (trifluoro-m-tolyl)-3(2H)pyridazinone (Sandoz-6706). Biochem Biophys Res Comm 48:16–22

Bassford P, Beckwith J, Berman M, Brickman E, Casadaban M, Guarente L, Saint-Girons I, Sarthy A, Schwartz M, Shuman H, Silhavy T (1978) Genetic fusions of the LAC operon: a new approach to the study of biological processes. In: Miller JH, Reznikoff WS (eds) The operon. Cold Spring Harbor Press, New York, pp 245–261

Bevan MW, Flavel RB, Chilton M-D (1983) A chimeric antibiotic resistance gene as a selectable marker for plant cell transformation. Nature (London) 304:184–187

Brickman E, Silhavy TJ, Bassford PJ, Suman HA, Beckwith JR (1979) Sites within gene LACZ of *Escherichia coli* for formation of active hybrid beta galactosidase molecules. J Bacteriol 139:13–18

Casadaban MJ, Cohen SN (1980) Analysis of gene control signals by DNA fusion and cloning in *Escherichia coli*. J Mol Biol 138:179–207

Casadaban MJ, Martinez-Arias A, Shapira SK, Chou J (1983) β-Galactosidase gene fusions for analyzing gene expression in *Escherichia coli* and yeast. Meth Enzymol 100:293–308

Cashmore AR (1983) Nuclear genes encoding the small subunit of ribulose-1,5-bisphosphate carboxylase. In: Kosuge T, Meredith CP, Hollaender A (eds) Genetic engineering of plants. Plenum, New York, pp 29–38

Cashmore AR (1984) Structure and expression of a pea nuclear gene encoding a chlorophyll a/b-binding polypeptide. Proc Natl Acad Sci USA 81:2960–2964

De Greve H, Dhaese P, Seurink J, Lemmers M, Van Montagu M, Schell J (1982) Nucleotide sequence and transcript map of the *Agrobacterium tumefaciens* Ti plasmid encoded octopine synthase gene. J Mol Appl Gen 1:499–511

Depicker A, Stachel S, Dhaese P, Zambryski P, Goodman HM (1982) Nopaline synthase transcript mapping and DNA sequence. J Mol Appl Gen 1:561–573

Helmer G, Casadaban M, Bevan M, Kayes L, Chilton M-D (1984) A new chimeric gene as a marker for plant transformation: the expression of *Escherichia coli* beta-galactosidase in sunflower and tobacco cells. Bio/Technology 2:520–527

Herrera-Estrella L, De Block M, Messens E, Hernalsteens J-P, Van Montagu M, Schell J (1983) Chimeric genes as dominant selectable markers in plant cells. EMBO J 2:987–995

Hiromi Y, Kuroiwa A, Gehring WJ (1985) Control elements of the *Drosophila* segmentation gene fushi tarazu. Cell 43:603–613

Horsch RB, Fry JE, Hoffmann NL, Eichholtz D, Rogers SG, Fraley RT (1985) A simple and general method for transferring genes into plants. Science 227:1229–1231

Jefferson RA, Kavanagh TA, Bevan MW (1987) GUS fusions: beta-glucuronidase as a sensitive and versatile gene fusion marker in higher plants. EMBO J 6:3901–3907

Kalnins A, Otto K, Rüther U, Müller-Hill B (1983) Sequence of the lacZ gene of *Escherichia coli*. EMBO J 2:593–597

Langley KE, Zabin I (1976) β-Galactosidase α complementation: properties of the complemented enzyme and mechanism of the complementation reaction. Biochemistry 15:4866–4875

Matsumoto S, Takebe I, Machida Y (1988) *Escherichia coli* LACZ gene as a biochemical and histochemical marker in plant cells. Gene 66:19–29

Miller JH (1972) Experiments in molecular genetics. Cold Spring Harbor Lab Press, New York

Nielsen DA, Chou J, MacKrell AJ, Casadaban MJ, Steiner DF (1983) Expression of a pre pro insulin-beta-galactosidase gene fusion in mammalian cells. Proc Natl Acad Sci USA 80:5198–5202

Odell JT, Nagy F, Chua NH (1985) Identification of DNA sequences required for activity of the cauliflower mosaic virus 35S promoter. Nature (London) 313:810–812

Ow DW, Wood KV, DeLuca M, de Wet JR, Helinski DR, Howell SH (1986) Transient and stable expression of the firefly luciferase gene in plant cells and transgenic plants. Science 234:856–859

Simpson J, Timko MP, Cashmore AR, Schell J, Van Montagu M, Herrera-Estrella L (1985) Light-inducible and tissue-specific expression of a chimeric gene under control of the 5′-flanking sequence of a pea *Pisum sativum* chlorophyll a/b-binding protein gene. EMBO J 4:2723–2729

Simpson J, Timko MP, Cashmore AR, Schell J, Van Montagu M, Herrera-Estrella L (1986) Photosynthesis-associated gene families: differences in response to tissue-specific and environmental factors. Science 233:34–38

Ullman A, Perrin D (1970) Complementation in beta-galactosidase. In: Beckwith JR, Zipser D (eds) The lactose operon. Cold Spring Harbor Lab Press, New York, pp 143–172

Ullman A, Perrin D, Jacob F, Monod J (1965) Identification par complémentation in vitro et purification d'un segment peptidique de la β-galactosidase d'*Escherichia coli*. J Mol Biol 12:918–923

Regulation of Photosynthesis in *Daucus carota* and *Arachis hypogea* Cell Cultures by Exogenous Sucrose

K.-H. Neuman, U. Groß[2] and L. Bender[1]

1 Introduction

It seems to be a general characteristic of mixotrophic and photoautotrophic plant cell cultures to perform CO_2 fixation via Rubisco and PEPC simultaneously (Yamada and Sato 1978; Bender et al. 1981; Seeni and Gnanam 1982; Herzbeck and Hüsemann 1985; Nato et al. 1985). As compared to PEPC in C_4 plants the K_m of the enzyme in cell cultures from C_3 plants is considerably lower, (800–1200 μM vs 50–120 μM, O'Leary 1982; Kumar et al. 1988). It has a molecular weight of 440 000 and consists of four identical subunits (Sato et al. 1988). Thus, the molecular weight is in the order of magnitude of the C_4 enzyme (Uedan and Sugiyama 1976). Also, in young and actively growing parts of leaves of C_3 (Aoyagi and Bassham 1986a) and C_4 (Aoyagi and Bassham 1986b) plants a PEPC of similar properties exists and even dominates in carbon fixation. During maturation of leaf tissue the activity and concentration of this enzyme does not change, whereas Rubisco in C_3 plants and also C_4-PEPC in C_4 plants increase during maturation to take the major role in carbon fixation. As reported earlier, a similar influence of developmental processes exists also in photomixotropic carrot tissue cultures (Kumar et al. 1983, 1984) and other cell culture systems (e.g. Nato et al. 1985).

During late log phase of cell division the activity of both carboxylases is almost equal with a slight dominance of Rubisco, whereas during stationary phase, i.e. in "mature" cultures, the activity of Rubisco strongly increases with a concomitant decline of PEPC activity. In this respect, cultured cells are apparently not different from other developing plant systems and therefore could serve as model systems to study the regulation of the development of the photosynthetic machinery. Such cultures are in many ways easier to handle than young leaf tissue of intact plants.

Besides the developmental stage of the cells, also some other factors exert an influence on the ratio of activities of these two carboxylases, e.g. light intensity (Nato et al. 1985), the pH of the nutrient medium (Herzbeck and Hüsemann 1985) or phosphorus supply (Bender 1985). For the latter two parameters, this conclusion was derived from ^{14}C labelling experiments. Another factor hitherto neglected is the sugar concentration of the nutrient medium which will be dealt with in the present chapter. For the experiments described a photomixotrophic carrot callus system characterized in earlier publications (for summary, see Bender et al. 1985; Neumann and Bender 1987) and a photoautotrophic *Arachis hypogea* cell sus-

[1]Institut für Pflanzenernährung, Justus-Liebig-Universität Gießen, Südanlage 6, D-6300 Gießen, FRG
[2]This chapter contains parts from the PhD. Thesis of U. Groß, Justus-Liebig-University

Primary and Secondary Metabolism of Plant Cell Cultures, ed. by W.G.W. Kurz

pension growing now for about 10 years on a sugar-free medium were employed. The original tissue of the *Arachis* system was isolated from mesophyll cells of cotyledons and at first cultured on an MS medium on which greening was also induced (Kumar 1974). In 1977 a subculture of this cell material was transferred onto a sugar-free medium (Neumann 1966) at ambient CO_2 concentration with transfer intervals of about 2–3 months. Some characteristics of this cell culture system were described earlier (Bender et al. 1981, 1985). Although sugar-free, this nutrient medium still contains 200 ppm casein hydrolysate, the usual vitamins (Neumann 1966) and some growth substances (2 ppm IAA, 50 ppm m-inositol and 0.1 ppm kinetin). In Table 1a and b some growth data of autotrophic *Arachis* cultures on liquid and on agar-solidified media are summarized and, as can be seen, the growth capacity realized on a sucrose-supplemented medium is by far not achieved by the photoautotrophic cultures at ambient CO_2. However, the relative growth rate (R) of photoautotrophic cultures at ambient CO_2 on liquid and on solid media are the same. An elevation of the CO_2 concentration increases the relative growth rate by a factor of 6 (Table 1b). The cultures in the experiments in Table 1b were harvested at a much earlier stage of the growth cycle which accounts for the

Table 1. Some growth data of an autotrophic *Arachis hypogea* cell culture strain

a. Photoautotrophic and mixotrophic growth of *Arachis hypogea* cultures grown at ambient CO_2 on liquid and agar solidified medium without sugar, or on agar medium with 1% sucrose. Liquid cultures: 15 ml nutrient solution in T-tubes on auxophyton. Agar cultures: 60 ml of medium in 250-ml Erlenmeyer flasks. $R = (DWt_n - DWt_0)/DWt_0$; t_n, time of harvest; t_0, time of inoculation. Data represent mean ± SE of 3 experiments

Medium type	mg	DW		DM (%)	
	t_0	t_n	R	t_0	t_n
Liquid, – sucrose, 21 days of culture	5.8	6.1 ± 1.0	0.05	3.9	5.7
Liquid, – sucrose, 115 days of culture	5.8	22.4 ± 2.0	2.90	3.9	1.3
Agar, – sucrose, 115 days of culture	15.5	58.0 ± 4.7	2.7	5.3	3.6
Agar, + sucrose, 115 days of culture	12.3	278.6 ± 18.0	21.7	4.1	3.7

b. Photoautotrophic growth of *Arachis hypogea* cultures grown in 60 ml sugar-free liquid medium at ambient or elevated CO_2 concentration in double tier flasks

	DW (mg)		R
Days in culture	0	43	
Ambient CO_2	27.4	31.1	0.1
2.34% CO_2	26.6	43.4 ± 4.5	0.6

c. Holoautotrophic growth of *Arachis hypogea* suspension cultures on an inorganic nutrient solution. Data represent means ± SE of 3 experiments per culturing time indicated (R, see Table 1a)

Days in culture	0	42	84	R_{42}	R_{84}
mg DW	4.5	6.1 ± 2.1	13.6 ± 5.1	0.4	2.0

low values of the relative growth rate. A subculture was transferred to a nutrient medium 28 months ago, consisting only of inorganic components. The cells continue to grow at ambient CO_2 concentration without loss of growth vigour on solid or liquid media (Table 1c) and this material will be called "holoautotrophic".

2 Results and Discussion

2.1 Carbon Fixation and Carbohydrate Metabolism of Photomixotrophic Cultures

The ultrastructural development of chloroplasts in carrot cultures was summarized earlier and comparative investigations on some aspects of the light reaction system indicated that these cultured carrot cells, freshly isolated from the secondary phloem of the tap root, are able to develop a fully functional light reaction system (Kumar et al. 1983, 1984) within about 2 weeks. In the present chapter some results on carbon metabolism will first be summarized, and then an attempt will be made to relate results on the metabolism of exogenous sugars to data on the rate of carbon fixed by photosynthesis.

The experiments on carbon metabolism described here were carried out 3 weeks after explantation and callus induction, and at that stage of the culture cycle (transition from log to stationary phase), the activity of Rubisço and PEPC is almost the same. After 15 s of $^{14}CO_2$ fixation in light the primary products of both carboxylases, PGA (Rubisco) and aspartate and malate (PEPC), are labelled with a slight dominance of the latter (Kumar et al. 1983; Bender et al. 1985).

At that stage PEP which should connect both carbon fixation pathways is not labelled, indicating that under mixotrophic conditions, at which the carbon acceptor could be derived from exogenous carbohydrates, carbon fixation by PEPC operates basically independent of the C_3-pathway. After 4 min of exposure to $^{14}CO_2$ heavy labelling could be detected in hexose phosphates followed by glucose and fructose. Some activity was now also detected in PEP. Carbon fixed by PEPC takes the C_4 position of the malate molecule, whereas C_{1-3} is derived from PEP (POPP et al. 1982). The determination of isotope distribution in malate (Table 2) indicates that after 4 min of fixation about 30% of the ^{14}C occupies the positions 1–3, whereas about 70% was detected in the C_4 position and was, therefore, derived from PEP carboxylation. This indicates the metabolic connection of the two

Table 2. Percent ^{14}C-distribution pattern in malate of *Daucus* and *Arachis* cell cultures ($NaH^{14}CO_3$)

	C_{1-3}	C_4
15 s Light		
Daucus	12.3	87.7
Arachis	20.1	79.9
4 min Light		
Daucus	28.3	71.7
4 min Dark		
Daucus	13.1	86.9

carboxylation pathways in the light, and a considerable contribution of C_4 fixation by PEPC to total carbon fixation of the cells at this stage of the cultural cycle. As was shown earlier light enhances the activity of PEPC five-fold (Bender et al. 1985; Bender 1987). Such a conclusion could also be drawn by comparing the ^{14}C fixation data in light and darkness (Bender et al. 1985).

In earlier publications we reported a strong depression of carbon fixation by a sucrose supplement to the nutrient medium (Neumann and Raafat 1973) during the application of radio-labelled carbon which could be again demonstrated by the experiments summarized in the present chapter. In carrot cultures (Table 3), ^{14}C incorporation into products of both carboxylating enzymes is depressed in the presence of sucrose, however, C_3 fixation seems to be more affected than fixation by PEPC (minus 43% vs minus 23%). Considering the decrease of PEP labelling due to the reduction of C_3 fixation, PEP carboxylation may even be increased following the sucrose supplement. An even stronger depression of carbon fixation following a sucrose supplement to the medium can be observed in photoautotrophic *Arachis* cultures if they are grown mixotrophically. In this system ^{14}C in C_3 fixation derivatives is drastically reduced while labelling of C_4 fixation products seems to be unaffected, coming incidentally to exactly the same value for ^{14}C incorporation in both treatments (Table 4).

Before some data on the influence of phosphorus on carbon fixation and on the metabolic breakdown of exogenous sugars is discussed, some other observations shall be considered. Whereas neither NADPH nor ATP derived from the light reaction system can be transported across the chloroplast envelope to the cytosol, organic acids like malate, oxoglutarate or oxaloacetate and amino acids like aspartate or glutamate apparently can do so freely in both directions (Heber and Heldt 1981). In the chloroplast oxaloacetate of cytosolic origin (and aspartate after deamination) can be reduced to malate by chloroplastic NADPH coming from the light reaction system. Then malate may leave the chloroplast to the cytosol where it can be either taken up by mitochondria to fuel the Krebs cycle or to be decarboxylated and oxidized by the NADP-malic enzyme to yield pyruvate and NADPH. Although the significance of this so-called dicarboxylate translocator for the export of light-generated reducing equivalents from the chloroplast to the

Table 3. The influence of sucrose (2%) on ^{14}C distribution following the application of $NaH^{14}CO_3$ to *Daucus* cell cultures (20 days of culture, 30 min $^{14}CO_2$, dpm × 10^4 g^{-1} dry wt)

	– Sucrose	+ Sucrose	
P-esters	73.7	20.9	
Neutral sugars	91.0	81.6	
Neutral AA	171.0	88.0	
Total	336.7	190.5	- 43%
Organic acids	222.0	196.0	
Acidic AA	283.0	192.0	
Total	505.0	388.0	- 23%
Total	841.0	579.0	- 31%

Table 4. The influence of sucrose (2%) on ^{14}C distribution following the application of $NaH^{14}CO_3$ to *Arachis* cell culture (4 min $^{14}CO_2$, dpm × 10^4 g^{-1} dry wt)

	– Sucrose	+ Sucrose
P-esters	2373	490
Neutral sugars	991	444
Neutral AA	248	232
Total	3612	1166
Organic acids	528	544
Acidic AA	248	232
Total	776	776
Total	4388	1942

cytosol was originally overestimated as compared to the P-translocator. (Giersch 1982), in special circumstances like cell cultures under mixotrophic conditions the contribution of this alternative route of export from the chloroplast should be considered. An indication for the operation of the dicarboxylate shuttle is the labelling ratio of malate to aspartate of > 1 in light and $\leqq 1$ in darkness. After an exposure of 21-day-old carrot cultures to $^{14}CO_2$ a ratio of about 1.5 can be observed with both $H^{14}CO_3$ or ^{14}C-fructose as substrates, indicating the possibility of the operation of this dicarboxylate shuttle (Bender et al. 1985). In the dark, this ratio comes to about 0.5. In agreement with this conclusion are also data on influences of DCMU on $^{14}CO_2$ fixation of *Arachis* cultures. Upon application of this inhibitor of electron flow through the two light reaction systems the malate/aspartate labelling ratio was reduced from 1.5 to 1 (Table 5).

The operation of the P-translocator requires P_i in the cytosol which is taken up stoichiometrically by chloroplasts in exchange for triose phosphates. Con-

Table 5. Distribution of ^{14}C in photoautotrophic *Arachis hypogea* cell cultures in light ($\pm 5 \times 10^{-6}$ DCMU) or darkness for 10 min prior to 10-s exposure to $^{14}CO_2$ (dpm × 10^3 g^{-1} dry wt)

	Light	Light + DCMU	Darkness
P-esters	550	51	0
Alanin	6	0	0
Glycine	11	0	0
Total "C_3"	567	51	0
Malate	45	26	13[a]
Aspartate	31	26	11[b]
Total "C_4"	76	52	24
Total	643	103	24
"C_3"/"C_4"	7.5	1.0	–

[a]Total anionic fraction.
[b]Total amphoteric fraction.

sequently, the concentration ratio of P_i/organic P in the cytosol regulates the extent of triose-phosphate export from the chloroplasts. Phosphorus deficiency should result in a decrease of triose phosphates leaving the chloroplast and in the chloroplast these triose phosphates and other assimilates should pile up, starch will accumulate, and after the storage capacity for starch in the chloroplast is exhausted, RUBP carboxylation will slow down due to feedback reactions and others. Although growth of the cultures is reduced considerably at P deficiency, $^{14}CO_2$ fixation is almost unaffected (Bender et al. 1985). As the ^{14}C distribution between products of PEP and RuBP carboxylation clearly indicated, at limited P supply C_3 fixation is decreased, whereas C_4 fixation shows a strong increase. Although many data are still lacking, based on the evidence presently available, it is hypothesized that at limited P supply the dicarboxylate shuttle serves as some kind of an anapleurotic reaction to compensate at least partly for a decreased export of reducing equivalents from the chloroplasts via the P-translocator.

To understand the mechanism of the negative influence of sucrose on carbon fixation, ^{14}C-fructose and ^{14}C-glucose were supplied to the carrot cultures and the ^{14}C distribution to a number of metabolites was followed for various periods of uptake (Table 6, Bender et al. 1987). Without going into all the details of the experiments, a picture as summarized in Fig. 1 emerges. Apparently, immediately after entering the cells the two hexoses are phosphorylated and converted into each other. A rather low percentage of these hexose phosphates is directly metabolized, whereas the bulk is used to synthesize sucrose as indicated by the accumulation of ^{14}C in the sucrose pool upon either ^{14}C-fructose or ^{14}C-glucose feeding. Sucrose might then again be hydrolyzed to glucose and fructose. These free hexoses seem to serve as storage pools for exogenous carbohydrates, which is in agreement with the high accumulation of these monosaccharides in cultured carrot cells (Table 7) as compared, e.g. to the various parts of the taproot (Table 8).

The phosphorylation of exogenous sugars commits phosphorus otherwise required in inorganic form to operate the P-translocator. Therefore, apparently some metabolically caused internal P_i deficiency may develop following the application of sucrose. Consequently, this export route for triose phosphates from chloroplasts may slow down and in due course C_3 fixation is reduced, whereas C_4 fixation and the operation of the dicarboxylate shuttle is less affected or may even be increased. This is just an hypothesis, other mechanisms are possible, e.g. the competition of some sugar breakdown products for active sites on enzymes common to pathways of assimilates and to derivatives of exogenous carbohydrates. Thus, further experiments are required.

2.2 Enzyme Activity and Enzyme Concentrations

All the results discussed till now were obtained using cell material precultured in sucrose-containing media growing photomixotrophically. The metabolic changes observed should be *short-term effects* implemented on a pre-existing enzyme system as it was established for mixotrophic nutrition. To study *long-term effects* induced by the prevailing nutritional regimes and which are associated with changes in the enzymatic apparatus of the cells, autotrophic *Arachis* cultures

Table 6. The distribution of ^{14}C from ^{14}C-glucose or ^{14}C-fructose after various durations of uptake by carrot tissue cultures (3 weeks in culture, dpm $\times 10^3$ g^{-1} dry wt; av. of 4 replicates)

Uptake (min)	PGA	P-Esters	Glu	Fru	Suc	Pyr	Gly	Mal	Citr	Succ	Asp	Glut	Glyc + Ser	Ala	$^{14}CO_2$	Total
^{14}C-Fructose																
4	0	16.5	0	571	0	0	0	8.4	10.2	0	14.1	5.3	0	0	0	625.5
30	0	26.7	97.8	1790	301	0	4.3	11.2	5.2	0	10.5	12.6	13.2	13.6	1.5	2287.6
360	5.2	36.9	752.5	3304	4883	5.2	13.0	137.5	81.3	27.3	43.7	122.9	73.5	84.2	134.0	9704.2
^{14}C-Glucose																
4	0	11.4	748.0	0	0	0	0	5.4	6.2	0	5.7	3.7	0	0	0	780.4
30	0	18.8	1653.0	104	294	0	4.5	13.2	7.4	0	12.6	6.8	11.2	9.7	1.6	2136.8
360	3.2	38.0	3992.0	647	4611	3.2	9.7	156.2	88.5	22.9	44.7	128.0	49.3	89.2	128.0	10010.9

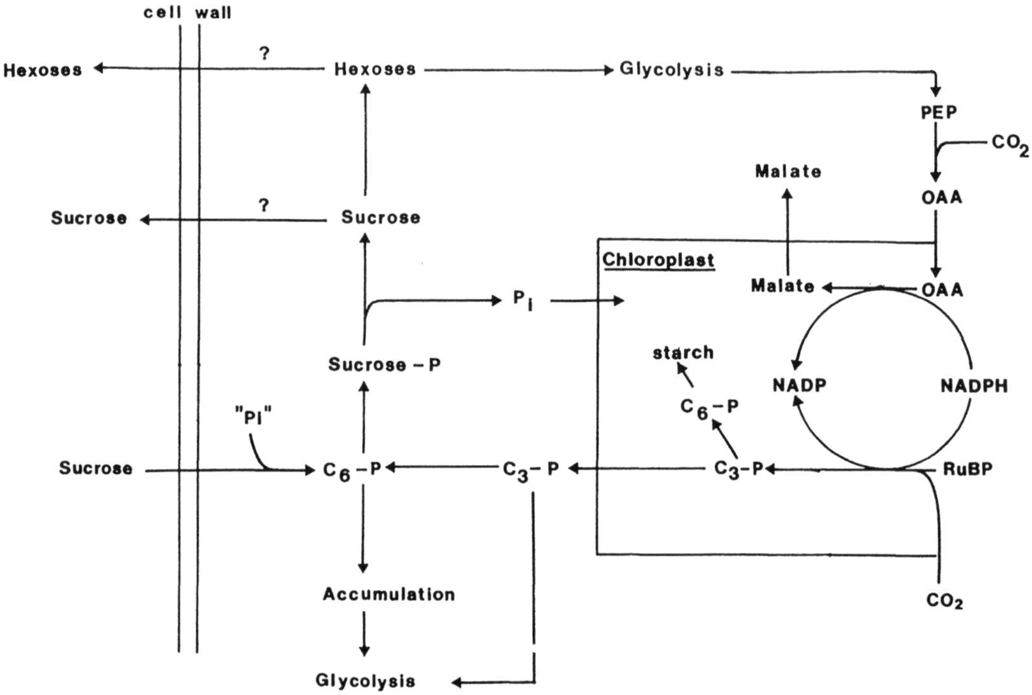

Fig. 1. Tentative scheme of metabolic interactions of photosynthates and sugar from the medium

Table 7. The concentration and the specific activity of free carbohydrates of carrot tissue cultures after exposure to ^{14}C-labelled glucose or fructose and to $NaH^{14}CO_3$ (6-h exposure, 21 days of culture)[a]

	Glucose			Fructose			Sucrose		
	a	b	c	a	b	c	a	b	c
^{14}C-glucose	130.4	399.2	3.1	118.6	64.7	0.55	21.8	461.1	21.2
^{14}C-fructose	130.4	75.3	0.58	118.6	330.4	2.8	21.8	488.3	22.4
$NaH^{14}CO_3$	130.4	327.0	2.5	118.6	231.0	1.9	21.8	30.0	1.4

[a] a = mg g^{-1} dry wt; b = dpm × 10^4 g^{-1} dry wt; c = specific activity dpm × 10^4 mg^{-1} sugar.

Table 8. The concentration of some free carbohydrates in various parts of the root of adult carrot plants (mg g^{-1} dry wt)

	Glucose	Fructose	Sucrose
Cortex	31.3	15.5	21.3
Cambial area	15.0	6.3	T[a]
Pith	18.1	9.7	22.9

[a] T = trace.

growing originally on a sugar-free nutrient medium were subcultured for various lengths of time in nutrient media supplemented either with 1 or 4% sucrose at the beginning of the experiments. At the transition from the log phase of growth into the stationary phase the activity of Rubisco and PEPC was determined and for Rubisco also a semiquantitative estimation of its concentration was attempted. As compared to the cultures in the sucrose-supplemented nutrient media, the growth performance of the cultures on a sugar-free medium is very low (Table 1) and therefore here the cultures were used for the investigation at the end of a subculture period before transfer into a new medium.

Whereas in the cells cultured in the sugar-free medium the activity of Rubisco dominates, at 1% sucrose PEPC activity was higher than that of Rubisco (Table 9). The activity level of PEPC remains quite the same, also at 4% sucrose with a slight insignificant decreasing tendency. At this sucrose concentration no activity of Rubisco could be detected in spite of several, though futile, attempts to concentrate the enzyme in the protein extract tested (Groß et al. in prep.). In tendency, influences of sucrose on $^{14}CO_2$ fixation by both enzymes in short-term experiments are reflected by activity changes of the two carboxylases, which should represent long-term effects.

In agreement with the activity data, the results by densitometric scanning of SDS-PAGE lanes (Table 10, Fig. 2a, b) indicate also a strong decrease of the concentration of Rubisco at increased sucrose concentrations in the medium. This was confirmed using the western blott technique (Groß et al., in prep.). To determine whether the influence of sucrose on Rubisco can be reversed, experiments were also conducted in which, after a preculture at 4% sucrose, the cells were transferred to a medium with 1% sucrose. As the data in Table 10 show, the negative influence of sucrose on the activity and the concentration of Rubisco is reversible. Although the ratio of the small and the large subunit of Rubisco in cultures grown on a sucrose-supplemented medium differs considerably from that in the autotrophic material, quite good agreement exists between the data derived from cells which remained in the 1% sucrose medium and those transferred from 4 to 1% sucrose. Similar experiments on the influence of sucrose on the activity and on the concentration of PEPC are presently in progress.

3 Conclusion

At present it looks as if short-term influences of sucrose on carbon fixation at a given complement of carboxylases are converted into a stable, though reversible, pattern of enzyme activity and as far as we know for Rubisco also of enzyme concentration. A more detailed picture describing the mechanism of regulation of the concentration and the activity of Rubisco and PEPC by sucrose will be attempted as soon as data on the concentration of PEPC at various sucrose regimes are available. Meanwhile an autotrophic cell line of *Daucus* cultures (which is also embryogenic) has been established which has been growing now for 18 months in a sugar-free medium at ambient CO_2, and which will be included in the study for comparison.

Table 9. The influence of sucrose on the activity of Rubisco and PEPC of autotrophic *Arachis* cultures (n kat mg^{-1} protein)

	Sucrose in nutrient medium (%)		
	0	1	4
Rubisco	1.19	0.88	Not detectable
PEPC	0.69	1.47	1.19

Table 10. The influence of various carbohydrate regimes on the specific activity and the concentration[a] (in percent of total soluble protein) of the two subunits of Rubisco of cultured *Arachis* cells

Preculture	Subculture	Spec. activity (n kat mg^{-1} protein)	LS	SS	LS/SS
0 Sucrose	0 Sucrose	1.20	10.4	5.7	1.8
1% Sucrose	1% Sucrose	0.92	7.4	2.8	2.6
1% Sucrose	4% Sucrose		Not detectable		
4% Sucrose	1% Sucrose	0.89	7.1	2.2	3.2

[a] Semiquantitative estimation by densitrometric scanning of the SDS-PAGE lanes. LS = large subunit; SS = small subunit.

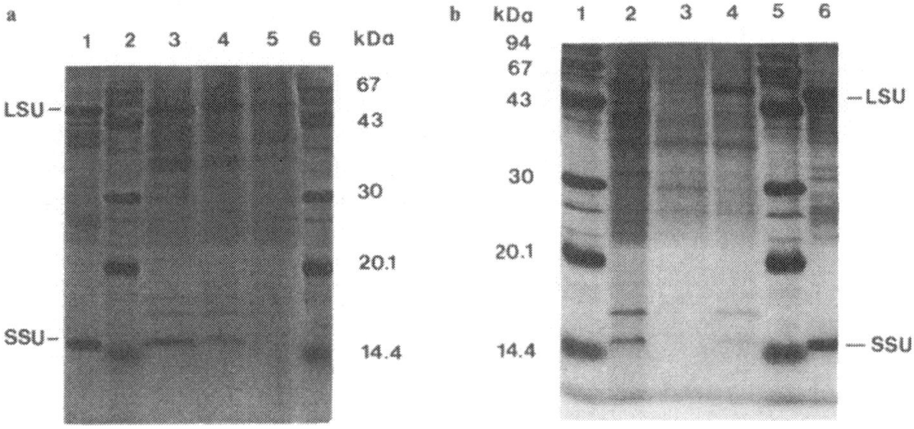

Fig. 2. a *1* Rubisco spinach; *2,6* calibration proteins; *3* Arachis without sucrose; *4* Arachis 1% sucrose; *5* Arachis 4% sucrose. **b** *1,5* Calibration proteins; *2* Arachis 1% sucrose from 1%; *3* Arachis 4% sucrose from 1%; *4* Arachis 1% sucrose from 4%; *6* Rubisco spinach

References

Aoyagi K, Bassham JA (1986a) Appearance and accumulation of C_4 carbon pathway enzymes in developing wheat leaves. Plant Physiol 80:334–340

Aoyagi K, Bassham JA (1986b) Appearance and accumulation of C_4 carbon pathway enzymes in developing maize leaves and differentiating maize A 188 callus. Plant Physiol 80:322–333

Bender L (1985) The influence of phosphorus nutrition on growth and CO_2-fixation of chlorophyllous carrot callus cultures. Z Pflanzenernähr Bodenkd 148:674–653

Bender L (1987) Fixation and metabolism of carbon dioxide by photosynthetically active carrot callus cultures. Can J Bot 65:1768–1770

Bender L, Kumar A, Neumann KH (1981) Photoautotrophe pflanzliche Gewebekulturen in Laborfermenten. In: Lafferty RM (ed) Fermentation. Springer, Berlin Heidelberg New York, pp 193–203

Bender L, Kumar A, Neumann KH (1985) On the photosynthetic system and assimilate metabolism of *Daucus* and *Arachis* cell cultures. In: Neumann KH, Barz W, Reinhard E (eds) Primary and secondary metabolism of plant cell cultures. Springer, Berlin Heidelberg New York Tokyo, pp 25–42

Bender L, Pauler B, Neumann KH (1987) On carbohydrate metabolism of cultured carrot root explants. Plant Cell, Tiss Org Cult 8:135–146

Giersch C (1982) Capacity of the malate/oxaloacetate shuttle for transfer of reducing equivalents across the envelope of leaf chloroplasts. Arch Biochem Biophys 205:246–259

Heber U, Heldt HW (1981) The chloroplast envelope: structure, function and role in leaf metabolism, Annu Rev Plant Physiol 32:139–168

Herzbeck H, Husemann W (1985) Photosynthetic carbon metabolism in photoautotrophic cell suspension cultures of *Chenopodium rubrum* L. In: Neumann K-H, Barz W, Reinhard E (eds) Primary and secondary metabolism of plant cell cultures. Springer, Berlin Heidelberg New York Tokyo, pp 15–23

Kumar A (1974) In vitro growth and chlorophyll formation in mesophyll callus tissues on sugar free medium. Phytomorphology 24:96–101

Kumar A, Bender L, Pauler B, Neumann KH, Senger H, Jeske C (1983) Ultrastructural and biochemical development of the photosynthetic apparatus during callus induction in carrot root explants. Plant Tissue Org Cult 2:161–177

Kumar A, Bender L, Neumann KH (1984) Growth regulation, plastid differentiation and the development of a photosynthetic system in cultured carrot root explants as influenced by exogenous sucrose and various phytohormones. Cell Tissue Org Cult 3:11–28

Kumar PP, Bender L, Thorpe TA (1988) Activities of ribulosebisphosphate carboxylase and phosphoenolpyruvate carboxylase and ^{14}C-bicarbonate fixation during in vitro culture of *Radiata* pine cotelydons. Plant Physiol (in press)

Nato A, Hoarau J, Brangeon J, Hirel B, Suzuki A (1985) Regulation of carbon and nitrogen assimilation pathways in tobacco cell suspension cultures in relation with ultrastructural and biochemical development of the photosynthetic apparatus. In: Neumann KH, Barz W, Reinhard E (eds) Primary and secondary metabolism of plant cell cultures. Springer, Berlin Heidelberg New York Tokyo, pp 43–57

Neumann KH (1966) Wurzelbildung und Nucleinsäuregehalt bei Phloem-Gewebekulturen der Karottenwurzel auf synthetischem Nährmedium verschiedener Hormonkombinationen. Phytohorm Organogenese 38:95–102

Neumann KH, Bender L (1987) Photosynthesis in cell and tissue culture systems. In: Green CE, Somers DA, Hackett WP, Biesboor DD (eds) Plant tissue and cell culture. Proc 6th Int Cong Plant Tissue and cell culture 1986, Minneapolis. Liss, New York, pp 151–165

Neumann KH, Raafat A (1973) Further studies on photosynthesis of carrot tissue cultures. Plant Physiol 51:685–690

O'Leary MH (1982) Phosphoenolpyruvate carboxylase: an enzymologist's view. Annu Rev Plant Physiol 33:297–315

Popp M, Osmond CB, Summons RE (1982) Pathway of malic acid synthesis in response to ion uptake in wheat and lupin roots: evidence from fixation of ^{13}C and ^{14}C. Plant Physiol 69:1289–1292

Sato F, Koizumi N, Yamada Y (1988) Purification and characterization of phosphoenolpyruvate carboxylase of photomixotrophically cultured green tobacco cells. Plant Cell Physiol 29(2):329–337

Seeni S, Gnanam A (1982) Carbon assimilation in photoheterotrophic cells of peanut (*Arachis hypogea* L.) grown in still nutrient medium. Plant Physiol 70:823–826

Uedan K, Sugiyama T (1976) Purification and characterization of phosphoenolpyruvate carboxylase from maize leaves. Plant Physiol 57:906–910

Yamada Y, Sato F (1978) The photoautotrophic culture of chlorophyllous cells. Plant Cell Physiol 19:691–699

Part V
Round Table Discussion

Cytodifferentiation and Compartmentation

K.-H. Neumann[1]

This round table discussion concentrated on five topics selected from this very broad field as related to the synthesis and storage of compounds generally designated as being part of so-called secondary metabolism. The first of these topics was the problem of relations between morphogenesis and secondary production, the second was concerned with the question: why do fast dividing cells not produce secondary metabolites, i.e. the correlation between cell division activity and cellular differentiation, and the third one was on the regulation of the expression of specific genes in specialized cells, which eventually led to a thorough discussion on the occurrence of "production" and "storage" cells. The last two topics were concerned with the distribution of enzymes of secondary metabolism between cellular compartments of which the plastids and the vacuoles were favored in the discussion, and on the active movement of enzymes as well as secondary metabolites between various compartments.

Since space is extremely limited, only a very condensed summary of ideas which emerged during this discussion of several hours can be attempted which necessarily also reflects personal ideas and experience. Generally, the production of secondary metabolites should be divided into two processes, i.e. the synthesis of these compounds and their storage, and to some extent the first may be regulated by the second. In intact plants both processes may occur in the same cell, production and storage may occur in different cell types connected by a transport system, and eventually the synthesis of secondary metabolites could require a series of specialized cells of which the last might be storage cells. In the intact plant and its various organs all these specialized cells will emerge in due course of species-specific development. It will depend on the species or on the substance or both as to which of these three models can be applied. Whereas in vitro in organized structures like hairy roots or, e.g. in somatic embryos from the torpedo stage onward, such specialized cells can differentiate, in cell suspensions, in cell clusters or callus cultures they could be either absent or present only in low numbers. One of the major prerequisites to an increase in the production of secondary metabolites in cell cultures, due to these models, should then be an increase in the population of such specialized cells according to the species and the compound to be produced. However, the knowledge of the ultrastructure and the molecular organization of such cells and the requirements to produce and accumulate them in cell culture is mostly lacking. If in cell suspensions synthesizing and storing cells are separated,

[1]Institut für Pflanzenernährung, Abt. Gewebekultur der Justus-Liebig-Universität Giessen, Südanlage 6, FRG

Primary and Secondary Metabolism of Plant Cell Cultures, ed. by W.G.W. Kurz

at least for hydrophilic compounds, precursors and products could possibly be transmitted through the nutrient medium.

In intact plants cells involved in the production of secondary compounds are rarely associated with primary meristems with a high cell division rate. Consequently, in cultures with fast dividing cells these specialized cells with enzyme complements and an ultrastructure suited for the production or storage of secondary metabolites are scarce or entirely absent. In many cases vacuoles serve as the storage compartment of secondary products and in fast dividing cells vacuoles, which in some instances seem also to function as some kind of lysosomes (generally absent in higher plants), are usually small and/or scarce. Let us assume that in individual cells the development of the organization required to produce and to store secondary products requires a certain length of the interphase between two cell divisions, then it becomes understandable why these fast dividing cultures with a very short interphase are unable to differentiate one or the other types of these specialized cells. One exception is represented by phytohormones like IAA or 2iP which can also be regarded as secondary products and which are synthesized in meristematic tissue in intact plants. These compounds occur in considerable concentrations also in cell cultures growing with a high cell division rate.

Surprisingly little seems to be known on processes of secondary metabolism at the various stages of the cell cycle and on correlations with the duration of one or the other of its phases. The duration of the cell cycle and at least the duration of the G1-phase can be influenced by temperature or the hormonal and nutritional supplement to the medium. Here, experiments with synchronized cultures are required in which the molecular and ultrastructural status concerning secondary metabolism is monitored in cell populations growing with variations in the duration of the cell cycle. Such experiments seemed to be "in the making" in the laboratory of one or the other of the participants.

Elicitors in in Vitro Cultures

F. Constabel[1]

1 Introduction

Induction of synthesis, accumulation, and release of secondary metabolites by elicitors is primarily a subject of phytopathology. The metabolites formed are plant-specific, in various plants structurally quite diverse, and generally show antimicrobial activity. They are referred to as phytoalexins. Isoflavonoids which occur in plants in response to elicitation by pathogens may be regarded as the model compounds of phytoalexins. Plant cell cultures can be employed to simulate the production of phytoalexins (Hahlbrock et al. 1988). Wolters and Eilert (1983) suggested to use plant cells cultured in vitro, specifically their formation of secondary metabolites upon elicitation for industrial exploitation. The results of subsequent experimentation in a number of laboratories vary from substantial enhancement of product formation to no measurable response, i.e. exposure of plant cells to elicitors did not increase the level of product accumulation over that obtained previously. As far as products resulting from elicitation of plant cell cultures furnished substantial yields of desirable compounds, elicitation in vitro has become a matter of much discussion. Focal points are (1) observations of product formation and industrial relevance, (2) the process of elicitation and aspects of genetic modification, and (3) re-elicitation.

Since the elicitors used may be biotic or abiotic in nature and may comprise artificial agents, and since the products accumulated may not have all the characteristic of phytoalexins, phytopathological terminology should be avoided, unless proven valid. One recommendation would be to simply refer to stress agents and stress products.

2 Observations of Product Formation

The synthesis and accumulation of secondary metabolites in cell cultures exposed to fungal homogenates and fractions thereof have widely been documented in the literature (Eilert 1988). The advent of hairy root cultures and the capacity of this material to furnish compounds which have previously been detected in in vitro systems in low concentrations, if at all, raised the question whether elicitation can enhance the accumulation of products and whether elicitation may be a means to

[1]National Research Council, Plant Biotechnology Institute, N.R.C., Saskatoon, Saskatchewan S7N OW9, Canada

Primary and Secondary Metabolism of Plant Cell Cultures, ed. by W.G.W. Kurz

enhance product release. Experiments with hairy roots of *Datura* showed no increase in tropane alkaloid levels upon exposure to heavy metal ions (Cu^{2+}), but did show the appearance of sesquiterpenoids. Also, such cultures released about 50% of the tropane alkaloids into the medium over about 40 h. Upon transfer of the roots to regular medium the product release ceased. After a period of time the cultures could be "milked" again. Success depended upon the developmental stage of the culture, use of a sufficient volume of medium, density of the root culture, and Cu^{2+} ion concentration (Rhodes). With chick pea cell cultures cadmium or better manganese ions stimulated flavonoid (phytoalexin) production. Here, again, ion concentration, duration of exposure to a stress agent, developmental stage of the culture, and peroxidase activity in the medium were seen to affect the process (Barz).

From an industrial as well as from a biological point of view it would be of interest to know whether secondary metabolites which result from elicitation, e.g. alkaloids, show antimicrobial/antibiotic activity as do phytoalexins. In many cases alkaloids have been seen as components of a plant's chemical defense system. When synthesized in the roots, alkaloids are transported to shoots and may provide protection from insects. Upon elicitation by microorganisms or wounding alkaloid levels in shoots may increase five times over that of controls (Wink). With opium poppy fungal infection may lead to loss of morphine, but thebaine and codeine remain unchanged. Whether benzophenanthridines, like sanguinarine, accumulate in response to elicitation remains to be confirmed, however, its antimicrobial activity has been demonstrated (Roberts). It would appear that the phytoalexin character of alkaloids which are synthesized and accumulated in cell cultures in response to elicitation has to be proven case by case. There clearly are cases where the natural phytoalexin is not the reaction product to elicitation with in vitro culture systems, it is rather a different metabolite. In *Capsicum frutescens* cell cultures, for example, it is not capsidiol, a phytoalexin, which is accumulated, but capsaicin which results from elicitation (Yeoman).

Designating sanguinarine in poppy cell cultures as a stress product, since this alkaloid was accumulated due to treatment with homogenates of fungal mycelia, confers a new consideration to the occurrence of sanguinarine in hairy root cultures not exposed to fungal elicitors. It would appear that hairy roots cultured in vitro are subject to stress, and that in general plant tissues cultured in vitro are subject to stress (Ellis).

Since some stress products found in in vitro systems are of industrial importance and, thus, substantiate the socio-economic relevance of research into the formation of such compounds, other responses by plant material to stress are being neglected. Some tissues may respond by lignification of their cell walls, for instance (Towers), others may show increased activity of alcohol dehydrogenase (Dixon).

3 Elicitation and Its Modification

The quality of elicitors may vary infinitely; they may be oligosaccharides or enzymes and the determination of an upper and lower limit of the molecular weight of such substances cannot be given. An elicitor, in order to be active, must be present in a minimum concentration and may be specific for a given tissue. Permeability of the cell wall would appear to be a factor (Ebel).

Once an elicitor is recognized by a cell a signal for gene activation is generated and transduced. Calcium appears to be involved, rather than cyclic AMP. Chelating calcium was seen to inhibit gene activation and phytoalexin production (Ebel).

Gene expression may be regulated by protein binding catalyzed by protein kinases. Characterization of respective binding proteins is being studied. Here, modification of the system could be achieved by enhancing reversible phosphorylation/dephosphorylation (Dixon).

Elicitation causes gene activation, then enzyme synthesis, then product formation. It may be of interest to know whether and to what extent these reactions are transient. There must be mechanisms which control the removal of the enzymes. The half-life of enzymes synthesized under elicitation may be different than those synthesized under non-stress conditions (controls). Is increased product formation the result of persistence of enzymes with a longer half-life? An analysis of enzyme turnover lies at the heart of such consideration (Dixon).

The sequence of enhanced enzyme activities in the synthesis of stress products must not be in accordance with the sequence as suggested by the biosynthetic pathway. In periwinkle cell cultures strictosidine synthase activity peaks prior to tryptophan decarboxylase in the formation of indole alkaloids. This finding may be explained by the high level of tryptamine in cell cultures at all times and so prior to elicitation. In seedlings with low levels of tryptamine tryptophan decarboxylase activity precedes strictosidine synthase activity as would be expected (De Luca).

4 Re-Elicitation

Re-elicitation of cell biomass in fermenters has been successfully demonstrated. Success, however, is restricted to using a very narrow range of elicitor concentration, duration of exposure to the elicitor, developmental stage of the culture, and recovery upon replacement of elicitor with regular medium. With regard to the developmental stage of the culture, it would be helpful to be assisted by markers (Tyler). Such markers may be found in conductivity data (Dixon). The question, however, is whether there is an optimum time for elicitation. In some cases, i.e. with heavy metals as elicitor, elicitation of cell cultures can be accomplished at all times (Barz).

Recovery of cell cultures after exposure to heavy metals as elicitors for purposes of re-elicitation would require a system which eliminates the effect of heavy metals quickly. Chelating agents may have to be employed. In the case of some aluminium-resistant plants it was shown that plants contain constitutive chelates. Such compounds may also contribute to recovery from heavy metal treatment of cells cultured in vitro (Yamada). This argument, however, cannot remain unchallenged, since the elicitation of cell cultures by heavy metal ions effects maximum phytoalexin production within 4–6 h, while the formation of peptides with chelating capacity may take at least double the time (Barz).

Industrial application of elicitation, enhanced by re-elicitation, will require quantitation and optimization beyond the present status and will encompass more factors then monitored at present, a fascinating rather than a depressing prospect.

References

Eilert U (1988) Elicitation: methodology and aspects of application. In: Constabel F, Vasil IK (eds) Cell culture and somatic cell genetics, vol 4. Academic Press, New York London pp 153–196

Hahlbrock K, Herrmann A, Hoffmann H, Kawallek P, Lozoya E, Scheel D, Schmelzer E, Schulz W, Somssich I (1988) Stress related genes in plants: structure and expression pattern in stimulated cell cultures as compared with intact tissue. 4th Int Congr Cell biology, Montreal, Abstr S47.1

Wolters B, Eilert U (1983) Elicitoren-Auslöser der Akkumulation von Pflanzenstoffen. Ihre Anwendung zur Produktionssteigerung in Zellkulturen. Dtsch Apotheker Z 123:659–667

Regulation of Metabolite Flow

V. De Luca[1]

When stimulated, plant cell cultures are capable of large changes in gene expression which results in altered metabolism. As demonstrated in this meeting, this property has permitted the isolation of inducible proteins by two-dimensional polyacrylamide gel electrophoresis and their identification by direct sequencing in a gas phase sequenator (Inzé, Sato). In addition, fungal elicitors have been added to the list of tools used to alter metabolism (Barz, Brodelius, Ebel, Eilert, Tyler), and both circumstantial and direct evidence indicates that this occurs as a result of de novo protein biosynthesis. The effects of elicitors seem to be widespread and several reports have demonstrated that a number of secondary metabolic pathways could be induced or repressed (several authors). Our ability to harness the biochemical machinery of the cell for production has, however, not been matched by an understanding of how these changes came about. The renewed interest in, and the apparent success of organ-specific secondary product synthesis by hairy root cultures (see Rhodes), reflects our lack of understanding of the biochemical processes involved and their regulation.

As recognized in previous meetings, it becomes apparent that our severe lack of knowledge about the expression of biochemical branch points, which connect primary and secondary metabolic pathways, has impeded the design of strategies to channel carbon, nitrogen and oxygen towards specific desirable end products. During the round table discussion, it was outlined that plant development controls gene expression and that metabolism is intricately associated with this developmental program. The expression of metabolic events is tied into a time frame which varies according to the needs of the organism and which is coordinated by the genome (Ellis). The point was raised that the stationary phase represents a physiological state induced by a specific nutrient limitation and that this state would vary based on which nutrient was limiting. This state or multiplicity of states would in turn define the type of gene expression obtained, whether it be in a plant cell culture or in an intact plant (Dougall).

These complex problems should be addressed at the biochemical and molecular levels (Inzé). The enzymes involved in secondary metabolite biosynthesis or degradation must be isolated, purified and characterized. The purified proteins should be utilized to produce antibodies and these antibodies can be used to isolate respective genes from complementary DNA and total genomic libraries. Cloned genes can be used to study gene expression as well as to isolate DNA binding proteins which may control tissue-specific and developmentally regulated

[1]Plant Biotechnology Institute, N.R.C. Saskatoon, Saskatchewan, Canada, NRCC 30263

Primary and Secondary Metabolism of Plant
Cell Cultures, ed. by W.G.W. Kurz

gene expression. Modification of regulatory sequences could alter the requirements for expression of a given metabolic pathway and thus permit, through the expression of previously silent genes, controlled production of metabolites in "non-differentiated" plant cell suspension cultures.

The technology to perform this research has become available and a few examples reported in this meeting (see text) indicate that secondary metabolism enzymes can be studied in this way. The rate-limiting step, however, seems to be at the stage of protein purification and the biochemistry of metabolic pathways. The other rate-limiting factor is the scarcity of researchers committed to plant biochemistry at the protein level. Expansion of research in this area, together with the utilization of molecular technology, promise to revolutionize our basic understanding of plant metabolism and in so doing will permit commercial exploitation of their chemical potential.

Bottlenecks Towards Commercialization of Product Synthesis by Plant Cell Cultures

W.G.W. Kurz[1]

Two major bottlenecks towards a wider industrial application of plant cell culture are product yield and productivity. Without significant progress within the biochemical, molecular biological, phytochemical, physiological and process technological sectors directed towards improvement of these two aspects, wide industrial acceptability of plant cell cultures cannot be expected.

Prolonged cultivation periods, the necessity of media exchanges to either promote cell propagation or product synthesis, the deposition of products within cell vacuoles, which forces the sacrifice of cells for product recovery, are but three aspects which make plant cell cultures poor competitors to microbial or synthetic processes for industrial purposes.

In order to get away from the shotgun approach of the past we now have to elucidate the biosynthetic pathways of the desired products, identify the enzymes involved and investigate/manipulate their regulation. Recombinant DNA technology is supplying the tools for identification and cloning of genes for specific regulatory enzymes, and combined with the elucidation of specific reaction sequences we are now able to take a more direct approach in regulating biosynthetic reactions. A major problem, however, is still the purification of some of the enzymes involved in secondary metabolism as many of these are of low abundance (De Luca).

On the technical side it is primarily reactor design which has to be adapted to the specific needs of plant cells and processes. No universal, optimal design can be expected to serve all needs.

The successful use of immobilized cell systems is dependent upon an effective, non-cell destructive product release which allows re-use of cell material. This is still in the developmental stage with no large-scale data available.

Cell line instability frequently observed has to be addressed much more vigorously than in the past. Hairy root cultures can grow at extremely high densities and because of the unique internal transport system within roots, mass-transfer problems inherent in high density cell culture systems do not necessarily apply in root systems. Furthermore, problems of genetic stability should be minor ones which make hairy root cultures good candidates for continuous culturing (Rhodes).

One possible way to avoid prolonged cultivation periods is offered by semicontinuous operations where certain portions of the culture are replaced by fresh

[1]Plant Biotechnology Institute, N.R.C., Saskatoon, Saskatchewan, Canada

Primary and Secondary Metabolism of Plant Cell Cultures, ed. by W.G.W. Kurz

medium on a regular basis. By choosing different ratios, different physiological stages can be obtained and optimal ones selected (Dougall).

Actively dividing cells usually do not produce secondary metabolites; one notable exception being *Coptis japonica* which produces berberine in large quantities while actively dividing. A closer investigation of this "anomalous" behaviour may give indications how to initiate a similar response in other cultures (Yamada).

Product stability within culture systems is still largely unresolved. Extracellular peroxidases and similar enzymes degrade products released into the media sometimes at rapid rates although these rates vary according to the environment in which the cells are growing. Cultures in bioreactors seem to be less affected than those in Erlenmeyer flasks (Barz). It has also been shown that by adding product-adsorbing resins to the culture, the amount of product within the medium could be decreased to unmeasurable levels and later eluted from the resin undegraded (Robins).

Variation was found with regard to elicitor sensitivity. The tolerance levels to elicitors in bioreactors are lower than in shake flasks indicating a greater stress factor in the former (Barz).

It is doubtful that elicitation will be useful in directing pathways or precursors towards one specific end product, however, a shift towards a number of structurally closely related compounds has been observed (Ebel).

Finally, socio-economic factors may need to be taken into consideration when calculating the economic feasibility of a plant cell culture process. This argument can be made when phytoproducts are obtained from plants of endangered species (Constabel). The case of cosmetic use of plant cell-derived shikonin is interesting. Here, the use of a natural compound in one product gave a company a market advantage over its competitors. This reflected positively on the sales of its other products (Robins).

Outlook for Future Research

M.J.C. Rhodes[1]

This workshop centred on the two major topics of the symposium, the study of primary metabolism in photoautotrophic cultures and the production of plant secondary metabolites using plant tissue cultures.

In photoautotrophic studies, the extension of the range of species in culture was recognized as a major need (Barz). With two exceptions, all the photoautotrophic cultures available are derived from C3-dicotyledonous plants: only one monocotyledon (asparagus) has so far been cultured photoautotrophically. The importance of extending this methodology to commercially important monocotyledons such as corn, rice and wheat and to plants with C4-photosynthesis was emphasized. Significant improvements in photoautotrophic culture techniques have led to cultures that grow at low (near atmospheric) levels of CO_2 and have levels of Rubisco similar to that of leaves, thus providing useful model systems for the study of metabolism in green tissues (Widholm). They differ from leaves in having relatively low levels of PEP carboxylase; the reasons for this need further investigation (Widholm, Barz). It is envisaged that photoautotrophic cultures will be used in basic studies of the unique features of the regulatory metabolism of green tissues such as the enzymology of Rubisco activation and in studies of chloroplast development (Widholm). However, it was also felt that these cultures will be used to investigate more applied problems such as in screening for new herbicides (Widholm) or in studies of plant diseases such as sun blotch viroid infection of avocado (Barz), in which important responses to infection occur in the chloroplast.

The discussion of secondary metabolism centred on the opportunities for developing more rational approaches to the improvement of the productivity of plant secondary metabolites in culture and, in particular, the contribution that the techniques of molecular biology might make to this area. Until recently, the approach used has been essentially empirical in nature, involving selection of elite explant material to initiate cultures, the manipulation of the medium constituents (particularly the plant growth substances) and of the physical environment of the culture. These factors, used in conjunction with screening and selection procedures which exploit the genetic and biochemical variation in cell cultures, have led to the isolation of high-yielding cell lines. However, the application of these approaches has led to uneven progress in the commercial exploitation of plant cell cultures; highly successful in a few cases, such as shikonin production, but unsuccessful in many others.

[1]Institute of Food Research, Norwich Laboratory, Norwich, England

Primary and Secondary Metabolism of Plant Cell Cultures, ed. by W.G.W. Kurz

A necessary precondition to a more rational approach is to develop greater knowledge of the routes of secondary pathways and of the main features of their regulation. The fundamental biochemical details of only a few secondary pathways, notably the biosynthesis of phenylpropanoids and of the isoquinoline alkaloid berberine, are known but knowledge of their regulation is limited. This regulation could be at the level of fine control, involving feedback control or reversible inactivation of enzymes, or by coarse control, involving transitory increases in transcription of specific genes and de novo enzyme synthesis. Work in culture on the phenylpropanoid pathway shows that de novo synthesis is the dominant feature of the regulation of this pathway, which may be induced by external factors such as UV light and fungal-derived elicitors (Dixon). However, it was pointed out that in this system a variety of other mechanisms including feedback inhibition, induction by intermediates and proteinaceous enzyme inhibitors also operate and that different regulatory mechanisms operate in different tissues (Dixon).

The opportunities for fine control were discussed and several potential systems have been described, notably feedback inhibition of PAL by cinnamate, of PAL by 3,4-dihydroxyphenyllactate in rosmarinic acid biosynthesis (Ellis) and of chalcone synthase by kievitone in ptercarpan formation (Dixon). Doubt was expressed as to whether these mechanisms operate in vivo, in view of the relatively high concentrations required in vitro. However, since these compounds act competitively, the concentrations of both substrate and inhibitor in the compartment in which the enzyme is held would have to be known before the importance of this type of control can be assessed (Widholm) and the possibility of genetically manipulating the target enzyme and returning it in a deregulated form to the tissue to enhance product formation could be considered.

Since it was widely agreed that the principal control of secondary pathways is at the level of transcription and de novo protein synthesis, substantial opportunities were recognized in returning key genes to cultures under the control of high-expression or deregulated promoters to increase flux through secondary pathways (De Luca, Inzé). To proceed in this direction, work is necessary at the biochemical level to identify in particular pathways the genes coding for enzymes which are underexpressed in culture and then to isolate these genes and to return them in a suitably modified form to the culture. The technology to isolate genes of secondary metabolism which may be expressed normally at only very low levels is available (Inzé, De Luca). Already the number of isolated genes available is increasing and at this meeting the work towards the isolation of the genes for tryptophan decarboxylase and the acetyl transferase of *Catharanthus* and hyoscyamine 6β-hydroxylase from *Hyoscyamus* have been described. The necessary vector systems with promoters of varying strength are available and, given the appropriate genes, the validity of this approach of attempting to eliminate limitations in the expression of secondary product pathways will be tested in the next few years.

The availability of a number of genes of a given metabolic pathway will offer additional opportunities in studying the overall regulation of such pathways at the genetic level. This approach involves study of the upstream sequences of the promoters of the secondary pathway genes. It is envisaged that a hierarchy of regulatory genes, producing specific proteins targeted to the upstream sequences of the individual genes, coordinates their expression (Inzé). Transformation of cells in the culture with a chimeric gene in which the promoter of the secondary metabolic gene is linked to a reporter gene such as β-glucuronidase (GUS) may help to elucidate this problem. Such an approach assumes that the reporter gene would reflect the expression of the pathway, since it would only be expressed when the regulatory genes and their protein products were operating. If cultures were raised from tissues carrying the construct, it would be possible to speed screening of factors such as medium constituents, physical conditions of the culture, etc. by rapid assay of the reporter enzyme. Another approach would be to transform cultures with a similar construct in which the promoter region was linked to a selectable marker (e.g. kanamycin or hygromycin resistance). Here, the presence of the high levels of the antibiotic would select cells in which the regulatory system, and hence the genes it controls, was constitutively expressed at a high level (Inzé). Knowledge of this regulatory system and of the promoters of the individual genes should lead to an understanding of the factors that control the intensity and duration of the coordinated expression of these pathways and this could be exploited to increase product formation in culture.

An important consideration, whether genetic or empirical approaches are made to improve secondary product formation, is the stability of expression of the improved phenotype during long-term culture. There was much discussion of the heterogeneity at the genetic level of some plant cell cultures (Rhodes, Robins). However, several speakers were able to point to their own experience of stability in cultures both in growth and in secondary metabolite formation over extended periods (Yamada, Dougall, Constabel). It was agreed that stability in culture is an important area in which more research is needed. This problem of general instability can be avoided by the use of organized cultures such as hairy roots, but even here the problem of the long-term stability of expression of introduced genes needs to be addressed.

Subject Index

SPRINGER NATURE

GPSR Compliance

The European Union's (EU) General Product Safety Regulation (GPSR) is a set of rules that requires consumer products to be safe and our obligations to ensure this.

If you have any concerns about our products, you can contact us on ProductSafety@springernature.com

In case Publisher is established outside the EU, the EU authorized representative is:

Springer Nature Customer Service Center GmbH
Europaplatz 3
69115 Heidelberg, Germany

Kolibri 360 GmbH
Ferdinand-Jühlke-Straße 7
99095 Erfurt, Deutschland
produktsicherheit@kolibri360.de